(Prospero's America)

Prospero's America

John Winthrop, Jr., Alchemy, and the Creation of New England Culture, 1606–1676

WALTER W. WOODWARD

Published for the Omohundro Institute of Early American
History and Culture, Williamsburg, Virginia, by the
University of North Carolina Press, Chapel Hill

The Omohundro Institute of Early American History and Culture is
sponsored jointly by the College of William and Mary and the Colonial
Williamsburg Foundation. On November 15, 1996, the Institute adopted
the present name in honor of a bequest from Malvern H. Omohundro, Jr.

© 2010 The University of North Carolina Press
All rights reserved

Set in Bembo by Tseng Information Systems, Inc.
Manufactured in the United States of America

Library of Congress Cataloging-in-Publication Data
Woodward, Walter William.
Prospero's America : John Winthrop, Jr., alchemy, and the creation
of New England culture, 1606-1676 / Walter W. Woodward.
p. ; cm.
Includes bibliographical references and index.
ISBN 978-0-8078-3301-8 (cloth : alk. paper)
1. Winthrop, John, 1606-1676. 2. Alchemists—New England—Biography.
3. Alchemy—New England—History—17th century. I. Omohundro Institute of
Early American History & Culture. II. Title.
[DNLM: 1. Winthrop, John, 1606-1676. 2. Alchemy—New England—Biography.
3. Famous Persons—New England—Biography. 4. History, Modern 1601—
New England. WZ 100 W792WA 2010]
QD24.W56W66 2010
540.1′1097409032—dc22
2009033899

The paper in this book meets the guidelines for permanence and durability
of the Committee on Production Guidelines for Book Longevity of the
Council on Library Resources.

The University of North Carolina Press has been a member
of the Green Press Initiative since 2003.

Frontispiece: John Winthrop, Jr. Oil on canvas by unidentified artist
of the school of Sir Peter Lely or William Dobson. [1634–1635].
Courtesy of the Massachusetts Historical Society

To Irene, Peter, Thomas, Michael, & Halley
A bookish family

Acknowledgments

While developing this study, so many people gave so generously of what they knew, who they were, and, in many cases, what they had that acknowledging my debt to them seems insufficient. My first and most heartfelt appreciation goes to Karen Ordahl Kupperman, who, for longer than either of us cares to remember, has been tutor, critic, counselor, and champion. My best decision when choosing to become a historian was to choose to apprentice under her guidance. She has been, from day one, the model mentor and intellectual guide. I would not be a historian, and this would not be a book, were it not for her unwavering support and damned good sense. T. H. Breen, too, has been a true friend and refiner of this project, from whom I have learned much, and learn still.

Other scholars, friends, and colleagues who helped midwife this book at various stages and in in many ways and whose advice I have valued if not always followed include William R. Newman and Richard S. Dunn, who read the original manuscript and made valuable suggestions for its improvement; Nick Bellantoni, Francis J. Bremer, Nancy Brennan, John Brooke, Kathleen Brown, Richard D. Brown, Steve Bullock, Cary Carson, Cornelia Hughes Dayton, Faith Davison, John Demos, Jorge Cañizares-Esguerra, Trudy Eden, Alison Games, Richard Godbeer, Robert Gross, Heather Kopelson, Kevin McBride, John Murrin, Robert Naeher, Susan Scott Parrish, Carla Pestana, Mark Peterson, Ann Plane, Shirley Roe, Guido Ruggiero, Bruce Stave, Fredrika J. Teute, Bruce D. White, Lisa Wilson, John Young, and Cynthia Van Zandt. All my colleagues at the University of Connecticut have also provided an inspiring environment in which to work, think, and share ideas. Gil Kelly showed me just how much a wise, careful, and talented copy editor can add to a book; working with him has been an enlightening and delightful experience.

A number of institutions generously funded the research upon which this book is based. I am grateful to the National Endowment for the Humanities, which funded a long-term fellowship split productively between two remarkable institutions: the Massachusetts Historical Society in Boston and the Huntington Library in San Marino, California. The American Antiquarian Society, in Worcester, Massachusetts, and the John Carter Brown Library, in Providence, Rhode Island, provided fellowships that measurably advanced this project. In each of these institutions, remarkably helpful and knowledgeable staff made research rewarding on both an intellectual and a personal level, providing experiences for which I am deeply grateful. Special thanks go to Peter Drummey. The Yale Center for Religion in American Life, in New Haven, Connecticut, also granted me a Pew Fellowship that facilitated both my research and writing.

The final and most important acknowledgment must go to my wife, Irene, who often held the fort while I wrote the book, and to my children, Peter, Thomas, Michael, and Halley, who grew into wonderful, grown-up human beings somewhere between researching Chapter 1 and writing Chapter 8. If this book is half as good as they are, it will be all I could possibly hope for.

Contents

Acknowledgments vii

List of Illustrations xi

Introduction 1

1: John Winthrop, Jr., and the European Alchemical Movement of the Early Seventeenth Century 14

2: The Republic of Alchemy and the Pansophic Moment 43

3: Founding a New London 75

4: Which Man's Land? Conflict and Competition in Pequot Country 93

5: Alchemical Vision Refined 138

6: "God's Secret": John Winthrop, Jr., Alchemical Healing, and the Medical Culture of Early New England 160

7: The Magus as Mediator: Witchcraft, Alchemy, and Authority in the Connecticut Witch-Hunt of the 1660s 210

8: "Matters of Present Utility": John Winthrop, Jr., the Royal Society, and the Politics of Intelligence in Restoration New England 253

Afterword 302

Index 309

Illustrations

Map

MAP 1. The New England World of John Winthrop, Jr. 96

Figures

FRONTISPIECE. John Winthrop, Jr.

FIGURE 1. *Monas Hieroglyphica,* by John Dee, Detail 34

FIGURE 2. *Monas Hieroglyphica,* by John Dee, Title Page 36

FIGURE 3. *Monas* Hieroglyphs 37

FIGURE 4. Ciphers 85

FIGURE 5. American Indian, by Charles Osgood 107

FIGURE 6. Medical Records, 1656 192

FIGURE 7. Winthrop Family Annotated Almanac 260

FIGURE 8. *A Funeral Tribute to . . . John Winthrope,* by Benjamin Tompson 304

Prospero's America

Introduction

The larger Atlantic world connections of colonization are now transforming Puritan studies. Colonial historians are rediscovering, although in new ways, something that Perry Miller noted more than two generations ago: New England's Puritans were continuing participants in a complex culture whose intellectual roots extended throughout Protestant Europe. This study adds another dimension to the discussion of this complex culture by demonstrating how one leading Puritan transferred Protestant alchemical beliefs and practices from the Old World to the New and how this Christian natural philosophy helped inform colonial expansion and influenced early colonial New England's culture.[1]

The life of John Winthrop, Jr., exemplifies the physical and intellectual links that spanned the Atlantic. Governor of Connecticut and son of the founding governor of Massachusetts, Winthrop was one of the most important men in colonial English America. Born in 1606, he was a political leader from his arrival in New England in 1631 until his death in 1676. He was also a leading undertaker of three new towns and the promoter of New England's first ironworks. Winthrop was a cosmopolitan intellectual and world traveler who journeyed through Europe and to the Middle East searching for knowledge of scientific mysteries. He was a successful suitor for a royal charter for Connecticut in the Restoration court of Charles II and a founding member of the Royal Society.

Like many natural philosophers of his age, Winthrop believed he lived in a time of special theological purpose, one in which God had elected to reveal again total knowledge of the natural and supernatural worlds. Such knowledge, once possessed by Adam, had been lost at the Fall, but it would be regained, many believed, through a process of research and discovery that would foreshadow Christ's Second Coming. Francis Bacon's call for an empirically based great instauration of knowledge and his utopian vision of a new Atlantis were rooted in this belief in providential intellectual renewal. So was

---

1. Perry Miller, *The New England Mind: The Seventeenth Century* (New York, 1939).

the fervor that greeted the appearance of the Rosicrucian manifestos, tracts of a supposedly secret group of European natural philosophers dedicated to restoring the "truth, light, life and glory" of the prelapsarian world.[2]

During the years when English Puritans were undertaking godly colonies in the Atlantic world, scientific reformers were working to synthesize Baconian empiricism and Rosicrucian millennialism into practical reform programs to improve world conditions in preparation for Christ's return. Winthrop was drawn to such religioscientific schemes. In his early twenties he began to study alchemy, the branch of natural philosophy many believed was the key to all understanding. Christian alchemists—those alchemists who believed God both influenced their quest for knowledge and intended their discoveries to be used to render Christian service to society—sought mastery over the natural world through the study and manipulation of the visible and occult forces permeating nature. While casual twenty-first-century observers often think of alchemy only as a vain and greedy quest to turn lead into gold, early modern practitioners thought of it as a legitimate, multidimensional science that could provide a profusion of benefits. Transmutation—turning lead into gold—was only one of alchemy's goals, and alchemy strove as much to achieve purity out of corruption as to generate any monetary value. Equally important was the search for the alkahest, the divinely granted elixir that would cure all diseases. Although Christian alchemists believed God granted only the most spiritually worthy adepts knowledge of such secrets, they also believed that, in the effort to attain them, alchemical practitioners were often given knowledge of lesser improvements with important practical benefits. Advances in medicine, mining, metal refining, husbandry, cloth dyeing, and military defense were common by-products of the chemical quest and important signs to alchemists that God was favoring their endeavors. Moreover, such discoveries could be instrumental in helping establish individual alchemists (not to mention whole societies) on a firm economic footing. The revenues from them could in turn fund the social amelioration many envisioned as part of their godly mission. This fusion of Christian quest for hidden knowledge with the simultaneous possibility of economic gain and utilitarian benefits helps explain alchemy's efflorescence among early-seventeenth-century European intellectuals and the profusion of programs that were advanced to perfect the world and hasten the millennium through alchemical study and its sequelae.

---

2. Francis Bacon, *The Great Instauration [1620]; and, New Atlantis [1627]*, ed. J. Weinberger (Arlington Heights, Ill., 1986); *Confessio Fraternitatis; or, The Confession of the Laudable Fraternity of the Most Honorable Order of the Rosy Cross, Written to All of the Learned of Europe*, trans. Thomas Vaughan (London, 1652), ed. Frances A. Yates, rpt. in Yates, *The Rosicrucian Enlightenment* (London, 1972), 256.

Following his emigration to America in 1631, John Winthrop, Jr., used alchemical knowledge as a foundation for Puritan colonization and economic development. Thereby, he and his alchemical associates helped shape New England culture in ways that have gone largely unnoticed by modern historians. This study examines five aspects of the alchemical practices of Winthrop and his associates to show the distinct and pervasive ways in which alchemical beliefs influenced colonial society.

The first two chapters explore the intellectual etiology of the occult alchemical philosophy to which Winthrop was attracted and his nearly lifelong participation in a pan-European network of alchemical practitioners who believed Christian alchemy could hasten the pansophic—that is, divinely sanctioned, knowledge-based—reformation of the human condition. It also locates these beliefs within the framework of evolving Puritan culture, to identify the points of concurrence and conflict that might occur between Christian alchemy and theology in Old and New England.

From the initiation of his alchemical studies in the 1620s, which he began with Edward Howes, his friend and fellow law student at London's Inner Temple, Winthrop was committed to the use of alchemy as a means of rendering Christian service and as a key to unlocking the hidden mysteries of nature. He and Howes became enthusiastic supporters of the Rosicrucian movement, which led Winthrop to undertake journeys to Europe and the Islamic world in search of alchemical knowledge. Among the important alchemical influences on Winthrop were Paracelsus, the English magus John Dee, the English physician and Hermetic theorist Robert Fludd, the utopian English alchemical entrepreneur Gabriel Plattes, and the great synthesist of the seventeenth-century universal reformation movement, Jan Comenius. Winthrop developed close relationships with alchemists throughout Protestant Europe and shared with them a vision of a world prepared for Christ's return through the alchemical recovery and deployment of practical scientific advances.

Winthrop's alchemical network reveals the importance of pan-European and transatlantic scientific alliances on New England colonization. Through relationships with Comenius, the circle of reformers such as Gabriel Plattes around the London-based German émigré Samuel Hartlib, alchemical friends in London, Hamburg, Amsterdam, New England, and the Caribbean, and in later years as a member of the Royal Society, Winthrop sought to use alchemy as a means of helping achieve the pansophic reformation of New England and the world while establishing the Puritan colonies on a sound and sustainable economic footing. For Winthrop, the goals of Christian reformation of the world and economic development were virtually synonymous. The pansophic goal of universal reformation framed an alchemical moral economy in which entrepreneurism in the service of Christian goals was not just justified; it was expected.

While Winthrop's commitment to Christian reformation was unwavering, his approach to religion was irenic. This was in keeping with both the Neoplatonic nature of his alchemical beliefs and the variety of sectarian interests—ranging from Dutch Reformed to Catholic—present in the Winthrop alchemical circle. It was not an indicator that Winthrop leaned toward radically enthusiastic religious movements such as Familism. Rather, Winthrop's irenicism reflected a Puritanism more in accord with mainstream members of the English Puritan movement such as Hugh Peter; William Fiennes, Lord Saye and Sele; and Robert Greville, Lord Brooke, than with the more narrowly defined Puritanism of Massachusetts Bay, and it became a factor fostering religious toleration elsewhere in the New England colonies.[3]

Chapters 3-5 focus on Winthrop's efforts, in the 1640s, to make New England a laboratory for alchemical transformation by creating a *new* London, where alchemists could collaboratively pursue scientific advances in agriculture, mining, metallurgy, and medicine. Several factors converged to make such a program seem both possible and necessary. The end of New England's Great Migration at the outbreak of the English Civil War had brought the region to the brink of economic collapse. The need to find some ways of stabilizing New England's economy coincided with the discovery of what appeared to be silver-bearing lead ore at a mine site in the interior of Massachusetts at the headwaters of the Pequot (now Thames) River watershed. Winthrop's return trip to England in 1641–1643 to find investors for a proposed New England ironworks and to confirm the silver potential of the lead ore occurred at a peak moment in the European pansophic movement. Samuel Hartlib's publication of Comenius's pansophic educational reform proposals, combined with Comenius's 1641 visit to England, filled many European alchemists with a fervor for universal reformation. Through the influence of Hartlib, Comenius, and Gabriel Plattes (whose *Macaria* outlined an alchemically centered program of economic development and societal improvement), Winthrop conceived of a plantation scheme that would involve exporting the silver-bearing lead ore through a port town to be established at the mouth of the Thames River watershed. The plantation, which Winthrop would call New London, would serve as an alchemical research center where improvements in agriculture, medicine, metallurgy, and other processes could be developed by a group of alchemical émigrés. Winthrop promoted this scheme to alchemical associates in England, Hamburg, and Amsterdam, and by the

---

3. On Winthrop and Familism, see Chapter 2.

time he returned to New England at least five alchemists had committed to joining him there. With Robert Child, an English alchemist who became an investor in both the ironworks and the lead mine, he began planning and implementing the alchemically based agricultural and industrial transformation of New England.

Two conflicts that had violently disrupted New England in 1637 reverberated with particular impact on Winthrop's new plantation. The aftermath of the Pequot War had destabilized already strained Indian relations in the proposed region of settlement. It had also created competing claims to the former Pequot lands between the colony of Connecticut (settled in the mid-1630s by dissatisfied Puritan emigrants from Massachusetts and Plymouth colonies) and its neighbor Massachusetts as well as among the colonies' former Indian allies. Winthrop's decisions to establish his new plantation in the heart of the conquered Pequots' territory and to serve as the self-appointed English protector of a surviving band of Pequots exacerbated these tensions and made his new plantation the focal point of serious intercolonial and intercultural conflict.

At the same time, the 1637 free grace controversy surrounding Anne Hutchinson had left in its wake still-unresolved questions about the limits of acceptable Puritan practice in New England. Winthrop, despite his personal adherence to Puritanism, recruited alchemists for his project who professed a broad range of confessional beliefs. The possibility of embracing such religious variety in New England caused some Puritans great concern, especially after Winthrop's alchemical partner Robert Child directly challenged Massachusetts's restrictive church membership and political enfranchisement policies. In the wake of Child's remonstrance, Connecticut's wariness of Winthrop's project sharply increased, and Massachusetts temporarily but harshly cracked down on some of the alchemists in its colony. The newly formed United Colonies of New England also acted in surprisingly harsh ways to curtail the ambitions of Winthrop's new plantation, even to the point of sanctioning Indian raids against the new settlement. Winthrop's pansophic errand into the wilderness became for a time a project that pitted colony against colony, Puritan against Puritan, science against religion, Indian against Indian, and, in the case of young Winthrop himself, father against son. While New England's anti-alchemical backlash was short-lived, it had a significant restraining effect on the start-up of the New London plantation.

Although for a time the New London venture appeared to be in jeopardy and most of the European alchemists ultimately chose not to emigrate, Winthrop persevered, pursuing an array of alchemically oriented projects and successfully developing a homegrown network of Puritan alchemical practitioners, who sought through alchemy to improve colonial and world living conditions. Although New London never fully realized Winthrop's vision, it nevertheless pointedly underscores the significance of alchemy in colonial New England settlement, economic development, intercolonial political relations, and intercultural diplomacy.

One of the ways in which Winthrop moved from a suspect newcomer to the beloved and perennially reelected governor of colonial Connecticut was through his administering alchemical medicines. Despite the occasional misgivings of some, most New Englanders came to have a deep appreciation for what they believed were the powerful curative effects of the alchemical medicines dispensed by Winthrop and other alchemical medical practitioners. Chapter 6 examines Winthrop's role as one of New England's most sought-after physicians and the importance that divinely derived alchemical medicines came to have in New England's highly providential medical environment. Alchemical medicines—derived from minerals and metals, unlike the herbal medicines that made up much of the colonial pharmacopoeia—were sought for both the powerful effects they produced on patients and the divinely sanctioned healing qualities they were believed to possess. Many alchemical medicines caused violent purgative reactions, which were interpreted as signs of the medicines' remarkable sanative qualities. Christian alchemists held that God had revealed these new medicines to them as a special gift for use in rendering Christian service to their fellow humans. In New England, which had a heightened sense of God's direct intervention into disease as a result of the selective contagion seen during the Indian epidemics, medical treatments deemed to be derived from direct divine sanction carried special significance. As one of New England's leading practitioners of alchemical medicine, Winthrop was inundated with medical requests. New London became a hospital town to which patients came from all over New England seeking cures for a host of medical conditions. To help meet the New England-wide demand for his medicines, Winthrop distributed them through a network of female practitioners, elite wives who incorporated Winthrop's color-coded packets of medicines into their own healing services. Both Winthrop and the elite wives distributed the medicines as a Christian service, without expectation of payment, a form of benevolence that distinguished them from the male and female doctors who healed for payment, acts of benevolence that further reinforced their families' status as community social and political leaders.

Winthrop's alchemical knowledge, coupled with his role as a political leader, gave him significant cultural as well as legal authority in the determination of witchcraft cases. Chapter 7 details Winthrop's forceful intervention to bring to an end the Hartford witch-hunt of the early 1660s. Prior to his becoming governor, the colony of Connecticut had been New England's fiercest persecutor of suspected witches. Between 1647 and 1654, seven people were tried for witchcraft in Connecticut and New Haven colonies. (New Haven would be absorbed into Connecticut in 1665.) All were convicted and hanged. Between 1655 and 1661, four more persons were prosecuted for witchcraft, and all were acquitted, thanks to the efforts of Winthrop. While he was in England seeking a royal charter for Connecticut, another witch-hunt broke out in Connecticut, and, by the time he returned, six women had been convicted of witchcraft, and four hanged. Immediately upon his return, Winthrop intervened forcefully to overturn witchcraft convictions and assure the safety of suspected witches. Winthrop's resistance to witchcraft convictions was based, not on a lack of belief in magic, but rather on the knowledge—confirmed by his study of alchemy, natural magic, and other occult philosophies—that manipulation of the occult was complex and difficult and that most charges of witchcraft were unfounded. With the assistance of the alchemist-minister Gershom Bulkeley, Winthrop helped create a definition of diabolical witchcraft that would end witchcraft executions in Connecticut permanently and help end them in all New England for more than a generation.

The final chapter again focuses on the transatlantic dimensions of alchemical culture, by examining Winthrop's election in 1662 as the first colonial fellow of the Royal Society. Because so many members of the newly chartered Royal Society held important positions within Charles II's agencies of colonial regulation—the Council for Plantations, the Board of Trade, and the Corporation for the Propagation of the Gospel—the Royal Society served as an important avenue of colonial political patronage. Winthrop's relationships with natural philosophers such as Robert Boyle, Sir Robert Moray, William Brouncker, and members of the Hartlib circle such as Samuel Hartlib and Benjamin Worsley were instrumental in helping him quickly secure a charter for Connecticut in 1662. He obtained a remarkably favorable charter as well, one that incorporated into the boundaries of Connecticut Colony all of New Haven Colony and almost half of Rhode Island.

The Connecticut charter represented Whitehall's first effort to implement in New England a new policy of centralizing and regulating the exercise of colonial authority. The New England colonies were an important part of the crown's plans for increasing the value of colonies through better regulation. Among the earliest, most-populated, and most religiously inclined plantations, New England was seen by both Whitehall and the members of the Royal Society as a prime venue for improving trade, commerce, and evangelization of the Indians. On the other hand, the Puritan colonies' reputation for religious intolerance, coupled with their traditions of relative political autonomy, made them a test case for the imposition of greater crown control.

In New England as elsewhere, the crown's knowledge of its plantations was abysmal. The most basic information about colonial governments, natural resources, populations, and productive potential was unavailable. For this reason, one of the crown's primary charges to all its colonial agencies was to collect relevant information about each colony. As a result of this knowledge imperative, the Royal Society, in addition to serving as a clearinghouse for scientific communication, acted informally as an essential intelligence-gathering agency for Whitehall.

The product that most clearly represented the fusion of political interest and natural philosophy was the natural history. Gerard Boate's 1652 *Irelands Naturall History* had set a standard for practical intelligence gathering. While he was in England, Winthrop was groomed by the members of the Royal Society to become the natural historian of New England. Winthrop valued his membership in the society, and he shared its pansophic vision. The society valued Winthrop's abilities as an alchemist and his expertise in mining and mineral matters. His unique status as a natural philosopher of *New England*, however, was what interested them most. Consistently, Winthrop was asked to report to the society about New England products, processes, and natural resources. When he returned to New England, it was with a set of specific instructions on what information he was obligated to report back.

As long as the imperial agenda of Whitehall and the territorial aspirations of Connecticut were in accord, Royal Society–Whitehall patronage was extremely advantageous to Winthrop. He took home to Hartford a charter for which he was lionized. Within a year of his return, however, his relationship to the same scientific patronage network became much more complex.

In June 1664 a fleet arrived in New England with royal commissioners sent to enforce liberty of conscience and oaths of allegiance and to settle colonial boundary disputes. In the same fleet came Richard Nicolls, military governor of New York under a new patent issued to James, duke of York. The York patent invoked imperial consolidation on a grand scale, incorporating into one patent all the land from Maine to the Delaware, including all of Connecticut's land west of the Connecticut River. Only the New England colonies (except, of course, for half of Connecticut) were unassimilated. The foundation for the York patent had been a short natural history of New England composed by Samuel Maverick, one of the royal commissioners. Polemical, biased against New England's Independents, and inconsistent in presentation, Maverick's account had nevertheless underscored the power of information in the service of empire.

Requests for Winthrop to provide a natural history of New England and information about New England's mineral wealth were consistent and often-strident features of every letter he received from the society. From the internal communications of Royal Society members, it is clear that obtaining a natural history was seen as a means of initiating consolidation of the colonies. Finding real evidence of mineral wealth could be the trigger that would initiate such consolidation.

To maintain the society ties he valued and the patronage ties he needed—so useful in dealing with the royal governors of New York who were making their own efforts to incorporate Connecticut into a consolidated government—Winthrop developed skillful strategies of cooperation and resistance. He ignored requests for natural histories (as long as he could) and substituted for them natural curiosities, which provided the color of New England but no valuable information that could be used in the way Maverick's or Boate's histories had been used: to undermine local autonomy. On mining matters, the inveterate promoter of New England mining schemes became a skeptic and cynic, arguing that New England had no readily accessible mineral resources. Winthrop provided astronomical information, information on processing, and natural wonders; any information that was not "a matter of present utility," he willingly contributed for the advancement of knowledge. Through these strategies he was able to maintain his membership in the Royal Society and to fight a stalling action against the imposition of greater imperial authority in Connecticut.

In each of the chapters in this study, alchemy, all but invisible to nineteenth- and twentieth-century historians of New England, plays a central role in one or more aspects of early colonial New England's cultural formation. Winthrop, as the foremost among the alchemists who participated in the development of New England, provides an important lens through which we can see an understudied and undervalued aspect of New England's crucial formative period. Winthrop and his associates had no doubt about the importance of alchemy to their world. Later historians, who from the other side of the Enlightenment came to view alchemy only as false magic and pseudoscience, separated the subject from its foundations in faith and utility. In doing so, they also banished alchemy from its central supporting role in the development of a self-consciously religious colonial society. This is not the first effort to refresh our understanding of alchemy's important presence in New England, though it is the first effort to locate New England alchemical study within its broader cultural context. As such, it is only a beginning, and the author is conscious that there is much more to learn about this surprisingly important topic. The author hopes this study will encourage continuing scrutiny and help make the once invisible world of New England alchemy highly visible once again.

> MIRANDA: O, wonder!
> How many goodly creatures are there here!
> How beauteous mankind is! O brave new world,
> That has such people in't!
> PROSPERO: Tis new to thee.[4]

ISSUES, CHALLENGES, AND METHODS

Two issues encountered during research made writing a book about John Winthrop, Jr., and his practice of alchemy particularly challenging. The first was the nature of the voluminous correspondence regarding Winthrop found in the Winthrop Papers; the second was Winthrop's unwavering commitment to the tradition of alchemical secrecy, of keeping the disclosure of alchemical information hidden from those who might be unworthy of receiving it or who might use such knowledge for improper purposes.

The Winthrop Papers are both a blessing and a curse. This multithousand-item collection of letters, account books, manuscripts, images, maps, and books at the Massachusetts Historical Society in Boston is a treasure trove of information concerning the Winthrops and their world, an almost unparalleled window into the details of the quotidian experience of one of colonial New England's most distinguished families. Through them, a careful student can reconstruct the lived experience of the Winthrop family with a depth impossible for any other seventeenth-century New England family. Because of the extensive nature of this collection, I was able to reconstruct with substantial accuracy important features of Winthrop's life as an alchemist: his network of alchemical correspondents both overseas and in America, the scope and relative importance of his alchemical entrepreneurial activities, the vital cultural significance of his alchemical medical practice, and the courtly tug-of-war he engaged in with members of the Royal Society over the natural history of New England. Yet, despite the wealth of information the Winthrop Papers provide, the gaps are infuriating. Many of the letters in the collection—especially, it seems, those from alchemical correspondents—are letters written *to* Winthrop in response to a missal *from* him that is no longer extant. We get responses to the ideas Winthrop expressed, follow-on thoughts, harsh critiques, or enthusiastic support, but what we don't get are the ideas them-

---

4. William Shakespeare, *The Tempest*, 5.1.

selves. We are often left to extrapolate Winthrop's beliefs, not from what he says or writes, but from what others say about them and how they react to them in writing. This is most often the case when it comes to figuring out the details of Winthrop's specific alchemical practices and experiments. Although he does occasionally reveal important specifics of his scientific practices or his philosophical beliefs, much of what we know is based upon secondhand information from those with whom Winthrop talked, worked, and experimented.

In part, this lacuna of personal information was intentional. Winthrop's own correspondence unequivocally shows that he believed in the importance of maintaining alchemical secrecy. The idea that information about alchemical experiments was too valuable and potentially too powerful to be shared with the unworthy was common among occult philosophers. Winthrop clearly preferred to communicate information about his experiments orally rather than through writing; he frequently mentions in letters to other alchemists that he has information too significant to put in writing but will share it in person when he and his correspondent are together. As a result, the Winthrop historian sifts through the evidence he has about the evidence he doesn't have and hopes he has correctly interpreted his clues.

Fortunately, between Winthrop's occasional comments about his own beliefs and his correspondents' comments about those beliefs, the marginalia found in his alchemical library, and a survey of his medical account books, a composite picture of Winthrop as an alchemist emerges, and that composite picture frames this analysis. Some may disagree with details of the interpretation advanced here, and future research may, and I hope will, help to clarify specifics, but I am confident that Winthrop himself would recognize his own views in most, if not all, of the positions I have attributed to him.

In describing Winthrop's alchemical philosophy, historians of science will see that I am employing a new term, *Christian alchemist*. I use this descriptor to clarify the nature of Winthrop's and many of his contemporaries' melding of religion and science and to distinguish it from two other widely advanced interpretations of early modern alchemy. The first of these interpretations has held that alchemy was primarily, if not exclusively, a spiritual pursuit, whose goal was less the transformation of matter than the transformation and purification of the soul. Elaborated in the psychoanalytic theories of Carl Jung and the cultural studies of Mircea Eliade, this interpretation of *spiritual alchemy* relegated the technological and productive motives behind alchemical research to insignificant status. What really mattered at the alchemical hearth, according to the most extreme interpretations of this view, was the psychic and spiritual transformation of the alchemist himself, not the physical experiment itself or its outcome. This in turn helped explain the obscure and enigmatic allegorical abstruseness of most alchemical texts. The extraordinary work of William Newman and Lawrence Principe has done much to undermine this long-held view of alchemical study. Not only have they shown that supposedly enigmatic alchemical texts actually encoded sophisticated chemical laboratory processes and experiments; they have traced the etiology of spiritual alchemy, not to early modern writers, but, rather, to Victorian-era interpreters of those writers, who imposed on the original texts fundamentally esoteric and occultist interpretations.[5]

In sharp reaction against the spiritual alchemy perspective, scholars have recently advanced new interpretive models stressing alchemy's utilitarian value and how it was both patronized and practiced for its economic value. Such studies dramatically downplay the presence of religious motivations behind alchemical research, often presenting it as a fundamentally secular enterprise.[6]

Christian alchemy, the term I am employing to describe the general alchemical philosophy of John Winthrop, Jr., and many of his alchemical associates, seeks a middle path between these two models. It fully embraces the concept that alchemy was pursued as a practical, utilitarian enterprise with the goal of economic gain but simultaneously recognizes that such pursuits neither ruled out nor contrasted with the pursuit of alchemy for religious ends. A reading of any representative sample of alchemical writings from the early modern period reveals that many, if not most, alchemical writers viewed God as an active agent in the alchemical quest and that God intended alchemical knowledge to be the province of pious practitioners who would be dedicated to using the fruits of their quest for godly ends. In this view economic gain from alchemical discovery was welcomed because it provided means through which godly works of reformation and social amelioration could be accomplished. Personal spiritual transformation might also follow from pious experimentation, but it, like economic gain, was not an end in itself. Christian alchemists sought through their experiments, discoveries, and application of

---

5. See, for example, C. G. Jung, *Alchemical Studies,* trans. R. F. C. Hull, The Collected Works of C. G. Jung, XIII (Princeton, N.J., 1967); Mircea Eliade, *The Forge and the Crucible: The Origins and Structures of Alchemy,* trans. Stephen Corrin (Chicago, 1978); Lawrence M. Principe and William R. Newman, "Some Problems with the Historiography of Alchemy," in Newman and Anthony Grafton, eds., *Secrets of Nature: Astrology and Alchemy in Early Modern Europe* (Cambridge, Mass., 2001), 385–431.

6. Pamela H. Smith, *The Business of Alchemy: Science and Culture in the Holy Roman Empire* (Princeton, N.J., 1994); Pamela H. Smith and Paula Findlen, eds., *Merchants and Marvels: Commerce, Science, and Art in Early Modern Europe* (New York, 2002); Harold J. Cook, *Matters of Exchange: Commerce, Medicine, and Science in the Dutch Golden Age* (New Haven, Conn., 2007).

alchemical knowledge to do God's work in the world, by engaging in practical activities that in no way ruled out making money from them. While critics might accuse them of greed on the one hand or radical religious enthusiasm on the other, in reality they operated within a context where economic gain and pious experimentation were considered compatible, natural, and productive. This, I believe, was the philosophical framework that informed the activities of John Winthrop, Jr., Robert Child, Samuel Hartlib, Johann Moraien, and most of their fellow alchemists of the early modern period. By using the term *Christian alchemists,* I do not mean to imply that all alchemists were Christians, or that all alchemists who were Christian held the views I have just described. It is, rather, a shorthand phrase intended to suggest that much of early modern alchemy was pursued for simultaneously practical, economically productive, and godly ends.

I have adopted certain conventions in the writing of this book. For dates, I have kept the day and month as listed on the original documents, even when the date was in Julian, or Old Style. I have, however, when necessary, adjusted the year to New Style to reflect the Gregorian calendar's New Year date of January 1 (against the Julian calendar's March 26). For example, a letter dated March 23, 1630, Old Style, in this study is dated March 23, 1631.

In transcribing quotations, I have in most cases retained the orthography and punctuation exactly as they were produced in the source from which I derived the quotation. The one consistent departure from direct transcription is in my automatic modification of the thorn, or *Y* words ("yt," "ym," and "ye") to their current equivalents "that," "them," and "the."

Finally, this study falls in among, without wholly resting within, several disciplinary approaches. It is not a history of science, though it hopes to contribute modestly to that field and is greatly informed by that discipline's vast literature on early modern science and alchemy. It is not an ethnohistory, though it hopes to offer a somewhat new understanding of the forces influencing inter- and intracultural relations in early New England. It is not a religious history, though it hopes to offer some insight into the religious impulses framing New England's early years. Nor is it a political history, though it hopes to add to our understanding of intercolonial and transatlantic political changes during the early colonial period. This book is a cultural history about John Winthrop, Jr., and the influence of alchemy on the creation of early New England's culture. As such, it touches on a number of disciplines, and dances, not always in perfect harmony, with several intellectual partners. I hope the reader agrees that the insights derived from the breadth of this study have made the interdisciplinary boundary crossing worthwhile.

(ONE)

John Winthrop, Jr., and the European Alchemical Movement of the Early Seventeenth Century

Today most historians of science view alchemy as an important contributing factor in the development of modern chemistry and experimental science. While they are still working out the exact nature of alchemy's contributions and the complex motivations leading early modern Europeans to pursue the alchemical quest, the generally positive current attitudes of historians toward alchemy differ markedly from the views prevailing only a generation ago. Then, and for a very long time before that, alchemy was lumped together with pursuits such as astrology, geomancy, Cabala, and other occult arts and dismissed as pseudoscience. A great deal of careful work by a generation of scholars less committed than their forebears to presenting scientific development as the march of progress and a victory of reason over superstition has helped secure the newfound respect for alchemy and its practitioners and for their role in the transformation of natural philosophy into modern science.

Greater recognition of the widespread practice of alchemy in early modern England has also helped to mute, if not fully resolve, long-standing debates about the relationship of alchemy to Puritanism and of Puritanism to the rise of modern science, especially among historians of seventeenth-century England. Whether modern science emerged from the spread of radical Puritan values or from the moderate, tolerant, and primarily Anglican beliefs of the post-Restoration period was a highly contentious issue that spilled over from sociology into history and the history of science during the 1960s and 1970s. It remained a heated source of scholarly debate well into the 1990s. In this debate, alchemists—often seen as closely aligned to radical Puritanism—were also seen as a drag on scientific development. Their theosophical, pseudoscientific pursuits were characterized as impediments to the real scientific progress being made by natural philosophers of more tolerant, latitudinarian religious beliefs. The reconceptualization of alchemy as useful and contributing, coupled with the recognition that it was practiced by English elites professing a very broad spectrum of religious beliefs, has not fully ended the argument, though it has substantially undermined its fundamental premises.[1]

Oddly, the long and contentious debate about alchemy, Puritanism, and science in early modern England had no influence on scholars of Puritanism in America. Convinced by an older historiography that alchemical study was at its core a diabolical form of Faustian conjuring, they conceived of Puritanism and alchemy—if they thought of the latter at all—in a false binary opposition: religion versus magic. Alchemy, these colonial scholars averred, was a form of magic. Since New England's Puritans had no tolerance for magic, it was assumed they had little tolerance for alchemy and its related practices.[2]

Important studies by William Newman and Patricia Watson have demonstrated the widespread presence of alchemical practitioners among New England's Puritan ministerial and medical elites, and Newman has underscored the oversimplification that derives from equating alchemy with magic. Yet, despite what would seem like mortal blows to the religion-versus-magic binary, that view still has considerable purchase. As a result, analysis of alchemy's and alchemists' roles in colonial settlement and of alchemy's influence on cultural formation in New England is still far from fruition.[3]

---

1. Stephen Cole, "Merton's Contribution to the Sociology of Science," *Social Studies of Science*, XXXIV (2004), 829–844; J. Andrew Mendelsohn, "Alchemy and Politics in England, 1649–1665," *Past and Present*, no. 135 (May 1992), 30–78; Margaret C. Jacob, *The Cultural Meaning of the Scientific Revolution* (Philadelphia, 1988), 82–83; Robert K. Merton, *Science, Technology, and Society in Seventeenth-Century England* (New York, 1970); Barbara J. Shapiro, "Latitudinarianism and Science in Seventeenth-Century England," in Charles Webster, ed., *The Intellectual Revolution of the Seventeenth Century* (London, 1974), 287; Christopher Hill, *The World Turned Upside Down: Radical Ideas during the English Revolution* (London, 1972); Hill, "Puritanism, Capitalism, and the Scientific Revolution," in Webster, ed., *The Intellectual Revolution of the Seventeenth Century*, 3; Max Weber, *The Protestant Ethic and the Spirit of Capitalism*, ed. Talcott Parsons (New York, 1958); R. H. Tawney, *Religion and the Rise of Capitalism* (New Brunswick, N.J., 1998).

2. Scholars of American Puritanism did not engage English debate on Puritanism and science, in part because of different historiographic conceptualizations of Puritan politics on different sides of the Atlantic. To English scholars, Puritan Independents were seen as members of the radical fringe. In New England, Independents controlled colonial governments in Massachusetts and Connecticut and have been considered the orthodox establishment.

3. John L. Brooke, *The Refiner's Fire: The Making of Mormon Cosmology, 1644-1844* (Cambridge, 1994); William R. Newman, *Gehennical Fire: The Lives of George Starkey, an American Alchemist in the Scientific Revolution* (Cambridge, Mass., 1994); William R. Newman and Lawrence M. Principe, *Alchemy Tried in the Fire: Starkey, Boyle, and the Fate of Helmontian Chymistry* (Chicago, 2002); Patricia A. Watson, *The Angelical Conjunction: The Preacher-Physicians of Colonial New England* (Knoxville, Tenn., 1991); Joyce E. Chaplin, *Subject Matter: Technology, the Body, and Science on the Anglo-American Frontier, 1500-1676* (Cambridge, Mass., 2001); Neil Kamil, *Fortress of the Soul: Violence, Metaphysics, and Material Life in the Huguenots' New World, 1517-1751* (Baltimore, 2005); Catherine L. Albanese, *A*

The English historical argument aligning alchemy with Puritan radicalism and the American historical argument that there was an antithesis between Puritanism and magic made it easy to overlook John Winthrop, Jr., and his fellow New England alchemists. Winthrop was a New England Puritan, a moderate Puritan, and an alchemist: as such, he was not easily located within the framework of either older debate. He was not a cultural anomaly, however. Winthrop was the leading exemplar of a group of Puritan Christian alchemists throughout seventeenth-century New England who helped shape the region's cultural, social, and political development.

Winthrop and his alchemical colleagues did not question whether alchemy was "scientific" or whether it was consistent with Puritan values. They understood alchemy to be a progressive, intellectual, immensely utilitarian but simultaneously spiritual undertaking of the utmost importance. They saw themselves as enlightened beings and lived in hope of achieving scientific advances of both immediate practical value and eternal importance. The alchemist's furnace was, to paraphrase Walter Pagel's description of natural philosophy, the place where grace from above met human aspiration for knowledge from below, a connecting link with divinity. It was also the place where real solutions to pressing current problems—economic, medical, agricultural, and metallurgical—might be realized at any time. The dual combination of potential spiritual illumination and practical problem solving and economic opportunity gave alchemy a unique role in New England's colonial cultural formation, one that produced substantial and lasting effects.[4]

The roots of Winthrop's alchemical beliefs may usefully be located in the challenge to medieval Scholasticism that began in Florence in the late fifteenth century. There, a group of Renaissance intellectuals led by Marsilio Ficino developed a new conception of humankind's relation to the cosmos, which, in contrast to the Aristotelian cosmology of the Scholastics, offered people newfound power to manipulate and control their natural environment. An essential component of this Neoplatonic movement was the translation and reinterpretation of works attributed to Hermes Trismegistus, an Egyptian magus—who is now known to be mythical, but who was believed by many early modern intellectuals to have been a real contemporary of Moses—who had prophesied the coming of Christ. Hermes was said to have possessed all knowledge, including the knowledge lost after Adam's fall from grace, and

---

Republic of Mind and Spirit: A Cultural History of American Metaphysical Religion (New Haven, Conn., 2007), 69–75.

4. Walter Pagel, "Religious Motives in the Medical Biology of the XVIIth Century," *Bulletin of the Institute of the History of Medicine*, III, no. 4 (April 1935) 97–312.

to have exercised godlike power through the use of astral magic, focusing the powers of stars and planets to reshape and accelerate nature's normal activities. The *Corpus hermeticum,* rediscovered and translated by Ficino, stressed humankind's ability to achieve dominion over nature through magical-religious communion with the cosmos. Through incantations, talismans, and such, the effluxes of the stars and the power of the planets, which operated upon all people and all things, could be harnessed to serve human ends. Hermeticism offered intellectuals steeped in the tradition of a divinely regulated cosmos new possibilities for interacting with and controlling the forces of nature.[5]

Although Hermes, because of his prophetic knowledge of Christ's appearance, was regarded as a divinely inspired ancient theologian, it was important for Renaissance Neoplatonism to connect the Hermetic tradition more closely to the Judaeo-Christian theological tradition. This was accomplished by Pico della Mirandola, a Florentine scholar who superimposed on the Hermetic belief in astral and natural magic a religious magic derived from the Hebrew Cabala. Cabala was a numerological magic based on the supposed hierarchies of heavenly angels, the twenty-two letters of the Hebrew alphabet, and the ten known names of God.[6]

In his 1533 *De occulta philosophia* Heinrich Cornelius Agrippa synthesized and extended the work of Pico and Ficino into a complete and integrated cosmology. With Agrippa, occultism, religion, and the utilitarian investiga-

---

5. Historians of science will readily recognize the roots of this etiology of Winthrop's alchemical beliefs in the delineation of the Hermetic tradition as outlined by Frances Yates. Although the Yates thesis has been subject to substantial critique and revision by a number of historians, others still recognize its continuing usefulness. In the case of Winthrop, the intellectual genealogy suggested by Yates remains apt, since he identified himself at various times as a Hermeticist, a proponent of the macrocosm-microcosm theory, sought repeatedly to make contact with the Rosicrucians, and was a self-professed disciple of John Dee. Brian P. Copenhaver delineates the complex origins, development, and division of both the Hermetic myth and the Hermetic texts in Copenhaver, trans., *Hermetica: The Greek "Corpus Hermeticum" and the Latin "Asclepius"* (Cambridge, 1992), xiii–lxi. Frances A. Yates, *The Rosicrucian Enlightenment* (London, 1972); Yates, *Giordano Bruno and the Hermetic Tradition* (London, 1964); Hermann Grieve, "Die christliche Kabbala des Giovanni Pico Della Mirandola," *Archiv für Kulturgeschichte,* LVII (1975), 141–161. On the critique of Yates, see John Clulee, *John Dee's Natural Philosophy: Between Science and Religion* (London, 1988); Brian Vickers, ed., *Occult and Scientific Mentalities in the Renaissance* (Cambridge, 1984), 1–55. On continuing usefulness of the Hermetic worldview, see Stuart Clark, *Thinking with Demons: The Idea of Witchcraft in Early Modern Europe* (Oxford, 1997), 157, 227–228; Lauren Kassell, *Medicine and Magic in Elizabethan London: Simon Forman: Astrologer, Alchemist, and Physician* (Oxford, 2005), 9–10.

6. Francis A. Yates, "The Hermetic Tradition in Renaissance Science," in Charles S. Singleton, ed., *Art, Science, and History in the Renaissance* (Baltimore, 1967), 258. On the further development of Christian Cabala by Johannes Reuchlin and Johannes Trithemius and its use by the English alchemist John Dee, see Deborah E. Harkness, *John Dee's Conversations with Angels: Cabala, Alchemy, and the End of Nature* (Cambridge, 1999), 157–194.

tion of nature became intertwined. The boundary between the natural and the magical was believed to be porous and diffuse. Mathematics, because of its abstraction and incorruptibility, was considered a kind of magic; mechanics, too, a realm affected by magical forces. The natural world itself was seen as active, animate, and psychic—natural objects, from stars to plants and animals, interacted and influenced one another's behavior.[7]

Renaissance Neoplatonic Hermeticism as developed by Ficino, Pico, and Agrippa provided a philosophy in which alchemical studies could flourish, but it did not give primacy to the practice of alchemy. That was done by a controversial, itinerant Swiss scholar-mystic-alchemist-physician named Philippus Aureolus Theophrastus Bombastus von Hohenheim. Paracelsus, as he came to be known, was raised by a father interested in transmutation and was apprenticed at an early age to serve in the Fugger mines near Villach, Austria. He was strongly influenced by the emerging artisanal culture of the sixteenth-century Holy Roman Empire, in particular the new emphasis naturalist artisans such as Martin Schongauer and Albrecht Dürer placed upon engagement with nature as the source of true knowledge. In sharp contrast to the centuries-old view prevailing among most European intellectuals that reliable knowledge was derived primarily from purely deductive reasoning, these late-Renaissance proponents of naturalism insisted that observation of and experiment with the natural world was the one reliable means for acquiring the deep knowledge God had encoded into all things. Many of these artisans considered alchemy—because it involved the experimental investigation into the properties bound within matter—both an ideal symbol of and method for engaging in this new investigation of nature. Paracelsus not only fully embraced the view that knowledge of God's creation was best obtained by direct investigation into nature; he developed an intense interest in alchemy and applied it to the medicine he practiced after receiving his medical education at Ferrara, Italy, between 1513 and 1516. By his early thirties, Paracelsus had incorporated Cabala magic and Hermeticism into a mystical chemical philosophy that emphasized alchemy and iatrochemistry—medical alchemy—a radical new approach to healing that stressed a chemical rather than herbal orientation to healing.[8]

---

7. Harkness, *John Dee's Conversations with Angels*, 157–194; Richard S. Westfall, "Newton and the Hermetic Tradition," in Allen G. Debus, ed., *Science, Medicine, and Society in the Renaissance: Essays to Honor Walter Pagel* (London, 1972), II, 184.

8. Ole Peter Grell, ed., *Paracelsus: The Man and His Reputation, His Ideas, and Their Transformation*, Studies in the History of Christian Thought, LXXXV (Leiden, 1998), 1–18; Pamela H. Smith, *The Body of the Artisan: Art and Experience in the Scientific Revolution* (Chicago, 2004), 59–95, 155–183; Paracelsus, *Selected Writings*, ed. Jolande Jacobi, (Princeton, N.J., 1979), xxxix–xlvi. Grell's volume

Paracelsus's chemical philosophy, with its pious insistence on discovering the divine keys to the workings of creation through the practical investigation of nature, provided the first truly innovative medical ideas in centuries, and knowledge of his works spread rapidly throughout Europe. By the later sixteenth century, Paracelsianism was a powerful and spreading intellectual movement. The key to its success lay in the way in which it wove so many different strands of Renaissance intellectual thought — mysticism, experimental investigation into nature, astrology, Hermeticism, magic, and alchemy — into an intensely utilitarian theosophy compatible with the rapidly expanding Protestant Reformation.[9]

Paracelsus's primary concern was with man's relation to God. He believed that unity and harmony in the universe were the intentional creation of God's wisdom and benevolence. The individual was a microcosm of the cosmos, center of a complex system of correspondences between the celestial and terrestrial worlds. Paracelsianism posited a world of sympathies and antipathies — of attraction and repulsion between objects and the unseen influences that emanated from them — that made possible astrology, alchemy, and the practice of all kinds of magic. Humans had the ability to gain power over the natural, celestial, and even the supernatural worlds, yet attaining that power was entirely dependent on the will of the Creator. The mission of the Paracelsian was to bring purity to a world enveloped in "a darkness that strives for light"; mankind was divinely charged with obtaining "the understanding and the fulfillment of the world."[10]

Such fulfillment would come about as the result of careful experimentation and firsthand observation of the natural world. Members of the plant and the mineral kingdoms were to be closely investigated to determine the celestial objects with which they had sympathetic correspondence; mathematics was to be studied to reveal the divine harmonies resonating between the macrocosm and the microcosm. Alchemy was the essence of this experimental quest for fulfillment, a material, psychic, and spiritual search for both practical knowledge and spiritual purification. The gradual transformation of

---

delineates the significant modern reconsideration of Paracelsus, in which he is seen less as a natural philosopher with clearly delineated ideas and more as an ambiguous figure posthumously co-opted by proponents of a variety of ideologies, approaches, and disciplines.

9. Smith, *Body of the Artisan*, 149–183; Allen G. Debus, *The English Paracelsians* (London, 1965), 36–39.

10. Paracelsus, *Selected Writings,* ed. Jacobi, xlvi, xlviii; Keith Thomas, *Religion and the Decline of Magic* (New York, 1971), 223; William R. Shea, "Trends in Interpretation of Seventeenth-Century Science," in M. L. Righini Bonelli and Shea, eds., *Reason, Experiment, and Mysticism in the Scientific Revolution* (New York, 1975), 3–4; Charles Webster, *The Great Instauration: Science, Medicine and Reform, 1626-1660* (New York, 1976), 286.

base matter, *materia prima,* into *materia ultima,* the purest form of matter, whose attainment represented the ultimate goal of the alchemist, was seen as the symbolic pattern followed by everything in creation. Paracelsians believed the earth was animate, and every plant, stone, animal, and mineral in it was striving to achieve its perfect state. The alchemist was one who, in a literal way, hastened imperfect metals on their path to purity. Paracelsus called alchemy "the highest and greatest *mysterium* of God, the deepest mystery and miracle that He has revealed to mortal man." Alchemists, through distilling the spiritual principle, the essence, from various minerals and species of plants, could recover the perfect knowledge of medicines once known by Adam but lost through his fall from God's grace and turn impure metals into precious gold and silver. The ingredient most useful in achieving *chrysopoeia,* the transformation of the five base metals (copper, lead, tin, iron, and quicksilver, or mercury) into gold, was the philosopher's stone. Attaining knowledge, through study and experiment, of how to produce this stone was one of the two great goals of alchemy. The other great goal of the alchemist, also attained by diligent and pious labor at the furnace, was the alkahest, an elixir, or universal solvent, that would cure all diseases. Paracelsus insisted that achieving either the philosopher's stone or the alkahest required, in addition to a great deal of technical knowledge about chemical processing, absolute purity on the part of the alchemist. Experiments simply would not work unless the adept had first purged himself of all vices. For that reason, the alchemical quest was undertaken by many as an arduous spiritual pilgrimage. It was, however, a spiritual pilgrimage only in part, and only in tandem with far more practical and potentially profitable goals.[11]

Never far distant from the spiritual side of the alchemical quest was its utilitarian imperative: the conviction that the search for and discovery of new chemicals, medicines, production processes, industrial technologies, and related inventions would help improve the human condition, even as they potentially returned a handy profit. The moral economy of Christian alchemy did not rule out gain, nor did it turn a blind eye to the economics of chemical discovery. It was assumed that products and processes that bettered human life could also produce economic returns that would support further experiment and possibly fund other efforts aimed at social amelioration. The profit mo-

---

11. Debus, *The English Paracelsians,* 18–19; Paracelsus, *Selected Writings,* ed. Jacobi, 215–219; Webster, *The Great Instauration,* 285; Paulo A. Porto, "'*Summus atque felicissimus salium*': The Medical Relevance of the *Liquor Alkahest,*" *Bulletin of the History of Medicine,* LXXVI (2002), 1–29; Thomas, *Religion and the Decline of Magic,* 269–272.

tive, far from being alien to the Christian alchemical quest, was an essential though subordinate component of it.

Historians of science have emphasized the economic drives that underpinned many early modern patrons' support for alchemists and their experiments. They have also shown that there were in early modern Europe a substantial number of technical alchemists—craftsmen and tradesmen who usually specialized in some particular aspect of chemical production—who seem to have pursued alchemy from purely economic motives. This new emphasis on the utilitarian and economic underpinnings of chemical production is a valuable corrective to older views that alchemy was at heart exclusively a quest for personal spiritual transformation. Sometimes lost in this new emphasis, however, is the way in which many alchemists were simultaneously pursuing both economic and Christian objectives. These alchemists did not lack a desire to create products of real value or to profit from them. Where Christian alchemists differed from their exclusively profit-oriented counterparts was in their belief that a quest for profit for its own sake was misguided. A Christian researcher's profits could, with God's blessing, be considerable, even huge, but whatever control of nature the alchemist gained from his practice was to be employed first and foremost in an effort to perform Christian service to society and further the positive and godly reformation of the world.[12]

Some Christian alchemists pursued godly goals exclusively through *philosophical* alchemy—research undertaken purely for the acquisition of new knowledge and spiritual advancement. This often involved conducting highly speculative chrysopoetic experiments aimed at attaining transmutation or the alkahest. Serving God through alchemy, however, was not limited to pursuing the alkahest alone. Many researchers combined the philosophical quest with more utilitarian experiments, such as efforts to perfect a dyeing process or improve a medicine. Even the most mundane alchemical product, discovered through pious experimentation and deployed benevolently, had a role to play

---

[12]. Much of the modern literature on alchemy emphasizes its utilitarian aspects. See Pamela H. Smith, *The Business of Alchemy: Science and Culture in the Holy Roman Empire* (Princeton, N.J., 1994); Deborah E. Harkness, *The Jewel House: Elizabethan London and the Scientific Revolution* (New Haven, Conn., 2007); Tara E. Nummedal, "Practical Alchemy and Commercial Exchange in the Holy Roman Empire," in Pamela H. Smith and Paula Findlen, eds., *Merchants and Marvels: Commerce, Science, and Art in Early Modern Europe* (New York, 2002), 201–222; Harold J. Cook, *Matters of Exchange: Commerce, Medicine, and Science in the Dutch Golden Age* (New Haven, Conn., 2007). In contrast, however, see Harkness, *John Dee's Conversations with Angels*, and Smith, *Body of the Artisan,* especially Smith's discussion of Johann Glauber, 165–177, where the links and tensions between commerce and piety are clearly delineated, if not fully resolved.

in fulfilling the Christian imperative to serve others and better the human condition.

In their efforts to serve pious ends, Christian alchemists were certainly attentive to economics. The alchemical hearth was, for them, very similar to one Puritan's description of New England: a place where "religion and profit jump together." Profiting from alchemy funded the ability to implement God's plan for improving human life on earth, which many considered a prerequisite to the anticipated return of Christ. In England, the Elizbethan court astrologer John Dee, whose work significantly influenced John Winthrop, Jr., promoted alchemical experimentation while casting astrological charts, conjuring and conversing with angels, and giving practical mathematical instruction to pilots and navigators going to the New World. From the records of his communications with angels, it is clear that Dee believed strongly in the supernatural origins and eschatological importance of the information revealed to him by his angelic communicants. This did not, however, deter him from selling the information they provided. The kind of world improvements alchemists such as Dee envisioned required massive economic investment, which they hoped could be funded, at least in part, through the advances produced by their alchemical research.[13]

While much of the alchemical literature printed in early modern England spoke of alchemy and alchemical medicine in simultaneously spiritual and practical terms, some alchemical writers expressed skepticism about the most zealous spiritual claims made by Christian alchemical authors. A number of alchemical patrons and many craftsmen also seem to have focused primarily on the economics of alchemical production, disregarding or minimizing whatever spiritual implications their efforts might have. Nevertheless, given the pervasive early modern belief in the interpenetration of the spiritual and mundane worlds, it is probable that the majority of early modern alchemists, and certainly all Christian alchemists, failed to draw sharp distinctions between the spiritual, philosophical, and economic motives behind alchemical research, thinking of them as intertwined and inseparable.[14]

---

13. Yates, "The Hermetic Tradition in Renaissance Science," in Singleton, ed., *Art, Science, and History*, 259; Harkness, *John Dee's Conversations with Angels*, esp. 215; William H. Sherman, *John Dee: The Politics of Reading and Writing in the English Renaissance* (Amherst, Mass., 1995); Nicholas H. Clulee, "John Dee and the Paracelsians," in Allen G. Debus and Michael T. Walton, eds., *Reading the Book of Nature: The Other Side of the Scientific Revolution* (Kirksville, Mo., 1998), 116–117.

14. Harkness, *The Jewel House*; Nicholas H. Clulee, "The *Monas Hieroglyphica* and the Alchemical Thread of John Dee's Career," *Ambix*, LII (2005), 210; J. T. Young, *Faith, Medical Alchemy, and Natural Philosophy: Johann Moriaen, Reformed Intelligencer, and the Hartlib Circle* (Aldershot, 1998); Jim Bennett and Scott Mandelbrote, *The Garden, the Ark, the Tower, the Temple: Biblical Metaphors of Knowledge in*

PARACELSIANISM, WHICH LAY at the heart of the Christian alchemical movement, posed a challenge to the Aristotelian Scholasticism that had dominated European intellectual thought for centuries. It rejected the Aristotelian theory of matter and stressed rather the animism and unity of the organic and mineral kingdoms. In medicine, for example, Paracelsianism rejected the time-honored humoral theories of Galen, posing instead the protomodern theory that illness was caused by disease astra—life forces unique to each disease—that settled in the body and attacked the human astra of the afflicted. These disease agents could be overcome only by counterastra, medicines whose active agents were similar to the disease they treated and therefore capable of neutralizing it. Such medicines were often the products of successful alchemical research. Paracelsus placed these new theories for overcoming old problems in a devoutly Christian context, which caused one early modern writer to speak of Paracelsian alchemy as a "Theologicall Phylosophy: Wherefore the New Birth is first to be sought for, and then all other Naturall Things will be added without much labour."[15]

As it evolved through the writings of many sixteenth- and seventeenth-century authors, Paracelsianism became a broad-ranging, multifaceted movement with great appeal to an age experiencing radical religious transformation and witnessing the discovery of new worlds. Many intellectuals embraced the new Hermetic and Paracelsian philosophies, and hundreds of alchemical works came from Europe's newly invented printing presses. On the Continent a school of alchemy developed, centered at the court of the Emperor Rudolph II, who made Hradcany palace in Prague a visual metaphor for the alchemical unity of the macrocosm and microcosm, where alchemists from throughout Europe enjoyed extensive patronage and power. Other Continental patrons of alchemical production included the duke of Saxony, the prince archbishop of Cologne, the landgrave of Hesse, the margrave of Brandenburg, the dukes of Braunschweig and Bavaria, Frederick II of Denmark, and,

---

Early Modern Europe (Oxford, 1998); Hereward Tilton, *The Quest for the Phoenix: Spiritual Alchemy and Rosicrucianism in the Work of Count Michael Maier (1569-1622)* (Berlin, 2003), 11.

15. Oswald Croll, "Discovering the Great and Deep Mysteries of Nature," in H. Pinell, ed. and trans., *Philosophy Reformed and Improved, in Four Profound Tractates* (London, 1657), 132; Paolo Rossi, "Bacon's Idea of Science," in Markku Peltonen, ed., *The Cambridge Companion to Bacon* (Cambridge, 1996), 25–46. William R. Newman argues that many alchemical writers were Aristotelian and that it is incorrect to assume that Aristotelianism was opposed to experiment; see "Alchemical and Baconian Views on the Art-Nature Division," in Debus and Walton, eds., *Reading the Book of Nature*, 91; Allen G. Debus, "The Chemical Debates of the Seventeenth Century: The Reaction to Robert Fludd and Jean Baptiste van Helmont," in Righini Bonelli and Shea, eds., *Reason, Experiment, and Mysticism*, 22–23; Webster, *The Great Instauration*, 285, 289, 330.

during his short life, Henry, prince of Wales, son and heir apparent of James I of England.[16]

These leaders patronized alchemists for a variety of expressed motives, sometimes principally philosophical or spiritual, sometimes economic, sometimes political, most often from a mixture of these. Politically, the Protestant alchemical quest for universal world improvement provided a useful counter to the Catholic universalism of the Hapsburg Empire (though alchemy was by no means an exclusively Protestant endeavor). Moreover, the technical productions of alchemists helped buttress the authority of rulers struggling to consolidate their states.

By 1600, London, too, had become an international center of chemical exploration. Practitioners from the Low Countries, France, Italy, Spain, Portugal, Scotland, and elsewhere came together in parish-centered neighborhoods often known for distinct practices, such as chemical distilling or medicine. Seventy-four alchemists are known to have practiced in Elizabethan London, a number that may significantly underrepresent the total number of practitioners. Within this burgeoning alchemical culture, a debate over Paracelsian ideas became the greatest polarizing issue.[17]

For all its promise, there were problems with Paracelsus's philosophical system, among them the inconsistencies frequently encountered in Paracelsus's writings and the obscurity and mysticism of Paracelsian alchemical tracts. From treatise to treatise, one could not be sure whether Paracelsus's views

16. Bruce T. Moran, *Distilling Knowledge: Alchemy, Chemistry, and the Scientific Revolution* (Cambridge, Mass., 2005), 103–104; Moran, "Patronage and Institutions: Courts, Universities, and Academies in Germany: An Overview, 1550–1750," in Moran, ed., *Patronage and Institutions: Science, Technology, and Medicine at The European Court, 1500-1750* (Rochester, N.Y., 1991); Smith, *The Business of Alchemy;* Carl Wennerlind, "Credit-Money as the Philosopher's Stone: Alchemy and the Coinage Problem in Seventeenth-Century England," *History of Political Economy,* XXV, suppl. (2003), 234–261.

Because of the diffuse ways in which the terms have been employed both historically and among present-day historians, some historians of science have criticized the use of both "Hermeticism" and "Paracelsianism" as descriptive terms. While recognition that both concepts were and still are subject to multiple interpretations is important, I believe the creative ambiguity and definitional imprecision inherent in them helps account for their long-term use and usefulness. When Winthrop identified himself to contemporaries as both a Hermeticist and a student of Paracelsian medicine, he clearly expected that to help define for them the nature of his scientific interests. Brian P. Copenhaver, "Natural Magic, Hermeticism, and Occultism in Early Modern Science," in David C. Lindberg and Robert S. Westman, eds., *Reappraisals of the Scientific Revolution* (Cambridge, 1990), 261–301; Steven Pumfrey, "The Spagyric Art; or, The Impossible Work of Separating Pure From Impure Paracelsianism: A Historiographical Analysis," in Grell, ed., *Paracelsus: The Man and His Reputation,* 21–52.

17. Deborah E. Harkness, "'Strange Ideas' and 'English' Knowledge: Natural Science Exchange in Elizabethan London," in Smith and Findlen, eds., *Merchants and Marvels,* 137–162.

were changing. Worse still, the mystical, allegorical, and metaphorical manner in which alchemical processes were described often made their meanings abstruse, even to the alchemists.[18]

This obscurity, common to many alchemical texts of the period, was intentional—a carefully encoded effort to keep the secrets of the most knowledgeable alchemists from the hands of the impure and vulgar, who might use such knowledge for exclusively self-serving and ungodly purposes. For many researchers, at least some of alchemy's appeal came directly from these enigmatic texts, because translating them and sharing their meanings became the special province of an informed and self-referential cognoscenti. Being able to demonstrate one's understanding of opaque alchemical allegories was a sign of being singled out by God as worthy of such knowledge, and a ticket into social circles that might otherwise have excluded one. "To write more about this mystery [the alchemical mystery] is forbidden and further revelation is the prerogative of the divine power," Paracelsus had said. "For this art is truly a gift of God. Wherefore not everyone can understand it."[19]

Not everyone agreed with this assessment. Historians have shown how the metaphorical obscurity of alchemical texts often actually represented carefully encoded descriptions of important chemical processes, but the allegorical texts' lack of transparency frustrated and baffled many readers, leaving alchemists open to attacks by skeptical critics. The attacks ranged from charges that alchemy was pure foolishness to assertions that alchemists engaged in deliberate obfuscation to more serious charges that the enigmatic texts concealed the writers' diabolical intentions. Thomas Erastus, the Swiss physician and theologian, claimed Paracelsus was a black magician in league with the devil. Georgius Agricola, a writer on mining and minerals, sarcastically asked why, with so many alchemists "straining every nerve night and day to the end that they may heap a great quantity of gold," there was not more evidence of success, since they should have already "filled whole towns with gold and silver." Alchemists responded with assertions that the critics of their failure to publicly display wealth had missed the point. Anyone seeking the secret of transmutation from motives of greed was, they insisted, destined to fail.[20]

---

18. Betty Jo Teeter Dobbs, *The Foundations of Newton's Alchemy: or, "The Hunting of the Greene Lyon"* (Cambridge, 1975); Steven Pumfrey, "The Spagyric Art," in Grell, ed., *Paracelsus: The Man and His Reputation*, 21–52.

19. Paracelsus, *Selected Writings*, ed. Jacobi, 144, 223.

20. William Newman has pointed out that secrecy also had practical benefits, including protecting trade secrets and increasing the social and intellectual status of the possessor, in *Gehennical Fire*, 42–48, 62–78. Georgius Agricola, *De Re Metallica* (1556), trans. Herbert Clark Hoover and Lou Henry Hoover (1912; New York, 1950), xxviii–xxix.

One of the main critics of seventeenth-century alchemy was the English philosopher Francis Bacon. Bacon attacked Paracelsian alchemy because he found it riddled with false theories and procedures but protected by its enigmatic language. Alchemy, astrology, and natural magic all were, as far as Bacon was concerned, philosophies that "hold much more of imagination and belief than of sense and demonstration." Scholars seeking to draw a clear demarcation between alchemy and modern science have often cited Bacon as one of the figures instrumental in marking the boundaries between the two. Modern work has shown, however, that this is a false dichotomy. Bacon's work was richly informed by alchemical authors, and the goal of his reform agenda was the same as the Paracelsians': to recover the *prisca theologia,* the Edenic past when humankind had dominion over nature and lived in harmony with God's will. Despite differences over methodologies, Bacon and the alchemists shared many fundamental assumptions. Both stressed the importance of experiment and observation of the natural world. Both wanted to free natural philosophy from the stultifying authority of the Scholastics; both stressed the utilitarian value of scientific investigation. Christian alchemists emulated the Baconian emphasis on minerals and mining as sources of practical knowledge, and, with the caveat that knowledge should be restricted to those empowered by God to understand it, they shared the Baconian desire for the free communication of ideas. Bacon admitted that alchemists and natural magicians pursued a desirable goal with regard to understanding nature; in fact, he listed the alchemist Sir Walter Ralegh as one of the English people he thought would and could help advance his ideas.[21]

Alchemy grew in importance in England throughout the first three quarters of the seventeenth century; it played an important role in English intellectual life throughout Winthrop's lifetime. The London minister Thomas Tymme wrote, *"Halchemy* should have concurrence and antiquitie with The-

21. James Spedding, Robert Leslie Ellis, and Douglas Denon Heath, eds., *The Works of Francis Bacon* (London, 1857), III, 503; P. M. Rattansi, "Alchemy and Natural Magic in Raleigh's *History of the World," Ambix,* XIII (1966), 133–134; Righini Bonelli and Shea, eds., *Reason, Experiment, and Mysticism,* 262. Stephen A. McKnight, "The Wisdom of the Ancients and Francis Bacon's New Atlantis," in Debus and Walton, eds., *Reading the Book of Nature,* 91–92, Newman, "Alchemical and Baconian Views on the Art-Nature Division," 90; Charles Webster, *From Paracelsus to Newton: Magic and the Making of Modern Science* (Cambridge, 1982), 58–59; Webster, *The Great Instauration,* 346, 384–385; Yates, "The Hermetic Tradition in Renaissance Science," in Singleton, ed., *Art, Science, and History,* 269–270; Harkness, *The Jewel House,* 211–253. Harkness recharacterizes Bacon's philosophical agenda not as an anti-Paracelsian effort to establish rational and transparent inquiry into nature—a turning point in a putative scientific revolution—but rather as an effort to wrest authority over scientific interpretation away from London's polyglot community of practitioners of vernacular science and make it the province of an educated elite.

ologie." For him, Creation was a chemical process, and so would be the Last Judgment. The utopian poet John Hall declared that alchemy "hath snatcht the keyes of Nature from the other sects of Philosophy, by her multiplied experiences," and the New Model Army chaplain John Webster called alchemy "the most abstruse and most excellent part of all Natural Philosophy," praising "the most noble of all Arts, the Transmutation of Metals."[22]

When Winthrop took up alchemy in London in the early 1620s, he was not a lone magus engaged in the mystical and solitary pursuit of knowledge, as sometimes characterized. He was, rather, a progressive intellectual participating in one of the advanced natural philosophical pursuits of his age. He shared the alchemical quest with many of the leading thinkers of his time, no small number of whom, importantly, were Puritans. For them, alchemy united the quest for useful knowledge with the quest for grace and focused both on the mission to render Christian service to a world reshaping itself in preparation for the return of Christ.[23]

EVER SINCE the start of the Tudor Reformation, and especially since the victory over the invincible Spanish Armada, Protestant England had felt a sense of prophetic destiny. Many English people believed they were living in the climax of human history and that England's role in the Reformation gave it a special mission of bringing about the millennium—the thousand-year period mentioned in Revelation during which Christ was to reign on earth.

A millenarian prophecy of particular importance to English Puritan intellectuals like Winthrop was the belief in a Great Instauration, a rapid increase in learning, that, in the days before the Second Coming of Christ, would restore the dominion man lost over nature at Adam's Fall. A verse from the twelfth chapter of Daniel was the basis for the prophecy: "But you, Daniel, shut up the words and seal the book, until the time of the end. Many shall run to and fro, and knowledge shall increase."[24]

For those inclined to see it, confirmation of this scriptural passage was omnipresent. The sixteenth-century improvements in navigation and the

22. Thomas Timme, dedication, in Iosephus Quersitanus [Joseph Du Chesne], *The Practise of Chymicall and Hermeticall Physicke, for the Preservation of Health,* trans. Timme (London, 1605), A-3; Bruce Janacek, "Thomas Tymme and Natural Philosophy: Prophecy, Alchemical Theology, and the Book of Nature," *Sixteenth Century Journal,* XXX (1999), 987–1007; Webster, *The Great Instauration,* 388; John Webster, *Metallographia; or, A History of Metals* . . . (London, 1671), 2, 11.

23. Brooke, *The Refiner's Fire,* 36–37.

24. Webster, *The Great Instauration,* 22–24, 326–329; McKnight, "The Wisdom of the Ancients and Francis Bacon's *New Atlantis,*" in Debus and Walton, eds., *Reading the Book of Nature,* 111–132; Dan. 12:4.

discovery of the new worlds were prima facie evidence that many indeed now ran "to and fro." The information and technological explosions that had resulted in the printing revolution, cannons, compasses, clocks, and the telescope signified a vast increase in knowledge. Even the heavens seemed to confirm prophecy in the 1604 appearance of new stars in the constellations Cygnus and Serpentarius. To many, these stars' appearance foretold the onset of the millennium. In early-seventeenth-century England, it was easy for many to believe that the seal on the book of knowledge had been broken open, that the Word was alive in the world.[25]

Bacon's 1620 *Instauratio magna*—a guide to intellectual regeneration, philosophical progress, and social planning—set the stage for the rise of a complex international reform movement. Bacon called on natural philosophers to join in a great collaborative, utilitarian enterprise to improve practical and experimental knowledge. This would reverse the Fall of man, restore man's dominion over nature, promote social betterment, and confirm the overwhelming power of God's providence. "It would be disgraceful to mankind," Bacon wrote, "if the regions of the Material Globe, i.e., of land and sea, and of the stars, should be immensely laid open and illustrated, while the limits of the Intellectual Globe were bounded by the discoveries and narrowness of the Ancients." The key to gaining the knowledge Bacon sought was for natural philosophers to adopt the utilitarian, craft-oriented approach of Adam, the first man. Adam was viewed as a hands-on, practical scholar, a gardener and naturalist who knew entirely the secrets of the natural world. Importantly, among the highest attainments ascribed to Adam by contemporaries were mining and metalworking; he was even credited with knowing the secret of transmutation. Through arduous and pious effort, natural philosophers could recover Adamic knowledge and improve the world.[26]

Bacon was not the only, or the first, early modern philosopher to call for the universal advancement of human knowledge. Other reformers, putting greater emphasis on alchemical knowledge and the relationship between world reform and Christian service, preceded him. On the Continent in 1614 a German tract appeared, ostensibly authored by a secret society of alchemists, fully embracing the mission of intellectual regeneration. The *Fama Fraternitatis* of the Brotherhood of the Rosy Cross called for a "Universal and General Ref-

---

25. Webster, *The Great Instauration*, 22–24, 326–329; Miguel A. Granada, "Kepler v. Roeslin on the Interpretation of Kepler's Nova: (1) 1604–1606," *Journal for the History of Astronomy*, XXXVI (2005), 299–319.

26. Webster, *The Great Instauration*, 24, 326; Francis Bacon, *The Novum Organon; or, A True Guide to the Interpretation of Nature*, ed. G. W. Kitchin (Oxford, 1855), 62.

ormation of the whole wide world," through the use of natural magic, Cabala magic, and Paracelsian alchemy:

> God in these latter days hath poured out so richly his mercy and goodness to mankind, whereby we do attain more and more to the perfect knowledge of his Son Jesus Christ and Nature . . . wherein there is not only discovered unto us the half part of the world, which was heretofore unknown and hidden, but he hath also made manifest unto us many wonderful, and never heretofore seen, works and creatures of Nature, and moreover [he] hath raised men, imbued with great wisdom, who might partly renew and reduce all arts . . . to perfection; so that finally man might thereby understand . . . why he is called Microcosmus, and how far his knowledge extendeth into Nature.[27]

The Rosicrucian movement, which emanated out of the Heidelberg court of Frederick V, the elector palatine, was possibly influenced by and certainly compatible with the research agenda pursued by John Dee, who had left England in 1583 to advance an alchemical-religious movement in Bohemia. The philosophy espoused in the *Fama,* the first Rosicrucian tract, incorporated alchemy, geometry, mathematics, Cabala and natural magic, and, above all, religious and spiritual illumination, all in the service of millenarian reform. The Rosicrucians underscored the purity of their alchemical intentions; they pursued alchemy, not in the cause of "ungodly and accursed gold-making," but in support of world improvement and Christian reformation. The speaker of the *Fama*, Christian Rosencreutz (Rosy Cross), made the distinction clear: "He [the alchemist] doth not rejoice that he can make gold . . . but is glad that he seeeth the Heavens open." The Rosicrucian manifesto announced the existence of a secret society of enlightened Christian alchemists, dedicated to knowledge, world reform, and good works.[28]

No one knows with certainty whether the Brotherhood of Rosicrucians really existed, but *belief* that they existed stimulated interest in alchemy and reform throughout Protestant Europe. Would-be alchemical reformers sought to identify and contact members of the brotherhood, to express admiration for their goals, and to ask to join their movement. The *Fama* had

---

27. *Fama Fraternitatis; or, A Discovery of the Fraternity of the Most Noble Order of the Rosy Cross,* trans. Thomas Vaughan (1652), in Yates, *The Rosicrucian Enlightenment,* 238, and xii, 30–40, 42, 236; Carlos Gilly and Friedrich Niewöhner, eds., *Rosenkreuz als europäisches Phänomen im 17. Jahrhundert* (Amsterdam, 2002); T. M. Luhrman, "An Interpretation of the *Fama Fraternitatis* with Respect to Dee's *Monas Hieroglyphica,*" *Ambix,* XXXIII (1986), 1–10.

28. *Fama Fraternitatis,* trans. Vaughan, in Yates, *The Rosicrucian Englightenment,* 250, and xii, 30–40.

proclaimed that, though the brothers intended to remain underground and incognito for a century, those desiring to join them should make their intentions publicly known. If the brotherhood deemed them worthy, they would be contacted.[29]

One of the English alchemists who tried repeatedly to make contact with the Rosicrucians was John Winthrop, Jr. He began his investigation into Christian alchemy while a student at London's Inner Temple, to which he was admitted in February 1624. He and a fellow student named Edward Howes enthusiastically embraced the Rosicrucian reform agenda. At the expense of their legal studies, the pair began to engage with the corpus of alchemical works coming from Europe and embarked on the pursuit of the alkahest and the philosopher's stone. Winthrop's interest in the medical aspects of alchemy, which became central to his service to New England's Puritan colonies, might also have been stimulated by his uncle Thomas Fones, a London apothecary with whom he lodged, whose daughter Martha became Winthrop's first wife in 1630. Winthrop found in alchemical culture an intellectual and Christian natural philosophy to which he could fully commit and through which he could seek knowledge and material gains while fulfilling his Christian duty to improve the world and serve others. In London, barely in his twenties, he embarked on a Christian scientific pilgrimage that would continue throughout his life.[30]

Mastering alchemy often involved physical pilgrimage. An adept-in-training was expected to acquire a full knowledge of the mundane as well as the hidden properties of minerals, plants, and animals. Only through careful observation and exploration of the natural world could an alchemist acquire such empirical knowledge. On the journey to perfection, he gained practical experience in locating ores, refining metals, processing minerals, and making medicines, skills considered especially useful in filling the Christian alchemist's obligations to serve others. The influential Danish Paracelsian Peter Severinus underscored the importance of such empirical pilgrimages:

> Sell your lands, your houses, your clothes and your jewelry; burn up your books. On the other hand, buy yourselves stout shoes, travel to

---

29. The Rosicrucians emerged in the writings of the mathematician Jungius, or Johann Valentin Andrae, who is the likely author of the *Chemical Wedding of Christian Rosencrantz* (1616). Yates, *Rosicrucian Enlightenment*, 66–67, 102.

30. Yates, *Rosicrucian Enlightenment*, 25–27. Winthrop's grandfather Adam had been admitted as barrister in the Inner Temple in 1584. Robert C. Black III, *The Younger John Winthrop* (New York, 1966), 23; Ronald Sterne Wilkinson, "John Winthrop, Jr. and the Origins of American Chemistry" (Ph.D. diss., Michigan State University, 1969), 14.

the mountains, search the valleys, the deserts, the shores of the sea, and the deepest depressions of the earth; note with care the distinctions between animals, the differences of plants, the various kinds of minerals, the properties and mode of origin of everything that exists. . . . Lastly, purchase coal, build furnaces, watch and operate with the fire without wearying. In this way and no other, you will arrive at a knowledge of things and their properties.[31]

Winthrop took the alchemical injunction to attain knowledge via pilgrimage seriously, and in 1627, at age twenty-one, he attempted to secure through family connections a purser's berth aboard a vessel bound for Turkey. The Islamic world was another primary source of alchemical knowledge central to the European alchemical tradition. It had been the region from which Europeans first received alchemical texts, through the translation of the *Seventy Books* attributed to Jabir Ibn Hayyan by Gerard of Cremona in the twelfth century. The importance of the East as a source of knowledge was underscored in the Rosicrucian tracts, which had announced that C. R. (initials for Christian Rosenkreutz), the "chief and original" member of the Rosicrucian order, had himself made a pilgrimage to the Arab world, where he "obtained much favour with the Turks." By means of this journey C. R. was said to have gained advanced knowledge of medicine, mathematics, and magic. Medical advances were, of course, primary goals of one branch of alchemical study, and many considered mathematics and natural magic essential to unlocking the keys to alchemical attainment. The fact that the Muslims were not Christian limited the knowledge the East could provide; from Islamic sources, it was not altogether pure. C. R. had noted that some of the information he attained was clearly "defiled with their religion," but the Arab world was nevertheless a wellspring of alchemical truths. The millenarian agenda of the Rosicrucians called for world unification, and, despite its Islamic foundations, as a Christian philosopher C. R. had found in the Arab world "more better grounds for his faith, altogether agreeable with the harmony of the whole world." Winthrop's desire to travel to Turkey was probably motivated by a hope to gain knowledge similar to that gained by Christian Rosencreutz, perhaps even to consciously emulate his quest.[32]

31. Petrus Severinus, *Idea Medicinae Philosophicae* . . . (Hague, 1660), 39, quoted in Debus, *English Paracelsians*, 20.
32. Black, *Younger John Winthrop*, 27–28; Robert C. Winthrop, *Life and Letters of John Winthrop, Governor of the Massachusetts-Bay Company at Their Emigration to New England, 1630*, 2d enl. ed. (Boston, 1869), I, 236–237; Samuel Eliot Morison et al., eds, *Winthrop Papers* (Boston, 1929–), I, 347n–348 (hereafter cited as *Winthrop Papers*). William Newman discusses the intellectual etiology of

Winthrop's initial attempt to secure a maritime position that would take him to Turkey was unsuccessful. Through family connections he did secure a position as captain's secretary in the fleet being sent for the relief of the Protestants at La Rochelle. This ill-fated, first-hand encounter with the fortunes and perils of war—the English army lost more than half of its invasion force to no avail—might have been the source of Winthrop's lifelong aversion to military conflict as an expression of government policy. The journey also advanced Winthrop's alchemical knowledge, for during this expedition he met Cornelius Drebbel and his son-in-law Abraham Kuffler, who served as explosives experts to the duke of Buckingham. Both men displayed the polymathic curiosity about the natural world common among seventeenth-century natural philosophers. Drebbel, an alchemical physician and inventor who had been a favorite in the court of Henry Frederick, prince of Wales, prior to the prince's death in 1612, fabricated a number of early microscopes, created a celebrated camera obscura, pursued perpetual motion, and even designed a submarine that was tested successfully in the Thames River in England. Winthrop's admiration for Drebbel can be seen in the comment he inscribed, long after the La Rochelle voyage, on the flyleaf of the alchemist Basil Valentine's book *Of Natural and Supernatural Things*. "This was once the booke of that famous philosopher and naturalist Cornel: Drebbel, wh. He Usually carried wt him in his pockett and after his death was given me by his son in law Mr. Abram Kuffler. John Winthrop." Abraham Kuffler and his brother Johann were successful practitioners of a number of alchemically related arts and became famous for their special cochineal textile-dying process. During a lifelong relationship, the brothers helped link Winthrop to many like-minded European practitioners.[33]

Winthrop returned from the La Rochelle expedition in November 1627, still eager to go to Turkey. The following June, having secured a position as the supercargo aboard the *London*, a merchant vessel of the Levant Company,

---

seventeenth-century alchemical theories in Newman, *Gehennical Fire*, 92–114; *Fama Fraternitatis*, trans. Vaughan, in Yates, *Rosicrucian Enlightenment*, 239–240. Winthrop would later claim to Edward Howes that the Rosicrucian's knowledge of alchemy was Arabic in origin. Edward Howes to John Winthrop, Jr., Feb. 25, 1640, *Winthrop Papers*, IV, 202.

33. Black, *Younger John Winthrop*, 33, 87; Wilkinson, "John Winthrop, Jr. and the Origins of American Chemistry," 15; Svetlana Alpers, *The Art of Describing: Dutch Art in the Seventeenth Century* (Chicago, 1983), 6–7; L. E. Harris, *The Two Netherlanders: Humphrey Bradley and Cornelis Drebbel* (Cambridge, 1961); Basilius Valentinus, *Von den natürlichen und übernatürlichen Dingen* (Leipzig, 1624), flyleaf, New-York Historical Society Library. An excellent account of Drebbel's and the Kufflers' contribution to cochineal dyeing is in Amy Butler Greenfield, *A Perfect Red: Empire, Espionage, and the Quest for the Color of Desire* (New York, 2005), 125–142. Greenfield's characterization of seventeenth-century alchemy is, however, highly inaccurate.

he departed for Constantinople. After visits to Leghorn, Pisa, and Florence, he was in the Turkish capital by mid-September, where he began seeking out alchemical contacts. For a time, he considered moving farther east, as the Rosicrucians' C. R. had done, but rumors of conflict led him to abandon a proposed journey to the holy city of Jerusalem. One of Winthrop's most productive encounters occurred in Venice the following winter, as the *London* was returning home. There, he encountered Jacob Golius, a Dutch mathematician and scholar philosopher who had just completed a tour of the Ottoman lands, acquiring Arabic and Persian manuscripts. During a monthlong quarantine because of illness, Winthrop and Golius developed a warm relationship through shared interests in the intellectual mysteries of the East. Additional stops on Winthrop's pilgrimage included visits to Padua, home of one of the best universities in Europe, where many alchemical doctors acquired their medical degrees, and the cosmopolitan crossroads of Amsterdam. Winthrop arrived back in London in August 1629, fourteen months after his pilgrimage had begun.[34]

Two weeks after his return, the Massachusetts Bay Company received its royal patent, and on October 20, John Winthrop, Sr., was selected to lead the company's migration of Dissenting Puritans to New England. Even as the younger Winthrop assisted in the preparations for his father's journey, helped sell off family property, attended to the varied needs of the New England company, and married his first cousin Martha Fones (on January 8, 1631), he and Edward Howes were once again engaged in the alchemical quest. A letter from Howes to Winthrop written at the end of March 1630 described in some detail the progress of an experiment aimed at perfecting the philosopher's stone.[35]

At about the same time, Winthrop began to display a special affinity for the English alchemist John Dee. Dee, whose mystical approach to experimental science has been linked by historians to both the origins of the Rosicrucian movement and the Arabic works of Avicenna (Ibn Sina) and Artephius, had a special interest in scientific exploration of the New World. He had given instruction and advice to pilots and navigators conducting exploratory voyages to North America. He also conjured angels to ask them of the success of a colony he proposed to establish there, which he intended to call Atlantis. Dee's "Praeface" to the English translation of *Euclid's Elements of Geometry* (1570) explained the link he believed existed between experimental science

---

34. Winthrop, *Life and Letters of John Winthrop*, 268, 270–271; Black, *Younger John Winthrop*, 35, 37–38.

35. *Winthrop Papers*, II, 227, IV, 155–156.

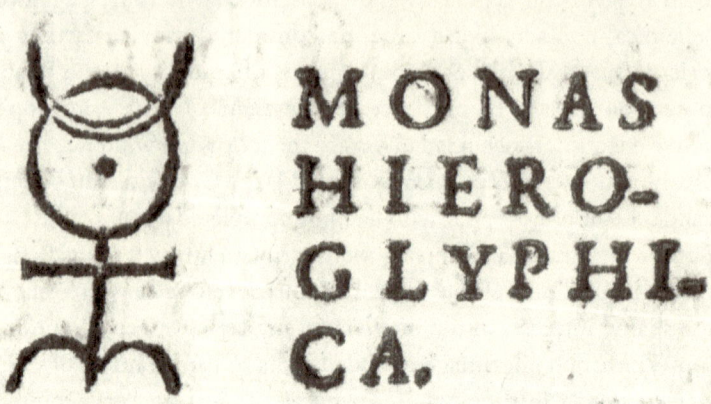

FIGURE 1. *Monas Hieroglyphica,* by John Dee. Detail of page 12. Antwerp, 1564. *Courtesy of the Massachusetts Historical Society*

and magical arts. Like Paracelsus, Dee believed that the experimental method reached truths other methodologies could not attain, because it investigated through actual experience and direct observation. But empirical investigation alone was not enough. By incorporating natural magic into experimentation, Dee believed natural philosophers could transcend the limits of direct observation and acquire knowledge of the future, past, and present. Dee developed an alchemical symbol, the *monas hieroglyphica,* which contained within it the "true cabala of nature." It became closely identified with the Rosicrucian movement and was adopted by several authors as a symbol of their commitment to the alchemical quest.[36]

Drawn entirely from a point, a line, and a circle, the most simple of all things, the monas, in Dee's eyes, epitomized all creation. The point represents the earth, and the circle the sun and the heavens. The semicircle represents the moon, and the dual semicircles at the bottom the sign of Aries, the first sign of the zodiac. The circular components relate to the heavens, and the cross to the Christian cross and the four sublunar elements (earth, air, fire, and water). The monas incorporated the alchemical symbols for the seven metals (lead, tin, copper, iron, mercury, silver, and gold), which corresponded to the seven "planets" (Saturn, Jupiter, Venus, Mars, Mercury, the moon, and the

---

36. On Dee's "true cabala of nature," see Harkness, *John Dee's Conversation with Angels,* 181–194.

sun, respectively). Dee thought the monas revealed a new and sacred symbolic language of nature, a "cabala of the real," and that it demonstrated both the unitary word of God and the integrated relationship between the terrestrial and celestial. The natural philosopher who could master the language of the monas would, Dee believed, gain access to the innermost secrets of the cosmos. According to Nicolas Clulee, "The *Monas Hieroglyphica* was a daring and inventive proposal for a symbolic language that had the power to reveal the divine plane of creation, to explain the workings of the material world in the principles of alchemy, and to assist the mystic ascent of the soul." Not everyone understood all the intended meanings of the monas, and people could and did read it selectively. Nevertheless, it became a symbol closely identified with both alchemy and the Rosicrucian movement.[37]

By 1631, the year he left to join his father in New England, young Winthrop was collecting works from Dee's personal library. He also adopted Dee's *monas hieroglyphica* as a personal mark inscribed next to his name within his alchemical texts and emblazoned on the crates of chemicals, instruments, and other supplies he had transported to America. It was a sign that identified to the knowledgeable his commitment to pursue passionately the search for alchemical knowledge in New England just as he had in the Old.[38]

It is essential to distinguish between natural magic and diabolic magic. The former involved using natural, divinely sanctioned means to manipulate the occult, that is, the unseen or hidden forces believed to influence change in the natural world. Diabolical magic, on the other hand, necessitated compacting with the devil—considered the most powerful of all magicians—to gain mastery over those same forces. While natural magic was a widely accepted component of early modern natural philosophy, diabolic magic was vehemently proscribed, an object of fierce civil and theological attack. Much of the work in conducting alchemical experiments involved the kind of technologically focused activities we associate with the modern laboratory: regulating fires, processing chemicals, purifying and compounding ingredients, measuring outcomes. Many alchemists augmented these technical processes

---

37. For deconstruction of the monas, I am indebted to Clulee, *John Dee's Natural Philosophy*, 65. Yates, *Rosicrucian Enlightenment*; Peter J. French, *John Dee: The World of an Elizabethan Magus* (London, 1972); Young, *Faith, Medical Alchemy, and Natural Philosophy*, 168; Clulee, "The *Monas Hieroglyphica* and the Alchemical Thread," *Ambix*, LII (2005), 197–215, esp. 197; Clulee, "John Dee and the Paracelsians," in Debus and Walton, eds., *Reading the Book of Nature*, 111–125; Peter J. Forshaw, "The Early Alchemical Reception of John Dee's *Monas Hieroglyphica*," *Ambix*, LII (2005), 247–269.

38. Ronald S. Wilkinson, "The Alchemical Library of John Winthrop, Jr. (1606–1676) and His Descendants in Colonial America, Parts I–III," *Ambix*, XI (1963), 33–51; Wilkinson, "John Winthrop, Jr. and the Origins of American Chemistry," 19, appendix, no. 11.

FIGURE 2. *Monas Hieroglyphica*, by John Dee. Title page. Antwerp, 1564.
Courtesy of the Massachusetts Historical Society

The monas is the symbol within the oval.

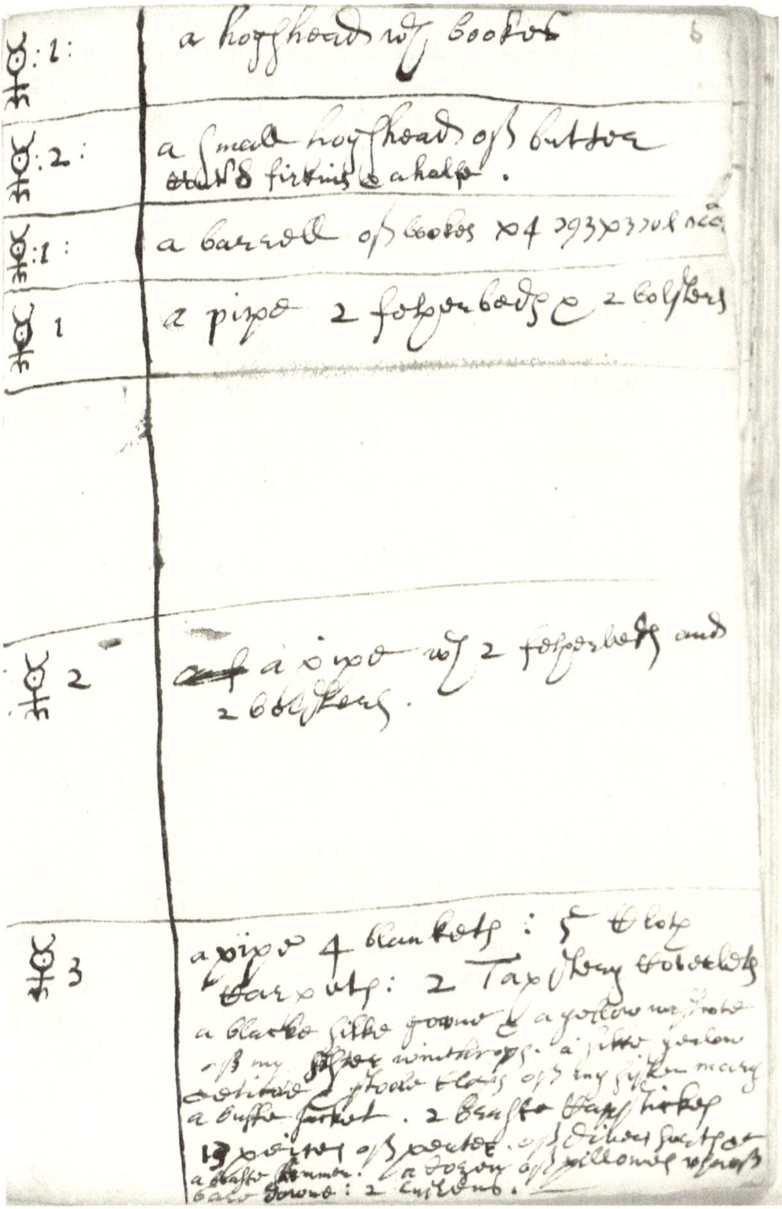

FIGURE 3. *Monas* Hieroglyphs. Memoranda Book of John Winthrop, Jr., 1631, unnumbered page. Manuscript, Winthrop Family Papers, Massachusetts Historical Society, Boston. *Courtesy of the Massachusetts Historical Society*

Winthrop adopted the monas as a personal symbol, here identifying crates of goods he shipped to America.

with natural magic: activities (for example, Cabalic prayer or astrologically sensitive timing of experiments) aimed at making use of occult forces to improve the outcome of their experiments. Such efforts were neither inherently suspect to contemporaries nor culturally unusual in the context of the times.[39]

In reality, the alchemical practices and philosophies of Winthrop as well as of the many other Puritans who practiced alchemy in New England during Winthrop's lifetime generally complemented rather than challenged the fundamental themes of New England Puritanism. Winthrop's efforts to alchemically manipulate the occult forces in nature by accelerating natural processes and manipulating the internal properties of elements might have been considered partially magical by some, but they were never considered diabolical. Rather, the practical benefits people believed they derived from Winthrop's alchemical skills made him an immensely useful and sought-after figure in America. Through his alchemical medicines alone, many New Englanders came to revere and depend upon the fruits of Winthrop's alchemical labors.

Christian alchemy, after all, was both a chemical and a spiritual undertaking whose practices mirrored and symbolized the spiritual objectives of the Puritan saint. Just as Winthrop sought the purification of metals at the alchemical furnace, the Puritan sought the purification of his or her soul in the crucible of God's judgment. For the alchemist himself, the effort to achieve transmutation was also an act of piety and an exercise in self-purification. Basil Valentine, whose works influenced Winthrop, described the state in which the quest for alchemical knowledge must begin:

> *Invocation* of GOD must be made with a certain Heavenly Intention, drawn from the bottom of a pure and sincere Heart. . . . For GOD will not be mocked . . . as Worldly Men . . . think: GOD, I say, will not be mocked, but the Creator of all things will be invoked with reverential fear, and acknowledged with due Obedience. . . . No impious Man shall ever be partaker of true Medicine, much less of the eternal Heavenly Bread.[40]

---

39. Some scholars, failing to distinguish between natural and diabolic magic, have considered alchemy a magical practice rejected by Puritans. Perry Miller claimed, "New England divines never had any tolerance for astrology, the philosopher's stone, or incantation, for any device by which men sought to escape the rules of nature or to circumvent the settled order of things." Miller's and other historians' similar views limited the appreciation for alchemy's significance in New England's cultural formation. Miller, *The New England Mind*, 227; see also Brooke, *The Refiner's Fire*, 36–37.

40. Basil Valentine, *His Triumphal Chariot of Antimony, with Annotations of Theodore Kirkringius,*

In accord with Calvinist theology, alchemical success could not be achieved solely through self-effort; it was completely dependent on the will of God. God alone decided who would receive alchemical knowledge. Benjamin Worsley, an alchemist who would play an important role in helping Winthrop obtain a royal charter for Connecticut in the 1660s, knew his experiments succeeded because "the Lord hath his seasons, and that it is not of him that wills, or of him that runnes, but of God only who in this as in more higher things enlightens whom he will." Alchemical adepts, conscious of the sacred nature of revealed information, adopted codes of secrecy and obscure symbolism to keep these divine gifts from falling into profane hands. Winthrop, in a letter to his English colleague Samuel Hartlib defending the alchemical practice of secrecy, left no doubt as to the source of alchemical discoveries or to the duty of the alchemist to safeguard the distribution of this sacred knowledge:

> God hideth from the unworthy vaine sinfull world many excellent things, which he knoweth they are very apt to abuse, turning all to pride and luxury and vanity, and such as God hath revealed such excellent secrets as might be of use to mankind they [the alchemists] feare the divulging of them least they should be turned to abuse by the vaine world.[41]

The concept of alchemical purification proved such an apt symbol for its spiritual counterpart that Edward Taylor, the Puritan minister-poet of Westfield, Massachusetts, wrote several poems employing alchemy as a metaphor for attaining grace. The belief in the macrocosm-microcosm relationship—that the visible and occult forces permeating all aspects of the cosmos interact with and upon each other—was homologous with the Puritan belief in providences, portents, and prodigies. The New England divine John Cotton promoted the experimental observation of nature expressly because of what nature could reveal spiritually: "God having usually made this world to be a mappe and shadow of the spiritual estate of the soules of men ... learne wee to discern the signes of our owne times."[42]

M.D., with the True Book of the Learned Synesius a Greek Abbot Taken out of the Emperour's Library, concerning the Philosopher's Stone (London, 1678), 2–3.

41. Benjamin Worsley to ———, Feb. 14, 1656, Hartlib Papers [42/1/5a], University of Sheffield; Young, Faith, Medical Alchemy, and Natural Philosophy, 233; John Winthrop, Jr., to Samuel Hartlib, "Some Correspondence of John Winthrop, Jr., and Samuel Hartlib," Massachusetts Historical Society, Proceedings, LXXII (1963), 58–62.

42. Randall A. Clack, The Marriage of Heaven and Earth: Alchemical Regeneration in the Works of Taylor, Poe, Hawthorne, and Fuller (Westport, Conn., 2000); Randall Anthony Clack, "The Phoenix Rising: Alchemical Imagination in the Works of Edward Taylor, Edgar Allan Poe, and Nathaniel Haw-

The Neoplatonic Hermeticism that provided a foundation for Winthrop's alchemy was based on the concept of recapturing primordial wisdom. Reacquiring the prisca theologia, the knowledge that been lost at Adam's fall, was at the heart of the alchemical agenda. This scientific goal mirrored the theological desire Theodore Dwight Bozeman has noted among New England Puritans to return to the "first, original and most perfect state" of the primordial church, when "the giftes of the spirite of wisdome, discretion, knowledge ... were poured forth more plentifully than ever they were, eyther before, or shall be after." In Bozeman's view Puritans eschewed newness; their goal was the restoration of prior perfection: "The crucial premise was that the first is best." By rejecting modern concepts in favor of ancient ones, Bozeman says, New England Puritans strived to escape the degeneracy they saw everywhere around them.[43]

Winthrop's alchemical reform program was also a significant aspect of millenarian Puritanism. Puritans came to New England as futurists embarked on a mission to usher in the millennium, as Sacvan Bercovitch has argued. The participants in the Great Migration of 1630 to 1640 crossed the seas to make the New World of regeneration a physical reality. They arrived with a clear vision of their errand and a "sense of themselves as a chosen people under a covenant." "So now it was needfull," the Massachusetts Puritan historian Edward Johnson wrote, "that the Churches of Christ should first obtain their purity, and the civill government its power to defend them, before Antichrist comes to his finall ruine: and because you shall be sure the day is come indeed, behold the Lord Christ marshalling these N[ew] E[ngland] people."[44]

Alchemy, too, shared in these millenarian expectations. The alchemical revival ushered in with the Rosicrucian manifestos was explicitly millenarian, and New England the theater in which Winthrop hoped to use his alchemical knowledge to help usher in the general reformation of the world. In the effort to unlock the hidden secrets of nature, New England offered special potential and had special needs. Bacon's prophetic Great Instauration seemed confirmed by the discovery of and the many subsequent discoveries in the New World. Previously unknown plants and animals, not to mention the vast stores of gold and silver that had been transported back to Europe in the Spanish flotas, underscored the scientific, medical, and mineral revelations that might be

---

thorne" (Ph.D. diss., University of Connecticut, 1994); John Cotton, *Gods Mercie Mixed with His Justice; or, His Peoples Deliverance in Time of Danger, Laid Open in Severall Sermons* (London, 1641), 118.

43. Theodore Dwight Bozeman, *To Live Ancient Lives: The Primitivist Dimension in Puritanism* (Chapel Hill, N.C., 1988), 14, 23, 35.

44. Sacvan Bercovitch, *The American Jeremiad* (Madison, Wis., 1978), 33, 79.

revealed to a faithful American magus. The English alchemist Noah Biggs directly equated alchemical discoveries with the wonders of the western lands: alchemical study prepared the intellect to penetrate to the deepest mysteries, "and maketh an investigation into the America of nature." Since God was the ultimate distributor of all alchemical knowledge, it was not unreasonable to believe that the consciously religious refuge in New England might be a place for special alchemical revelations. There, the mystery of transmutation might be revealed to the worthy, mineral treasures be discovered by the knowledgeable. More pragmatically, in new plantations struggling to achieve stability the need for the practical application of knowledge about medicines, metals, and minerals was inordinately great.[45]

New England also offered benefits to alchemical development that Old England lacked. First, there were abundant land and natural resources. New England was a large-scale laboratory in which to test alchemical production schemes that required space or natural resources unavailable—or too expensive for experimental use—in England. America was also an ideal environment in which to experiment with new methods for iron foundries, tar production, mineral extraction and processing, and a range of other alchemically based agricultural improvements. The American climate offered opportunities that the English climate precluded—Winthrop would, for example, test several methods of extracting salt from seawater through a heat evaporation process unthinkable in England. In America, alchemically derived innovations could be vetted, local resources exploited, and an infant colonial economy placed on a firm footing.

In the late summer of 1631, at the head of a family group that included his wife, stepmother, and a younger stepbrother and stepsister, John Winthrop, Jr., set sail aboard the *Lyon* for New England. His preparations for emigration reflected his belief that New England would be a laboratory for alchemical research and development. Before embarking from England, he gathered and had carefully packed for shipping alchemical glassware and chemicals and a barrel full of alchemical texts. He established a network of alchemical com-

---

45. Francis Bacon *The Great Instauration [1620]; and, New Atlantis [1627]*, ed. J. Weinberger (Arlington Heights, Ill., 1986); Susan Scott Parrish, *American Curiosity: Cultures of Natural History in the Colonial British Atlantic World* (Chapel Hill, N.C., 2006); William Wood, *New Englands Prospect*, ed. Alden T. Vaughan (Amherst, Mass., 1977), 37; Raymond Phineas Stearns, *Science in the British Colonies of America* (Urbana, Ill., 1970), 74, 80, 147; Gabriel Plattes, *A Discovery of Subterranean Treasure* . . . (London, 1679), 14; Webster, *The Great Instauration*, 16, 36, 146, 394; Noah Biggs, *Mataeotechnia Medicinae Praxeos: The Vanity of the Craft of Physick; or, A New Dispensatory* (London, 1651), 57; Allen G. Debus, "Paracelsian Medicine: Noah Biggs and the Problem of Paracelsian Reform," in Debus, ed., *Medicine in Seventeenth Century England* (Berkeley, Calif., 1974), 40.

municants with whom he could share information and from whom he could request books and supplies. Arriving in Massachusetts in November 1631, Winthrop lost little time in setting up an alchemical furnace at his father's house in Boston. His texts told him that the millenarian reformation of the world would be accomplished through a comprehensive series of utilitarian improvement projects. Where better to test them than in a virgin wilderness already being transformed by godly colonists?[46]

---

46. "Accounts of John Winthrop, Jr.," *Winthrop Papers,* III, 1, and Bill of John Steward, Jr., to John Winthrop, Jr., July 22, 1631, 45–46.

{ TWO }

The Republic of Alchemy and the Pansophic Moment

On the November 1631 day that John Winthrop, Jr., stepped ashore to the welcoming salutes of cannon fire and musket volleys from the Bay Colony's trainbands, he began a career of colonial leadership that would see him become one of the most important figures in all English America. Twenty-five years old and the firstborn son and namesake of Massachusetts's governor, Winthrop was destined by birth for colonial preferment and position. His affable, entrepreneurial personality, intercultural sensitivity, political savvy, and scientific knowledge helped him parlay that preferment into positions of Atlantic world eminence. Over the next half century, Winthrop would found three colonial towns, serve as a Bay Colony assistant for nearly two decades, govern the colony of Connecticut for eighteen years, secure that colony a charter from the Restoration court of Charles II granting it virtual independence, found several New England iron foundries, serve as physician to nearly half the population of Connecticut, and become a founding member of the Royal Society. Alchemical knowledge and philosophies factored, often essentially, into each of these accomplishments. To understand John Winthrop, Jr., as an English colonial leader, it is essential to understand how his thinking about alchemy developed and how he deployed alchemy as a strategy for colonial development during a particular moment in the history of both alchemy and English colonization. To understand that is to see, in turn, the development of colonial New England, and especially Connecticut, from a new and revealing perspective. Long considered, if at all, a bucolic backwater of the colony to its north, Connecticut herein takes center stage as a place upon which many early modern Europeans focused great attention and from which a conflicting set of transformations came to be expected.[1]

Throughout his years in New England, and never more than during his

---

1. John Winthrop, Nov. 4, 1631, in Richard S. Dunn, James Savage, and Laetitia Yeandle, eds., *The Journal of John Winthrop, 1630–1649* (Cambridge, Mass., 1996), 60.

first years as a colonist, Winthrop's correspondence to and from Europe was punctuated with alchemical communications: orders for books and alchemical apparatus and discussions of experiments, alchemical movements and adepts, receipt of ore samples, and requests for chemicals useful in assaying metals. From first arrival Winthrop expected the broad-ranging technological aspects of alchemical natural philosophy to play a major role in establishing the Puritan project on a sound economic footing. To that end, he undertook a variety of tasks that afforded him the opportunity to explore the coastal regions and interior of New England, assaying the local physical environment for potential metal or mineral ore deposits. In addition to serving on the colony Court of Assistants, he managed the Bay Company's fur trade business and in the spring of 1633 moved north of Boston to found the town of Agawam (now Ipswich). His requests for chemicals during this period suggest the kinds of practical ventures he envisioned. He ordered ingredients useful for soap- and glassmaking, sulfur for use in gunpowder manufacture, sandiver for assaying metals, and other minerals with which to conduct alchemical experiments. He also ordered a variety of medicines while continuing his research into the panacea alkahest. Even at this early stage of his career, Winthrop probably provided medical care in the communities in which he resided, though it is unlikely that he practiced medicine to nearly the extent he was to do in later years.[2]

The books obtained for him by Edward Howes, with whom the alchemical partnership formed earlier at the Inner Temple continued to grow, show that Winthrop's interests melded the occult aspects of alchemy and related occult sciences into his practical chemical pursuits. Among the texts he acquired were Petrus Galatinus's *De arcanis catholicae veritatis,* a work on the mystical Cabala, a volume by the Bristol alchemist Samuel Norton on the philosopher's stone, and two volumes of the work *Utriusque cosmi mairoris scilicet et minoris metaphysica, physica atque technica historia* by Robert Fludd. Fludd, a Kentish-born member of the Royal College of Physicians, had published the first English defense of the Rosicrucian movement in 1616. Howes had strongly recommended Fludd to Winthrop as an alchemist whose works were "well drest for your Pallate."[3]

---

2. Ronald Sterne Wilkinson, "John Winthrop, Jr. and the Origins of American Chemistry" (Ph.D. diss., Michigan State University, 1969), 25–32; Edward Howes to John Winthrop, Jr., Nov. 9, 1631, in Samuel Eliot Morison et al., eds., *Winthrop Papers* (Boston, 1929–), III, 54–55, Frances Kirby to John Winthrop, Jr., Feb. 26, 1634, 150–152, Henry Jacie to John Winthrop, Jr., June 12, 1633, 126–128 (hereafter cited as *Winthrop Papers*).

3. Edward Howes to John Winthrop, Jr., Mar. 29, 1634, *Winthrop Papers,* III, 157–159.

Unlike the Scholastic philosophers who argued that God was an efficient but removed cause of events in the natural world, Fludd argued that God was literally active in every quotidian event. "It is his reall Spirit that filleth all things," Fludd wrote, "and not any accidental vertue, as is falsly imagined by some." True wisdom was to be found through the examination of God's two books of revelation—the book of scripture and the book of creation, that is, the natural world. Since man was a representation in miniature of the great world around him, Fludd suggested that a true Christian philosopher might learn most about the natural world by focusing his efforts on determining the proportional relationships linking the outer world, the heart, and the soul. Alchemy was the best science with which to gain this knowledge, since the effort to achieve chemical discoveries inevitably amalgamated the study of nature, mathematical analysis, and piety. In his emphasis on harmonies and proportions, Fludd reflected both the mystical mathematics of Pythagoras and the magical mathematical focus of John Dee, who like Fludd believed that abstract mathematical proportions contained the key to divine operations in the world. The presence of eleven of Fludd's works in the surviving Winthrop family library suggests Howes was correct in his opinion that Fludd's philosophy closely reflected Winthrop's own intellectual predispositions.[4]

Letters to Winthrop from Edward Howes (more than twenty-five surviving from Winthrop's first decade in New England) demonstrate the fusion of pragmatic economics and spiritual intentions that characterized the Christian alchemy Winthrop practiced in America. Howes, whose relationship with Winthrop was at one time so close that Winthrop called him his "alter idem"—another self—was an important source of information and advice on a wide range of topics, from piety to killing wolves. While helping nurture Winthrop's growth as an alchemist, Howes provided literature and counsel on many subjects pertaining to colonial development: how to manufacture large-bore military ordnance, how to make huge boiling pans from New England's abundant supply of wood, the "orderinge of silkworms" for use on the region's mulberry trees, the location of the Northwest Passage (up the Hudson

---

4. Robert Fludd, *Mosaicall Philosophy: Grounded upon the Essentiall Truth or Eternal Sapience, Written First in Latin, and Afterwards Thus Rendered into English* (London, 1659), 16; Allen G. Debus, *The English Paracelsians* (London, 1965), 115; Wilkinson, "John Winthrop, Jr. and the Origins of American Chemistry," appendix, 443–444. The Winthrop family library represents multiple generations of Winthrop alchemical texts, so it is not safe to assert that John Winthrop, Jr., acquired all these books. Nevertheless, all of the texts discussed herein were published before or during the years under discussion. On the book of nature, see James J. Bono, *The Word of God and the Languages of Man: Interpreting Nature in Early Modern Science and Medicine*, I, *Ficino to Descartes* (Madison, Wis., 1995), 123–166.

River), and establishing relations with native people. Howes recognized the need for "a chief pillar to the new sion" to apply knowledge against a broad spectrum of human activities, and he fully encouraged Winthrop's commitment to do so.[5]

When Winthrop returned briefly to England in the fall of 1634 following the death of his first wife, Martha, he and Howes renewed their joint alchemical quest, attempting to make contact with the Rosicrucians. The person they believed to be a member of that underground brotherhood was a certain Dr. E or Ever, whom David Como has identified as Dr. John Everard, a minister whose "alchemical divinity" had a deeply Hermetic focus and whose religious inclinations tended toward an extremely antinomian Puritanism that embraced aspects of Familism. Everard was a committed alchemist, an associate of Fludd, and the celebrated translator into English of the *The Divine Pymander of Hermes Mercurius Trismegistus* in 1650.[6]

Howes and Winthrop visited Everard before Winthrop returned to New England in July 1635. At that meeting, Winthrop attempted to query him regarding conflicts in the theories of various alchemical authorities but met with little success. Howes, who visited Everard two or three times after Winthrop's departure, also got "but small satisfaccion" about the same queries. He believed Everard "hath some preiudicate conceipt of one of us, or both." Howes implied that the prejudice was against Winthrop, for he noted, "I must confesse he seemed verie free to me, only in the maine he was misticall." Howes reported that Everard's response to his reiteration of Winthrop's earlier queries was to issue what was essentially a pro forma response among alchemists: that God holds the key to the mysteries of the alchemical quest and that, when he willed, he could grant Winthrop the knowledge that would resolve the inconsistencies he encountered in his alchemical treatises. "This he said," Howes told Winthrop, "that when the will of God is you shall knowe what you desire, it will come with such a light, that it will make a harmonie amonge all your authors, causing them sweetly to agree, and put you for ever out of doubt and question." When asked about how to contact the Rosicru-

---

5. Edward Howes to John Winthrop, Jr., Feb. 25, 1640, *Winthrop Papers*, IV, 202–203, Nov. 9, 1631, III, 54, Mar. 7, 1632, III, 66, Mar. 26, 1632, III, 72, Mar. 26, 1632, III, 73, Apr. 3, 1632, III, 75, Apr. 20, 1632, III, 76–77, July 1632, III, 85, Nov. 23, 1632, III, 94–95, Nov. 24, 1632, III, 96–97, June 5, 1633, III, 124, Apr. 18, 1634, III, 164.

6. Ibid., Feb. 25, 1640, IV, 202–203; David Raymond Como, "Puritans and Heretics: The Emergence of an Antinomian Underground in Early Stuart England" (Ph.D. diss., Princeton University, 1998), 1–214; Como, *Blown by the Spirit: Puritanism and the Emergence of an Antinomian Underground in pre-Civil-War England* (Stanford, Calif., 2004), 415–425.

cian brotherhood, Everard was close-lipped. "To discerne the fratre scientiae," Howes wrote, "I cannot as yet learne of him."[7]

Alchemy was a natural philosophy whose metaphorical texts could be interpreted as espousing a wide range of ideas, both scientific and theological. Christian alchemy was many things to many people, and, depending on which alchemical texts or interpretations of those texts one adopted, it could reflect a wide range of confessional beliefs, ranging from the most orthodox and commonplace to the most radical and extreme. Just as there was no single accepted interpretation of Calvinism, there was no single accepted interpretation of Hermeticism, Paracelsianism, Rosicrucianism, Christian Cabala, or many other fundamental concepts informing Christian alchemy. Although alchemy appealed to and was practiced by Puritans with a broad range of theological perspectives, some very vocal defenders of New England's Puritan orthodoxy viewed alchemy as potentially subversive. Most New England Puritans, most of the time, accepted the alchemists' practicing in their communities without reservation. Nevertheless, at various moments—usually during periods of religious tension or social or economic unrest—latent concerns about alchemy's compatibility with the prevailing Puritan orthodoxy surfaced, and at those times a significant number of Puritans cast a wary eye on the alchemists among the godly—even, on occasion, one as highly placed, respected, and politically important as Winthrop.[8]

Striking similarities have been noted between the chemical and spiritual illumination sought by alchemists and the perfectionist theology advocated by the antinomian religious sect known as the Family of Love, whose views had a strong influence on the theology of John Everard. Founded by the charismatic Dutch reformer Hendrik Niklaes in the 1540s, the Family of Love was noted

---

7. Edward Howes to John Winthrop, Jr., Aug. 21, 1635, *Winthrop Papers,* III, 206. In his work on the "antinomian underground" of seventeenth-century England, David Como has suggested that Everard's spiritual views were the draw that attracted Howes and Winthrop to visit him. In this interpretation, the two men approached Everard primarily seeking spiritual rather than chemical enlightenment, because of Everard's highly illuministic and antinomian theological beliefs. Such a reading would have significant implications for interpreting Winthrop's subsequent career in New England, since it would imply that he held the same highly enthusiastic religious views that provoked the violent reaction against Anne Hutchinson and her followers during the free grace controversy in Massachusetts only two years later. This in turn would make it difficult to explain his lifelong leadership roles in both of the North American Puritan colonies. Everard's alchemical reputation, rather than his radical antinomianism, was most certainly the attraction for Winthrop. Once it became clear that the cleric would speak only mystically about his purported alchemical knowledge, Winthrop rapidly lost interest in him. Como, *Blown by the Spirit,* 419.

8. Robert M. Schuler, "Some Spiritual Alchemies of Seventeenth-Century England," *Journal of the History of Ideas,* XLI (1980), 293–318.

(and reviled by many) for its conviction that true believers would experience a full personal indwelling of God, thus freeing them from dependence on oversight by an established church or even the scriptural commands of the Bible. At the same time, Familists, as members of the Family of Love were called, practiced Nicodemism, holding that it was acceptable to appear to conform outwardly to the requirements of any locally established church in order to avoid censure or persecution. The homologies between the expressed goals of many Christian alchemists and Familists—both sought to receive direct divine inspiration, and both aimed at restoring to the world a kind of prelapsarian purity—coupled with the Nicodemism that allowed Familists to masquerade as a member of any denomination have led historians to posit that a host of early modern natural philosophers had Familist leanings, including John Dee and Edward Kelly, Bernard Palissy, Jakob Böhme, Charles Lécluse (Carolus Clusius), John Everard, Edward Howes, and even Winthrop.[9]

Despite the compatibilities between Familism and Christian alchemy, one need not impute a commitment to Familism or any other form of antinomian radicalism to explain why Winthrop visited John Everard in 1635. His motives, to obtain alchemical knowledge and to make contact with the Rosicrucian brotherhood, are made clear in the correspondence between him and Howes. It was Everard's reputation as a Rosicrucian that drew the two men to seek him out, and in that regard they found the fiery divine a bit of a disappointment.

From private correspondence, it is clear that both men found little in Everard's belief or alchemical practices to their liking. A year after Winthrop had returned to New England, Howes wrote of Everard dismissively. "The Dr. I have not seene since last Sommer; I doubt all is not gold that glisters like it, and he that would learne to distinguish, may pay too deare for his knowledge." Three years later, when Howes wrote of Everard again, it was also in dismissive terms. Describing to Winthrop an Arabian alchemical philosopher named Dr. Lyon, Howes noted that he was "the best of the Rosicrucians that ever I mett with all, farre beyond Dr. Ever[ard]." For Winthrop and Howes, Everard was apparently neither a model alchemist nor a model theologian.[10]

---

9. Deborah E. Harkness, *John Dee's Conversations with Angels: Cabala, Alchemy, and the End of Nature* (Cambridge, 1999), 154–156; Neil Kamil, *Fortress of the Soul: Violence, Metaphysics, and Material Life in the Huguenots' New World, 1517–1751* (Baltimore, 2005), 199–207; Harold J. Cook, *Matters of Exchange: Commerce, Medicine, and Science in the Dutch Golden Age* (New Haven, Conn., 2007), 94–96, 107–108; Como, *Blown by the Spirit*, 219–265.

10. Edward Howes to John Winthrop, Jr., June 21, 1636, *Winthrop Papers*, III, 272, Feb. 26, 1640, IV, 202–203.

Winthrop's willingness to seek alchemical information from Everard as well as from other Christians of a variety of sectarian persuasions reflects Neoplatonic views widely held in both alchemical and mainstream Puritan circles. Belief that God had not fully revealed himself to any one Christian, but, rather, had given many different sects imperfect understandings of his perfect unity was commensurate with millenarian views holding that a flawed and corrupt world was groping its way toward perfection. As Christians sought an ever more perfect understanding of God and improvement of the world, it was wise to allow other Christians, within limits, to work toward their own Christian destiny and attain their own imperfect but perhaps revealing vision of godliness. By sharing these imperfect understandings of truth with one another, in an atmosphere of tolerance, the Christian community writ large could inch its way toward the perfect understanding all sought. Tolerance of religious diversity among Protestants would hasten the world toward its ultimate fulfillment. Such views of the state of Christian progress energized many of the important reform movements of the seventeenth century: John Dury's efforts to reunify the Protestant churches, the pansophism of Jan Comenius, and many elements of the early modern alchemical revival.

Many, if not most, alchemists believed that God granted his secrets to faithful researchers without regard to sectarian belief. This assumption helped men like Winthrop establish networks of communication that bridged sectarian distinctions and spanned the globe. From the Neoplatonic perspective, Massachusetts, through its policies limiting the honest exploration of religious belief and by giving churches control over the Bay Colony government, could be seen to be limiting the Puritan movement rather than defending it. Many of Massachusetts's staunch supporters, including perhaps even Winthrop, felt this to be the case. But, if Winthrop criticized Massachusetts's restrictive religious policies, he did so privately and as a Puritan Neoplatonist, not as a radical enthusiast.[11]

---

11. Karen Ordahl Kupperman, "The Connecticut River: A Magnet for Settlement," *Connecticut History*, XXXV (1994), 53; Keith Thomas, *Religion and the Decline of Magic* (New York, 1971), 227, 270–271; Brian Vickers, ed., *Occult and Scientific Mentalities in the Renaissance* (Cambridge, 1984), 1–55; Stephen Clucas, "Samuel Hartlib's *Ephemerides*, 1635–59, and the Pursuit of Scientific and Philosophical Manuscripts: The Religious Ethos of an Intelligencer," *Seventeenth Century*, VI (1991), 33–55; Mark Greengrass, Michael Leslie, and Timothy Raylor, eds., *Samuel Hartlib and Universal Reformation: Studies in Intellectual Communication* (Cambridge, 1994); Betty Jo Teeter Dobbs, *Alchemical Death and Resurrection: The Significance of Alchemy in the Age of Newton* (Washington, D.C., 1990); Edward Howes to John Winthrop, Jr., May 12, 1640, *Winthrop Papers*, IV, 241–242.

The once widely believed premise that English alchemical practice was synonymous only with Puritan radicalism is no longer credible. J. Andrew Mendelsohn, among others, has shown that alchemy was an ecumenical chemical theology. "The full spectrum of its politics . . . included

WHILE IN ENGLAND conducting experiments with Howes and seeking contact with the Rosicrucian brotherhood, Winthrop came in contact with a group of Puritan grandees whose association with the Providence Island Company had kept them engaged in matters associated with colonization throughout the 1630s: William Fiennes, Viscount Saye and Sele; Robert Greville, Lord Brooke; Sir Nathaniel Rich; Richard Knight; John Pym; and others. Fearing the worsening climate for Puritans in England, these leaders and their associates, who included Sir Richard Saltonstall, Sir Arthur Heselrige, John Humphrey, George Fenwick, and Edward Hopkins, were preparing to expatriate en masse to a refuge in America. They developed an ambitious scheme to settle a new Puritan colony along the Connecticut River that would be more tolerant of religious diversity than the Puritans of Massachusetts. They contracted with Winthrop to be the "Governour of the river Connecticut in New England" for one year and tasked him with establishing a fortification at the Connecticut River's mouth and seeing that houses "as may receave men of qualitie" were built within its walls. Winthrop accepted the appointment in July 1635, the day before he married his second wife, Elizabeth Reade, stepdaughter of the Puritan minister Hugh Peter, and within a few weeks the couple sailed for New England.[12]

They arrived in New England in October, and within a month Winthrop had sent an advance party to lay the groundwork for the new plantation. The next spring he journeyed to Connecticut to take charge of the Saybrook operation. Changing affairs in England, however, made the migration of grandees less likely. Winthrop's father wrote to him, "The gentlemen seem to be discouraged in their design here." Winthrop returned to Boston, sat as an assistant at the September meeting of the Court of Assistants, and then returned to the first town he had founded, which was by then sufficiently populated, fenced, and framed with timbered structures to have replaced its Indian name of Agawam with the civilized English name of Ipswich.[13]

Winthrop's reaction to the religious crisis surrounding Anne Hutchinson and her followers the following year must be inferred; no documents indicate his position on the issue. Important, *perhaps,* is the fact that he was not present

---

Anglicans, Puritans and sectaries, Royalists, Parliamentarians, and Levellers." "Alchemy and Politics in England, 1649–1665," *Past and Present,* no. 135 (May 1992), 37–38.

12. Kupperman, "The Connecticut River: A Magnet for Settlement," *Conecticut History,* XXXV (1994), 56; "Agreement of the Saybrook Company with John Winthrop, Jr., July, 1635," *Winthrop Papers,* III, 198–199; Wilkinson, "John Winthrop, Jr. and the Origins of American Chemistry," 43; Robert C. Black III, *The Younger John Winthrop* (New York, 1966), 85–90.

13. John Winthrop, Nov. 3, 1635, in Dunn, Savage, and Yeandle, eds., *The Journal of John Winthrop,* 161; John Winthrop to John Winthrop, Jr., June 10, 1636, *Winthrop Papers,* III, 268–269.

either to defend or to condemn her when she was tried in November 1637, although as a colony assistant he could have participated and probably was expected to participate. Undoubtedly, his absence was a source of concern or consternation to his father, who, as governor, led the prosecution. The younger Winthrop's reluctance to join in Hutchinson's prosecution is one of the historical markers indicating the ways he differed from his namesake. The differences between father and son—which never became a source of public controversy and only rarely boiled up in private communications—illuminate the ways in which alchemical culture produced an alternative culture in New England significantly different from that associated with Massachusetts's and Connecticut's early colonial leadership.

Although the issues in the free grace controversy surrounding Hutchinson were complex, at the heart of the crisis was a debate whether Christians received grace as a direct and immediate infusion of the spirit into the believer by God, or whether saints' awareness of the presence of grace in their lives developed more slowly, through the course of living a Christian life. The former view, held by Hutchinson and her supporters, led them to accuse many of the Bay Colony's ministers and church members of not being true Christians, but only observers of meaningless laws and rituals. The Hutchinsonians' belief in the direct infusion of grace, on the other hand, led critics to attack them as enthusiasts, Familists, and antinomians. These two positions became focal points that polarized the Massachusetts churches in 1636 and 1637, creating potential civil as well as religious unrest.[14]

The elder John Winthrop was at the very center of this controversy. Like his son, he believed in the value of a religious toleration that would allow the godly to work out their salvation through an open exchange of points of difference in religious matters. As Francis Bremer has written of the early years of New England settlement, "Unity, rather than uniformity, was the hallmark of the New England struggle for godliness." The elder Winthrop's range of tolerance, however, was more circumscribed than his son's, and it in no way extended to anyone espousing positions he viewed as Familist. After seeking assiduously and at significant cost to his own reputation to get the contending factions to moderate their attacks on one another, he determined that Hutchinson and her brother-in-law the Reverend John Wheelwright were indeed Familists, and he strongly supported, if not led, the successful effort to have them banished from the colony and their followers suppressed. The

---

14. The crisis is well summarized in Francis J. Bremer, *John Winthrop: America's Forgotten Founding Father* (New York, 2003), 275–300; the best complete account is Michael P. Winship, *Making Heretics: Militant Protestantism and Free Grace in Massachusetts, 1636-1641* (Princeton, N.J., 2002).

younger Winthrop does not seem to have agreed with the strict position of his father, as his absence from Hutchinson's trial and his discreet silence on the issue suggest. He would observe the same policy of absenting himself from a trial in which his own views differed from those of his father over a somewhat similar issue in 1646, when his friend and alchemical partner Robert Child challenged the authority of the Bay magistrates over the Bay's restrictive church membership and enfranchisement policies.[15]

Given his Neoplatonic view of Christian understanding, it is most probable that Winthrop believed both Hutchinson and Child were entitled to liberty of conscience, and to a degree his father found unacceptable. Throughout his life, the younger Winthrop maintained friendly relations with people holding a wide variety of religious opinions, from the ultrafundamentalist Roger Williams, to the Quaker William Coddington, to the Catholic alchemist and natural philosopher Sir Kenelm Digby. Among English natural philosophers he had a reputation for being an advocate of religious tolerance. The English alchemist and onetime surveyor general of Ireland Benjamin Worsley, when asking the promoter Samuel Hartlib to arrange an introduction to Winthrop, urged Hartlib to tell Winthrop "that in all things relating to publicke good, Just Lyberty of Conscience and any sort of ingenuus kinde of improvement, he will finde Mr. Worsley, as you beleeve, according to [his] owne hearts desire."[16]

Yet, despite his religious tolerance, Winthrop's absence from Hutchinson's and Child's trials might have reflected ambivalence rather than overt disapproval of the proceedings. Where the younger Winthrop seems to have drawn the line on liberty of conscience, at least later in life, was its exercise in a manner that provoked public unrest. Religious zealotry, when publicly targeted at the dominant religious settlement, was destabilizing. Those who sought to foment political unrest to gain religious advantage acted in ways that exceeded the boundaries of the personal exercise of liberty of conscience, and they therefore had to be prepared to face the consequences of their actions. After he was governor of Connecticut, Winthrop personally interceded with Massachusetts to urge it not to execute Quakers. Nevertheless, when the Quaker William Coddington harangued him to further loosen Connecticut's relatively moderate policy toward the sect, Winthrop firmly rejected his demand. Personal religious beliefs were one thing, aggressive advocacy an altogether different matter.[17]

15. Bremer, *John Winthrop*, 278.
16. Extract, Benjamin Worsley to ———, with a Message for Winthrop, undated, Hartlib Papers [33/2/27A], University of Sheffield.
17. John Winthrop, Jr., to William Coddington, Sept. 6, 1672, MS, Winthrop Family Papers, 5.75, Massachusetts Historical Society, Boston; Black, *Younger John Winthrop*, 185–186.

A document in Winthrop's handwriting, which recounts the position on religious liberty adopted at the 1659 Synod of Savoy, probably reflects Winthrop's personal attitudes toward liberty of conscience as much as it describes the approach he took as a public official toward sectarians.

> Although the magistrate is bound to encourage, promote, and protect the professor and profession of the Gospell and to manage and order civill administration in due subservience to the interest of Christ in the World . . . yet in such differences about the doctrine of the Gospell, or Waies of the Worship of God, as may befall men exercising a good conscience, manifesting it in their conversation, and holding the foundation, not disturbing others in their way of worship that differ from them, there is no warrant for the magistrate under the Gospell to abridge them of their liberty.

Christians could believe what they wanted to believe, but they could not seek to impose those beliefs on their neighbors or publicly attack those whose beliefs differed from theirs.[18]

INSTEAD OF PARTICIPATING in the trial of Anne Hutchinson, Winthrop remained in Ipswich, where he continued his alchemical studies. Correspondence with Edward Howes and John and Abraham Kuffler shows that Winthrop's interest in the philosopher's stone continued unabated. He sought details from the Kufflers about their own alchemical progress and secured from them a volatile and dangerous chemical water, which he had shipped to New England in a chest surrounded by five or six pecks of salt to reduce the risk of breakage. Winthrop also continued to aggressively apply his specialized knowledge to projects that would support New England's economic development.[19]

From the earliest English efforts to colonize northeastern North America, promoters had insisted that New England's surest and fastest path to economic stability would come from the fishing trade. Shortly after Winthrop's return to the Bay, his father-in-law, Hugh Peter, had gone "from place to place labouringe bothe publicly and privally to rayse up men to a public frame of spirit, and so prevayled as he procured a good summe of money to be raysed to sett on foote the fishinge businesse." Developing a substantial New England fishery depended, among other things, on having an ample supply of salt, to cure the

---

18. Black, *Younger John Winthrop*, 186; MS, Winthrop Family Papers, 1.161.
19. Francis Kirby to John Winthrop, Jr., Apr. 10,, 1637, *Winthrop Papers*, III, 385, Abraham Kuffler to John Winthrop, Jr., June 12, 1639, IV, 122.

fish in preparation for shipment. In the absence of a sufficient domestic supply, salt had to be imported at high cost. Winthrop, no doubt at his father-in-law's urging, decided to employ his chemical skills in establishing a New England saltworks. He was considering the venture as early as 1636, for in March of that year he received a letter from his aunt Lucy Downing offering to invest in the venture. Not until the summer of 1638, however, after he received concessions of land from the town of Salem, did he begin the project in earnest. He built a salthouse near the junction of the Bass and Danvers rivers and hired an employee to conduct the saltmaking operation. By 1639, he and his wife were resident at the site, and in a letter to Jacob Golius (the scholar of Arabic texts he had met on his journey to the Middle East) he explained that the saltworks was part of his continuing effort to expand and usefully employ his specialized knowledge. He told Golius that he was developing inventions needed by the Bay Colony, one of which was a fast method to extract salt from seawater. He related these efforts to his interests in the Hermetic philosophy, medicine, optics, and the mathematics of the superior and inferior harmonies.[20]

The value placed on salt in the thinking of many natural philosophers demonstrates how even mundane processing activities such as saltmaking could become imbued with profound metaphysical significance. Salt assumed a dominant role in the thinking of many who embraced the macrocosm-microcosm theory. Robert Fludd believed the life force was a divine salt, which he had extracted from wheat. Johann Glauber used scripture to show the primacy of salt above all other substances. Christ called his disciples the salt of the earth, and Christ himself, Glauber said, was a pure and divine salt. Salt's importance as a preservative and its value as a fertilizer demonstrated that it must have within it the life force itself. Glauber believed salt emanated from the sun's fire and passed down through air into seawater and then into the earth, bringing life and fecundity with it. The philosopher Jan Comenius used salt as an example of how alchemical research had been able to clarify the mysteries of scripture. Alchemists had, he argued, discovered God's reason for forbidding sacrifices to be made without salt.[21]

---

20. On the potential of a New England fishery, see Walter W. Woodward, "Captain John Smith and the Campaign for New England: A Study in Early Modern Identity and Promotion," *New England Quarterly*, LXXXI (2008), 91–125; John Winthrop, January 1636, in Dunn, Savage, and Yeandle, eds., *Journal of John Winthrop*, 164. On the importance of salt in New England, see Stephen Innes, *Labor in a New Land: Economy and Society in Seventeenth-Century Springfield* (Princeton, N.J., 1983), 234, 296; Lucy Downing to John Winthrop, Jr., Mar 6, 1637, *Winthrop Papers*, III, 369, John Winthrop, Jr., to Jacob Golius, Nov. 20, 1639, IV, 155–156; Wilkinson, "John Winthrop, Jr. and the Origins of American Chemistry," 50–55.

21. J. T. Young, *Faith, Medical Alchemy, and Natural Philosophy: Johann Moriaen, Reformed Intelligencer, and the Hartlib Circle* (Aldershot, 1998), 166–167.

If you know the true natural properties of salt, the mystic sense of this ordinance cannot escape you. It is because salt above all other material creatures is tenacious of its essential quality, and for this reason is, as it were, incorruptible (for it neither decays itself nor even allows things on which it is sprinkled to decay). And even if it is forced to change its external form—as may befal it when it is brought into contact with things of an opposite nature—for it is dissolved and dispersed in water and broken up by fire—still it does not change or lose its own nature. (For the water which dissolves the salt is compelled to drink in the quality or nature of the salt; and the salt when broken and reduced to atoms by fire, is still the same. Must not the belief, then, be held that by this symbol, God through the example of stones and fire enjoins constancy upon his worshippers?[22]

Saltmaking, which Winthrop pursued on several occasions, was simultaneously a commercial venture, a metaphysical exploration, and a source of scriptural exegesis, as were many of the utilitarian ventures of early modern alchemical projectors. Winthrop pursued the saltmaking operation at Salem nearly three years, suspending it just before his return to England in 1641. Little is known about the economics of the operation. Though some historians have assumed that the venture was closed because it had proved unprofitable, the eagerness with which Salem merchants later sought to support another Winthrop saltmaking venture suggests it showed, at the very least, great potential. In any event, the demand for domestically produced salt remained strong, and Winthrop continued his experiments to find cost-efficient means of salt extraction throughout his life. Winthrop's initial venture into industrial technology might have achieved marginal success, but it set the stage for greater projects ahead.

WINTHROP'S ENTREPRENEURIAL RESPONSE to the collapse of New England's economy at the onset of the English Civil War is noteworthy. The colonial economy that had developed in New England in the 1630s depended on a steady influx of immigrants. When the flow of immigrants suddenly stopped in the early 1640s as England itself moved into civil war, the resulting economic crisis in New England was immediate, severe, and sustained. Winthrop, through the knowledge of metals and processing he had gained from alchemical study, was able to conceive bold, imaginative solutions to New England's

---

22. John Amos Comenius, *The Way of Light* (1668), ed. and trans. E. T. Campagnac (Liverpool, 1938), 118–119.

crisis. Years of observation and assaying ores had persuaded him that exploiting New England's mineral wealth would do much to help restore economic stability. Moreover, he had the status, knowledge, connections, and entrepreneurial skill to shape that belief into schemes that would attract overseas investment. From 1641 to 1643, he returned to Europe to seek investors and capital for a New England ironworks. His arrival in England coincided with that of Jan Comenius, whose approach to uniting science and theology in the service of world improvement was to help shape Winthrop's goals.[23]

In a period of great eclecticism in natural philosophy—with "perhaps as many syntheses as there were authors"—Jan Comenius was among the most syncretic thinkers of his time. His vision of world reformation through a collaborative and systematic effort to identify the fundamental principles underlying all knowledge was the grand unified theory of its day. Pansophism, the possession or profession of universal knowledge, had roots reaching back to the Platonic ideal. It had been revived in the early modern period by an array of utopian thinkers and reformers, including Tommaso Campanella, Giordano Bruno, J. V. Andrae, Francis Bacon, Johann Heinrich Alsted (Comenius's professor at Herborn in Nassau), and Peter Laurenberg, all of whom (except perhaps Laurenburg) influenced the Comenian elaboration of the pansophic idea.[24]

Comenius called Bacon and Campanella the "Philosophiae restauratores gloriosos," but he sought to put realistic foundations under their utopian constructions. Because works such as Bacon's *New Atlantis,* Campanella's *Citta del sole,* Andrae's *Christianopolis,* and the Rosicrucian *Fama Fraternitatis* were presented as utopian tracts or secret manifestos, they were easily dismissed by

---

23. Richard S. Dunn, *Puritans and Yankees: The Winthrop Dynasty of New England, 1630-1717* (Princeton, N.J., 1962); E. N. Hartley, *Ironworks on the Saugus: The Lynn and Braintree Ventures of the Company of Undertakers of the Ironworks in New England* (Norman, Okla., 1957), 47-48; Samuel Eliot Morison, *Builders of the Bay Colony* (Boston, 1930), 269-288; Bernard Bailyn, *The New England Merchants in the Seventeenth Century* (New York, 1964), 47-49; Black, *Younger John Winthrop,* 110, 117.

24. The quotation is by Allen G. Debus in "The Chemical Debates of the Seventeenth Century: The Reaction to Robert Fludd and Jean Baptiste van Helmont," in M. L. Righini Bonelli and William R. Shea, eds., *Reason, Experiment, and Mysticism in the Scientific Revolution* (New York, 1975), 44; Daniel Murphy, *Comenius: A Critical Reassessment of His Life and Work* (Dublin, 1995), 20; Frank E. Manuel and Fritzie P. Manuel, *Utopian Thought in the Western World* (Cambridge, Mass., 1979), 205-221, 243-412; Tommaso Campanella, *La Città del Sole: Dialogo Poetico / The City of the Sun: A Poetical Dialogue,* trans. Daniel J. Donno (Berkeley, Calif., 1981); Frances A. Yates, *Giordano Bruno and the Hermetic Tradition* (London, 1964); Johann Valentin Andreae, *Christianopolis: The Ideal of the Seventeenth Century,* trans. Felix Emil Held (New York, 2007); Francis Bacon, *The Great Instauration [1620]; and, New Atlantis [1627],* ed. J. Weinberger (Arlington Heights, Ill., 1986); Johann Heinrich Alsted, *The Beloved City; or, The Saints Reign on Earth a Thousand Yeares* . . . , trans. William Burton (London, 1643).

critics as fantasies or impractical futuristic speculations. Bacon's *New Atlantis,* for example, when first published posthumously in 1627, included a disclaimer from Bacon's secretary, William Rawley, that Bacon had presented a model "more vast and high than can possibly be imitated in all things." Comenius believed such presentations undermined the author's serious intentions and inhibited meaningful reform. His *Labyrinth of the World,* written in 1623 to reflect his disillusion at the carnage of the early years of the Thirty Years' War, had been a satire of the utopian form. The English reformer Samuel Hartlib shared Comenius's view. "Two great faults are commonly committed by writers of utopia projects," Hartlib argued. "Either they propose their matter so that every body can perceive them to bee absolutely impossible to put in praxis . . . or they give false rules which should not at all be practiced, as in Campanella, that of viewing one another naked before they marry."[25]

Comenius sought to create a pragmatic plan that would give substance to the general reformation movement. Rather than relying on secret brotherhoods, as the Rosicrucians did, or on isolated and fantastic utopian communities, Comenius envisioned the creation of a publicly acknowledged, collaborative network of Christian natural philosophers who would take on the mission of improving the world for the coming of Christ. In works published by Hartlib and in the "Via lucis," a manuscript Comenius composed while in England in 1641 and 1642, Comenius outlined a reform program that led many to believe pansophic reform was a viable approach to effecting worldwide economic, technological, theological, social, and educational reform. The relative pragmatism of his pansophic outline, especially in comparison to the excesses of the utopias, helps account for the enthusiasm that greeted his ideas.[26]

Comenius sought to engage natural philosophers in a multidisciplinary, worldwide quest for knowledge. He chided those who sought only the truths

---

25. Johann Amos Comenius, *Pansophiae praeludium,* in *Opera didactica omnia* (1657) (Prague, 1957), I, part 1, 442, quoted in Manuel and Manuel, *Utopian Thought,* 317; W. Rawley, "To the Reader," in Bacon, *Great Instauration; and, New Atlantis,* ed. Weinberger, 36; Comenius, *The Labyrinth of the World and the Paradise of the Heart,* trans. Matthew Spinka (Ann Arbor, Mich., 1972); Francis A. Yates, *The Rosicrucian Enlightenment* (London, 1972), 156-170, esp. 161; Samuel Hartlib, "Ephemerides," 1640, Hartlib Papers [30/4/57a-57b].

26. Robert Fitzgibbon Young, whose positivist rejection of the mystical and occult dimensions of Comenius's work leads him to underestimate the value such a work could have for seventeenth-century natural philosophers, posited a different interpretation. He called the *Way of Light,* the tract Comenius penned while in England to promote the pansophic program, "ill-drawn and discursive," presenting a "rather nebulous and inconsequent account, tinged with chiliasm, of the scope and aims of the projected college." Young, ed. and trans., *Comenius in England: The Visit of Jan Amos Komenský (Comenius) . . .* (London, 1932), 6.

of a single science or discipline without simultaneously seeking the unifying divine principles that ordered all things. By focusing exclusively on the limited illumination gained from a discrete subject, he argued, Copernicus, William Gilbert, Campanella, and many others had gained only partial insights, which led them into numerous errors and contradictions. Ultimate knowledge could be derived only through the comparative analysis of discoveries made in a plethora of disciplines, coupled with an intense exploration of the Holy Scriptures to determine the divine principles that applied universally to them all. These principles, Comenius noted, were as harmonious as the macrocosm and microcosm, they were uniformly applied throughout the cosmos, and they would ultimately, following a long-term, worldwide education and reform program, be totally accessible to pious inquisitors.[27]

Comenius called for comprehensive educational reform. Pansophia demanded new pedagogical methods, new schools, even a new and universal language. The core of the program was to be a College of Light, whose elite members would be stationed throughout the world, pursuing knowledge, sharing discoveries, and improving society. The goal of pansophia, Comenius wrote, is "nothing in fact less than the improvement of all human affairs, in all persons and everywhere." Through pansophism, humanity would share "as a common possession, all the arts and sciences and the mysteries and treasures of wisdom." This knowledge, and the utilitarian benefits that derived from that knowledge, while central goals of the pansophic program, were not ends in themselves; they were merely the means through which pansophism would achieve its ultimate objective: the conversion of all non-Christians and the unification of all Christian sects in preparation for Christ's Second Coming.[28]

Comenius believed the pansophic educational program provided an especially effective means for converting the American Indians, who had to be won to Christ as a prerequisite to the millennium. Conversion of the natives required, most Protestants believed, distancing them from the unruly states of nature in which the English had found them and leading them to embrace civilized European social customs and labor practices. Without civility, the moral reforms required of those to be regenerated would be impossible to

---

27. Johann Amos Comenius, *The Great Didactic of John Amos Comenius, Now for the First Time Englished*, ed. M. W. Keatinge (London, 1896), 63.

28. Comenius, *The Way of Light*, ed. and trans. Campagnac, 6, 210–211. Comenius wrote the manuscript of the *Via lucis* while in England in 1641–1642, where it sparked great enthusiasm among Comenius's supporters. It circulated in manuscript until 1667, when it was printed in Amsterdam, with a new Comenius-penned dedication to the recently formed Royal Society. The quotation is from that dedication.

maintain, and conversion would be either impossible to achieve or of short duration. Some Protestant thinkers, including the influential Puritan minister and Cambridge professor Joseph Mede (Mead) thought the task so daunting that all conversion attempts were doomed:

> Concerning our Plantation in the *American* world, I wish them as well as anybody. . . .
> And though there be but little hope of the general Conversion of those Natives in any considerable part of that Continent; yet I suppose it may be a work pleasing to Almighty God and our Blessed Saviour, to *affront* the Devil with the sound of the Gospel and Cross of Christ in those places where he had thought to have reigned securely and out of the dinne thereof; and though we make no Christians there, yet to bring some thither to vex and disturb him, where he reigned without check.[29]

Comenius was more optimistic. Through appropriate educational approaches, the Americans could be freed from their ignorance and brought willingly to Christ. Pansophia was the key "whereby the hearts of all men may be opened unto all things; and universal assent won to the truth; likewise victory over whatever is of darkness from sheer lack of knowledge, such as prevails among those races."[30]

At a time when Europe was being savaged by religious war and sectarian conflict, Comenius's Neoplatonic belief in a single unified Christian religion free from sectarian conflict and uniting all people found many supporters, not only for its ideal of Christian unification but also for its programmatic approach to advancing knowledge and promoting technological innovation. Through the efforts of Samuel Hartlib, pansophia gained widespread public attention. Hartlib, a polymath reformer from Elbing (now in Poland), had come to England in the 1630s and was closely associated with John Dury's efforts to unify Europe's Protestant churches. Hartlib learned of Comenius's pansophic program from Moravian scholars visiting London. Comenius, himself a Moravian minister, had already acquired an international reputation as an educational reformer through a widely acclaimed text he had written promoting a new method for teaching Latin. Hartlib wrote him asking for details about his pansophic program, and Comenius replied with an outline of the

---

29. Joseph Mede, *The Third Book of the Works of the Pious and Profoundly-Learned Joseph Mede, B.D.* (London, [1664]), 980.

30. Comenius to Samuel Hartlib, June 5–15, 1647, reprinted in Robert Fitzgibbon Young, *Comenius and the Indians of New England* (London, 1929), 17–18.

project, which Hartlib hurried into print in 1637. This Latin work, *Conatuum comenianorum praeludia*, and a subsequent English translation of the pansophic scheme helped secure Comenius's reputation as Europe's leading proponent of the quest for universal wisdom and reform.[31]

Johann Tassius, professor of mathematics at Hamburg, wrote: "Every corner of Europe is filled with this pansophic ardor. If Comenius were to do no more than stimulate the minds of all men in this way he might be considered to have done enough." Another admirer claimed that humankind had received no greater benefit than pansophia since the light of God's word. Robert Greville, Lord Brooke—a prime force in two New World Puritan colonizing ventures, including the plantation Winthrop helped establish at Saybrook—proclaimed enthusiastically that, although truth was at present fragmented into "particular rivulets," "that learned, that mighty man Comenius doth happily and rationally endeavour to reduce all into one." Like his patron Brooke, Hartlib was caught up in the Comenian vision, and in the early 1640s he put in motion a scheme he hoped would lead to the implementation of the pansophic agenda on a global scale. He urged Comenius to journey to England, where Hartlib would use his extensive patronage network to seek support for the pansophic initiative. "Come, come, come," Hartlib urged a somewhat reluctant Comenius. "It is for the glory of God: deliberate no longer with flesh and blood!"[32]

---

31. Comenius's claim, "Christ, whom I serve, knows no sect," would resonate with particular strength with Winthrop, who had avoided participation in the persecution of Anne Hutchinson and her supporters in Massachusetts and was later to plead for mercy for Quakers there. Black, *Younger John Winthrop*, 105–106; Dunn, *Puritans and Yankees*, 106–107; Johann Amos Comenius, *Janua linguarum reserata, sive, omnium scientiarum et linguarum seminarium / The Gate of Languages Unlocked; or, A Seed-plot of All Arts and Tongues: Containing a Ready Way to Learn the Latine and English Tongue*, trans. Thomas Horne and John Robotham (London, 1647); Anthony Milton, "'The Unchanged Peacemaker': John Dury and the Politics of Irenicism in England, 1628–1643," in Greengrass, Leslie, and Raylor, eds., *Samuel Hartlib and Universal Reformation*, 95–117; Young, ed. and trans., *Comenius in England*; G. H. Turnbull, *Hartlib, Dury, and Comenius: Gleanings from Hartlib's Papers* (Liverpool, 1947); Turnbull, *Samuel Hartlib: A Sketch of His Life and His Relations to J. A. Comenius* (London, 1920). Hartlib might have learned of the pansophic program as early as 1632 and began collecting money to publish the pansophic tracts in 1634 (Young, ed. and trans., *Comenius in England*, 34). Hartlib followed Johanne Amoso Comenio and Samuel Hartlibius, *Conatuum comenianorum praeludia* . . . (London, 1637), with Johann Amos Comenius and Samuel Hartlib, *Pansophiae prodomus* (London, 1639), and John Amos Comenius, *A Reformation of Schooles, Designed in Two Excellent Treatises . . .* , ed. Samuel Hartlib (London, 1642).

32. Tassius quoted in Keatinge, introduction, in Comenius, *The Great Didactic*, ed. Keatinge, 31; Turnbull, *Hartlib, Dury, and Comenius*, 30; Robert [Greville], Lord Brooke, *The Nature of Truth* . . . (London, 1640), quoted in Karen Ordahl Kupperman, *Providence Island, 1630–1641: The Other Puritan Colony* (Cambridge, 1995), 225. The German scholar Joachim Hübner, a colleague of Hartlib's, wrote Comenius in December 1639, telling of plans for actually establishing the universal college

One of the appeals of pansophism, especially to those who viewed their age as a crucial premillenial moment, was the manner in which Comenius elaborated the relationship between empirical research and theology. The investigation of nature was the principal means by which most of the pansophic reforms, including the religious ones, would be accomplished. Comenius held that the natural world was "the first and greatest book of God." Studying nature for clues to the divine archetypes in nature was crucial, for the world functions as "nothing but a lower school into which we are sent before we can be promoted to the Heavenly Academy." Science, Comenius taught, would clarify scripture, and scripture would in turn unveil the divine, uniform principles applicable to all science.[33]

> A true knowledge of the world of Nature will be a key to the mysteries of the Scriptures. . . . For Scripture teaches us in general what is the origin of created things, and by what power they are sustained, and what end they are at last to have. But actual and particular created things reveal the true meaning of individual passages of the Scriptures, in which mysteries are offered to us under a veil or covering.[34]

Comenius believed reason functioned fully only when it was attuned to religious truth. "This, I say, is the Glory of Reason and of the Senses, to yield to the words of God, to believe incredible things when they are revealed, to attempt impossible things when they are enjoined, to hope for things invisible, when they are promised." Pansophism held up natural philosophy as the key to unlocking the mysteries of both God and nature.[35]

By subordinating empirical knowledge to scriptural truth and arguing that the ultimate goal of science was to ascertain the divine harmonics underlying all natural processes, Comenius implicitly granted primacy to Christian alchemy and other spiritually infused occult studies as the most productive areas of scientific investigation. Belief in the irreducible linkage between knowledge and piety was an axiom of Christian alchemy, as was Comenius's support of the macrocosm-microcosm analogue. Some skeptical intellectuals

---

and soliciting advice on the teachers that needed to be recruited. Murphy, *Comenius*, 23; "Komenský's Description of the Development of His Plan for an Encyclopaedia and a Great College for Scientific Research, and of His Visit to England in 1641–42," in Young, ed. and trans., *Comenius in England*, 25.

33. An example of how pansophia expected the natural world to illuminate scripture is seen in the way Comenius's description of the natural properties of salt revealed why God had forbidden the Israelites to make sacrifices without using salt. Comenius, *The Way of Light*, ed. and trans. Campagnac, 5, 14, 118–119.

34. Ibid.

35. Ibid., 15.

such as Descartes claimed that Comenius had confused philosophy with theology. Proponents of Christian alchemy such as Winthrop and Hartlib, however, viewed Comenius's synthesis as both welcome and essential.[36]

Comenius held that ameliorating the problems of human society globally was a long-term process that would be achieved through the cumulative impact of four universal programs, each of which could be implemented immediately: universal education for all men and women, the creation and distribution of universal books, the development of a universal language, and the establishment of a universal college. Public education was needed "not that all men should become learned . . . but that all men may be made wise unto Salvation." Comenius did not advocate a leveling of the social order. The upper echelons of enlightenment would remain the province of learned elites. "We do not mean that it is our desire that mechanics, rustics and women should devote themselves heart and soul to books," Comenius noted. "Those who have high position shall know their position and learn how to rule, and those who have a subordinate position may know theirs and learn to obey." With rudimentary education, however, the lower orders would play their own role in the general reformation, and they "will look up to the learned because they will remark in them the full flame of that light of which they have the sparks, and will rejoice and take pleasure in those sparks."[37]

Universal books, containing the condensed essence of all knowledge, would reveal the "very marrow of eternal truth . . . showing how all things proceed backward and forward from a single principle." Comenius called for the collaborative composition of three universal books: *Pansophia,* which would contain the essence of knowledge and reveal the unifying divine principles ordering reality; *Panhistoria,* history of the development and operation of these principles in specific disciplines; and *Pandogmatica,* which would review the competing theories that had been propounded over time regarding various things, and whether they were true or false. A single, worldwide

---

36. Jaromír Červenka, *Die Naturphilosophie des Johann Amos Comenius* (Prague, 1970); Yates, *Rosicrucian Enlightenment,* 156–170, esp. 168. Betty Jo Teeter Dobbs characterized Hartlib and the group of natural philosophers with whom he was associated as participating in but moving away from the more mystical and spiritual approaches to alchemy. Stephen Clucas has detailed the ways in which Dobbs's characterization of the Hartlib circle as rationalist-mechanistic was rather exaggerated (Dobbs, *The Foundations of Newton's Alchemy; or, "The Hunting of the Greene Lion"* [Cambridge, 1975], 62–80; Clucas, "The Correspondence of a Seventeenth Century 'Chymicall Gentleman': Sir Cheney Culpeper and the Chemical Interests of the Hartlib Circle," *Ambix,* XL [1995], 146–170). What the Hartlib circle best represents is "the simultaneous presence of occult and nonoccult" tendencies in Renaissance science discussed by Vickers (Vickers, ed., *Occult and Scientific Mentalities in the Renaissance,* 17).

37. Comenius, *The Way of Light,* ed. and trans. Campagnac, 128, 130–132.

language, a type of pansophic Esperanto to be taught as part of education, "would be the most appropriate means for reconciling [people] to each other and their concepts of things to the truth."[38]

Most important from the standpoint of implementing the program was the creation of the Universal College, the intellectual manufactory that would provide the "executants or workmen, who with God's help will carry the schemes which have been excellently thought out to completion." The possibility of creating the College of Light as a first step in the pansophic endeavor energized Hartlib and the patrons with whom he was associated, and it made Comenius the focus of much attention during his stay in England. Even before his arrival, John Gauden, the future bishop of Worcester, recommended Comenius to the members of Parliament as one who had "laide a fair design and foundation for the raising up of a Structure of Truth, Human and Divine, of excellent use to all mankinde." Comenius was favored by grandees such as Lord Brooke, the wealthy Kentish landowner Sir Cheney Culpepper, the parliamentarian John Pym, and the noted prelates Archbishop James Ussher and John Williams, bishop of Lincoln. Celebrated intellectuals, among them Theodore Haak, Joachim Hübner, and the mathematician John Pell, joined Hartlib in promoting the Comenian program. A proposal was brought forward to transform the College of Chelsea, founded in 1607 as a Protestant center for the campaign against Catholicism, into the site of the Universal College. Political unrest in England and rumors of insurrection in Ireland, however, led to numerous delays. While he waited in London throughout the summer of 1641 and into 1642, Comenius penned the *Via lucis,* a more detailed account of the overall pansophic program, and he helped promote the pansophic scheme to those who came in contact with the circle surrounding Hartlib and Dury.[39]

Winthrop arrived in England the same month as Comenius, and, in addition to meeting Comenius himself, Winthrop was during the next two years repeatedly in the company of other natural philosophers caught up in the pansophic moment. Both in England and on the Continent—in 1642 he traveled to Hamburg, Amsterdam, and the Hague pursuing alchemical and industrial knowledge—he found philosophers engaged in enthusiastic efforts to help initiate pansophic reforms.

One of them was Gabriel Plattes, who, like Winthrop, believed that alchemy was the key to understanding nature and the most useful tool for find-

---

38. Ibid., 139, 145–146, 186.
39. Ibid., 167; Comenius, *The Great Didactic,* ed. Keatinge, 45; Young, ed. and trans., *Comenius in England,* 43.

ing the means to improve the human condition. Plattes had come to the attention of Samuel Hartlib after two of his tracts, on improving mining and agriculture, were published in 1639. Shortly after Winthrop arrived in London in October 1641, Plattes anonymously published, probably with Hartlib's support, a utopian tract titled *A Description of the Famous Kingdome of Macaria, Shewing Its Excellent Government,* which presented a series of economic reform proposals directed to the newly assembled Long Parliament. These proposals reveal just how closely the vision of universal reformation was linked to early modern entrepreneurial projecting proposals.

Central to Plattes's argument was the proposition that establishing God's kingdom on earth would be achieved through massive economic development activity, much of which would be focused on the kinds of technological innovations generated by alchemists. Plattes saw clearly the economic potential of scientific knowledge and was particularly interested in advancing the development of metallurgy, husbandry, mining, and other alchemy-related industries, leading to the production of, among other things, precious metals and the development of new medicines. The pansophic agenda, while utopian in its ultimate goals, became, in the application of men such as Plattes and Winthrop, rooted in a very down-to-earth entrepreneurial pragmatism and an understanding that the ideals of universal reformation of the world could be brought about only through the successful implementation of a continual series of practical and economically successful improvement projects.[40]

Macaria, the name of the ideal kingdom Plattes's treatise was urging Parliament to emulate, was to be administered by a series of councils (the Council for Fishing, the Council for Trade by Land, the Council for Trade by Sea, and the Council for New Plantations), each with authority over a specific aspect of the nation's economy. Central to the success of the kingdom was the "Colledge of experience," a counterpart to the Comenian College of Light, dominated by its laboratory, where "all such as shall be able to demonstrate any experiment for the health or wealth of men, are honourably rewarded at the publicke charge." Despite *Macaria*'s use of the utopian genre, it was clear that Plattes believed the ideal society was attainable in the present—and was in fact a necessary condition for the millennium—through a series of systematic reforms that could be inaugurated immediately by Parliament and continually improved upon by natural philosophers. Plattes's reform scheme, with its coupling of utopian ideals and promotion of technological development, was tailor-made for Winthrop, who had come to London to seek funding for a

---

40. Charles Webster, *Utopian Planning and the Puritan Revolution: Gabriel Plattes, Samuel Hartlib, and "Macaria"* (Oxford, 1979), 25–32.

technology-based improvement scheme for New England of the type Plattes and Hartlib were encouraging for England itself. Charles Webster suggests Plattes might well have included Winthrop as one of the prospective founders of his proposed college. Whether or not he did, Winthrop was impressed with both Plattes and his alchemical knowledge. Years after he had returned to New England, he asked Hartlib whether Plattes had ever followed through on his offer to demonstrate transmutation to the members of Parliament.[41]

Another Hartlib associate, Robert Child, was also a reform philosopher, whom Winthrop had previously met in New England and with whom he was already engaged in alchemical planning. Child, a medical doctor, had acted as Winthrop's alchemical book buyer after he had returned from New England, attempting to find and forward the spagyric texts Winthrop requested from across the Atlantic. Child, like Plattes, was interested in agricultural reform and mining and had written to Winthrop announcing his intention to return to New England and possibly join with him in a mining venture while also establishing vineyards and experimenting with raising silkworms. Child and Winthrop reconnected in London, and, before Winthrop returned to New England, Child had become an investor in Winthrop's proposed ironworks and was promoting other joint projects the two would pursue back in America.

On the Continent, Winthrop established more relationships with members of the republic of alchemy. In Hamburg, a center for German intellectual development and a key link in the communications network of the Hartlib circle, Winthrop became friends with Johannes Tanckmarus, a doctor connected to the movement associated with the Paracelsian theologian Jakob Böhme. Winthrop might have studied alchemical medicine with Tanckmarus and with Paul Marquart Schlegel, another doctor and former professor of botany, anatomy, and surgery at the University of Jena. He met the Hebrew scholar, mathematician, and poet Johann von Rist, who years later remembered Winthrop's remarkable abilities with minerals. Winthrop also used his journey to expand his collection of scientific texts, which included at least one manuscript from the library of John Dee, a tract on perpetual motion by Thomas Norton.[42]

---

41. Gabriel Plattes, *A Discovery of Infinite Treasure* . . . (London, 1639); Plattes, *Macaria* (London, 1641), in Charles Webster, ed., *Utopian Planning and the Puritan Revolution*, 83; Webster, *The Great Instauration: Science, Medicine, and Reform, 1626-1660* (New York, 1976), 48–49; Manuel and Manuel, *Utopian Thought*, 325-326; John Winthrop, Jr., to Samuel Hartlib, Dec. 16, 1659, Hartlib Papers [40/3/2a].

42. Paul Marquart Schlegel to John Winthrop, Jr., September 1649, *Winthrop Papers*, V, 364, Johannes Tanckmarus to John Winthrop, Jr., October 1642, IV, 361; Johann Rist, *Die alleredelste*

A Hartlib circle member whose sphere of influence encompassed both Hamburg and Amsterdam was Johann Moraien, a Dutch Reformed minister, physician, and alchemist, then intensely interested in Comenius's pansophic scheme. He was friends with Abraham and Johann Kuffler, and it might have been through these old friends that Winthrop first came to know Moraien. Like Winthrop, the affable and well-respected Moraien was dedicated to demonstrating the ethic of public-spirited Christian alchemical service. In Amsterdam, he provided regular medical clinics, taking voluntary contributions. His knowledge of chemical medicines was extraordinary; a Paracelsian, he claimed to know forty-two different preparations of medicines derived from the mineral antimony. His medical philosophy influenced Winthrop's subsequent medical practice. After he returned to New England, Winthrop too began to hold public medical clinics and became famous for an antimonial preparation called rubila, the most sought-after medication in the Winthrop vade mecum.[43]

Moraien shared Winthrop's twofold commitment to pursuing alchemical discoveries for both utilitarian and Christian goals. He attempted, for example, to improve land that he had inherited with his wife, using the agricultural model advocated by Plattes. At the same time, he felt alchemical advances should serve the pansophic agenda of working toward the universal reformation of the world. Moraien was not averse to profit, but he believed the motive driving any discovery must be to serve the good of many. To that end, he steadily—and at times in cooperation with the Kufflers—pursued the chrysopoetic secrets of transmutation and the alkahest. Comenius's patron Laurens de Geer funded some of Moraien's experiments at transmutation. At one time Moraien was certain that he personally had transmuted antimony to gold. On another occasion he announced that he had discovered the *ludus paracelsi,* one of the great Paracelsian arcana, a stone that would dissolve stones.

Moraien became interested in optics while a minister in Cologne and came to believe that, through the use of lenses, sunlight could be concentrated into a material from which the "spiritus mundi," the life force animating all elements of the cosmos, could be extracted. He also claimed to have distilled a "universal menstruum" by attracting a salt of nature (probably dew) from the night air and distilling it for thirty days. Moraien's enthusiasm about alchemical achievements, John Young has noted, was occasioned less by a drive for

---

Tohrheit der gantzen Welt: Vermittelst eines anmuhtigen und erbaulichen Gespräches welches ist diser Ahrt die dritte und zwahr eine Märtzens-Unterredung (Hamburg, 1664); Stephen Clucas, "Samuel Hartlib and the Hamburg Scientific Community," MS, 1998; John Winthrop, Jr., to Samuel Hartlib, Dec. 16, 1659, Hartlib Papers [40/3/2a].

43. Young, *Faith, Medical Alchemy, and Natural Philosophy.*

profit than by the thrill of discovery and the fact that transmutation confirmed the pansophic worldview. Both provided certainty that humankind "could indeed comprehend the universal, harness cosmic forces, and discern the true pattern, the divine method, underlying Creation itself." Moraien had met Descartes, who was a critic of pansophism and Comenius, and, not surprisingly, he distrusted Descartes's reliance on rational deduction without experiment or scriptural inspiration.[44]

Moraien's belief in the unitary nature of divine will was reflected in his religious practices. Like Winthrop, he was both irenic and latitudinarian, prepared to accept a wide range of confessional expression. As a Reformed minister in Catholic Cologne, Moraien had personally experienced the repressive treatment of the Reformed minority by the Catholic majority there. In 1627, after enduring great persecution, he had been allowed to resign his post and move to Nürnberg, where he raised funds for Reformed sufferers in the Palatinate. There he subscribed to two documents put forth by John Dury, urging tolerant and collaborative relations between the English churches and the German Reformed churches. Having experienced sectarian persecution, Moraien became a strong advocate of the kind of religious tolerance often found in the republic of alchemy. Moraien and his wife Odilia served as godparents to two of the children of the Catholic chemist Johann Glauber.[45]

Despite the difference in their ages—Moraien was fifteen years older than Winthrop—the similarities of temperament, interests, and attitudes led to a lasting friendship between the two men. Winthrop not only emulated aspects of Moraien's medical practice upon his return to New England; he displayed the same tolerant disposition in religious matters, as best seen, perhaps, in his efforts to urge Massachusetts to moderate its persecution of Quakers. Like Moraien, Winthrop attempted numerous alchemical projects promoting the public good and worked collaboratively with others to succeed at transmutation. He also became increasingly interested in optics.

Decades after he returned to Connecticut, Winthrop would mention Moraien in his correspondence with other natural philosophers and inquire after the older man's welfare. Theodore Haak announced Moraien's death to Winthrop in 1671 by telling him "of the losse of many very special friends, in severall parts, and especially of that dear and worthy friend of ours Mr. Moriaen."[46]

44. Ibid., 11, 22, 229–230.
45. Ibid., 20–25.
46. John Winthrop, Jr., to Samuel Hartlib, Dec. 16, 1659, Hartlib Papers [40/3/2a]; Samuel Hartlib to John Winthrop, Jr., Sept. 3, 1661, Massachusetts Historical Society, *Proceedings*, 1878 (1879), 214, and Theodore Haak to John Winthrop, Jr., June 22, 1670, 247–248; John Winthrop, Jr., to Theodore Haak, July 1668(?), Winthrop Family Papers, Mss.5.163.

Through Moraien, Winthrop met another chemical philosopher who had a lasting influence on him. Johann Rudolph Glauber was a former apothecary at the court of Hesse-Darmstadt. He had arrived in Amsterdam in 1640 declaring himself a disciple of Paracelsus, who "had but a few Equals in true Genuine *Philosophy, Medicine,* and *Alchymy.*" Glauber defined himself as an apothecary, merchant and hermeticist; he wholeheartedly endorsed the macrocosm-microcosm theory and combined experiment with scriptural exegesis. He made significant advances in chemical technology—especially furnace design. He also discovered sodium sulfate, a white crystalline powder with cathartic effects, which in its mineral form is called mirabilite but is still commonly known as Glauber's salt.[47]

As an alchemist, Glauber displayed pious commitment to universal reformation and mercantile entrepreneurism. In Amsterdam, he sold a number of chemical products—such as pesticides, fertilizers, and preservatives. He had great faith in the chemical improvement of nature, and his 1656 work *Des Teutchlandts Wolfahrt; oder, Prosperitas Germaniae* would describe a comprehensive alchemically based program to help Germany recover from the devastation of the Thirty Years' War. Like Plattes's *Macaria,* it called for taking maximum advantage of improved mining techniques, agricultural improvements, and medical innovations, all of which were to be implemented by the application of advanced alchemical knowledge.[48]

At various times Glauber claimed to have perfected transmutation. One of the first books of Glauber's authorship owned by Winthrop combined useful practical descriptions of his furnaces with accounts of both *aurum potabile* and the *spiritus mundi. Aurum potabile* was an elixir made from gold, which was supposed to have astonishing curative properties. "Therefore forasmuch as my Potable Gold is . . . a concentrated Fire, and reduced into a liquid form, and all its whole Essence may be said to resemble nothing else but some tender penetrating Fire, yet void of flame, every one may readily conjecture what it is helpful for, and what use it is of in Medicine." Glauber asserted it could also transmute all metals, albeit at a cost higher than the cost of gold. Glauber's un-

---

47. Christopher Packe, trans., *The Works of the Highly Experienced and Famous Chymist, John Rudolph Glauber: Containing Great Variety of Choice Secrets in Medicine and Alchymy* . . . (London, 1689), 125.

48. Webster, *The Great Instauration,* 386–387. Pamela Smith, noting the combination of pious intention and market orientation displayed by Glauber, has seen him as embodying a new conception of a union between scholarly and artisanal traditions as shaped by the exchange economy of the seventeenth-century Netherlands. Smith, *The Body of the Artisan: Art and Experience in the Scientific Revolution* (Chicago, 2004), 165–177. On Glauber's reception among members of the Hartlib circle, see Allen G. Debus, "Palissy, Plat, and English Agricultural Chemistry in the Sixteenth and Seventeenth Centuries," *Archives internationales d'histoire des sciences,* XXI (1968), 84–85.

selfconscious blending of patently utilitarian instructions with esoteric philosophical analysis helped him gain a significant following, though at the time Winthrop was in Amsterdam he was as yet little known. Winthrop believed him to be a great alchemist and collected a substantial portion of the more than forty works Glauber ultimately wrote.[49]

While on the Continent, Winthrop shared ideas and hours and possibly experiments with Moraien, Glauber, the Kufflers, Tanckmarus, Schelegel, Rist, and others about whom little is known: Augustinus Petraeus and men named Haberfeld, Saarbrugh, and Caesar. What these contacts, considered as a group, reveal is that the community of natural philosophers of which Winthrop considered himself a member was not exclusively English, but pan-European and, ultimately, transatlantic. There was constant circulation of ideas between Europe and England, and especially between the northern European states and England during the early seventeenth century. Alchemical culture—the issues raised by the provocative theories of Paracelsus, Comenius, Glauber, Dee, Fludd, Jan Baptista van Helmont, and others, combined with the experiments and discussions that followed from those experiments—provided a lingua franca and a worldview that united intellectuals across Europe and across the Atlantic. Throughout his life, Winthrop sought to remain a part of this broad world of ideas and to benefit from the intellectual ferment a multinational scientific community could provide. It was the existence of this world of intellectual exchange that made schemes like pansophism appear viable, even in the face of crises such as the Thirty Years' War and the English Civil Wars. Winthrop's belief in the potential of alchemy was as much a creation of the Continent as it was of England. What Winthrop had experienced among the members of the republic of alchemy in both places, during the pansophic moment of 1641 and 1642, was a sense that profound change in the human condition was imminent and that alchemical philosophers had a duty to lead that transmutation.

Moraien exemplified both the pan-European circulation of ideas and the sense of expectation. As one of the most active correspondents of the Hartlib circle, he served as a conduit for both information and people between the Continent and England. His home was often a way station for natural philoso-

---

49. Packe, trans., *The Works of the Highly Experienced and Famous Chymist, John Rudolph Glauber*, 213; Ronald S. Wilkinson, "The Alchemical Library of John Winthrop, Jr. (1606–1676) and His Descendants in Colonial America, Part IV," *Ambix*, XI (1966), 139–186; Wilkinson, "John Winthrop, Jr. and the Origins of American Chemistry," 444–446; Bruce D. White and Walter W. Woodward, "'A Most Exquisite Fellow'—William White and an Atlantic World Perspective on the Seventeenth-Century Chymical Furnace," *Ambix*, LIV (2007), 285–298.

phers visiting Europe, and not infrequently he funneled promising European scientific prospects over to Samuel Hartlib to seek patronage or placement among the English. Because of the bidirectional nature of the Hartlib correspondence, there is a tendency to focus on the trans-Channel idea exchanges only. Yet, it is clear from the letters that, like Hartlib himself, Moraien was a hub through which the Hartlib circle's communications extended throughout Europe.

Moraien was undoubtedly the source of Winthrop's introduction to William Boswell, the English agent at The Hague, whom Winthrop visited in an effort to recover books and manuscripts (especially the manuscript he had acquired from John Dee's library) that had been captured by privateers during shipment from Hamburg to Amsterdam. Boswell, a Royalist, was also a correspondent of the Hartlib circle. Among the numerous letters in the Hartlib papers in which he is mentioned, there is a Boswell-authored paper describing a series of observations to be recorded by ships sailing south of the equator, aimed at determining longitude and latitude. Boswell, who, like Winthrop, was a devotee of and collector of manuscripts associated with John Dee, proved unable to help Winthrop recover the Dee manuscript, though he did help Winthrop recover one of Winthrop's own manuscripts. Boswell also provided Winthrop a letter of introduction to the English agent at Brussels. This introduction might have led to Winthrop's exposure to the innovative Walloon method of iron refining, which was subsequently employed at the New England ironworks.

Given the extensive and pan-European nature of his personal communications network, one wonders whether Moraien might also have played a role in connecting Winthrop to John Becx, a Dutch Reformed alien resident of London who became one of the primary investors in Winthrop's New England ironworks.[50] He might also have facilitated a meeting between Winthrop and Comenius, though it is likely the two men had met earlier in London. When they met, Comenius and Winthrop discussed the pansophic program and its potential application in America. Comenius would later say that the motivation for his 1641 journey to meet Hartlib was "propagation of the Gospel unto the nations of the world, and in particular the sowing thereof so happily made then in New England." For Winthrop, the Comenian agenda was a pragmatic extension of the principles of universal reformation he had endorsed since

---

50. Black, *Younger John Winthrop*, 115; John Winthrop, Jr., to Samuel Hartlib, Dec. 16, 1659, Hartlib Papers [40/3/2a]; "Memo Re Observations On Sea Voyages, Sir William Boswell," n.d., Hartlib Papers [71/16/4A–5B]; Hartley, *Ironworks on the Saugus*, 11–12, 64–66. On Boswell and Dee, see Julian Roberts and Andrew G. Watson, eds., *John Dee's Library Catalogue* (London, 1990), 66–67.

his early efforts to seek out the Rosicrucians. Winthrop, like so many others, seems to have become caught up in the pansophic moment, and particularly to have become enthusiastic about using New England as a pansophic laboratory.[51]

One of the curious eddies in the history of early colonial New England is the assertion, made by Cotton Mather on two separate occasions, that during his 1641 journey to Europe Winthrop offered Comenius the presidency of Harvard College. (I discuss the unlikelihood of this offer, and some alternative offers Winthrop might have made, in the next chapter.) More significant than whether Comenius received an invitation to come to New England (which, in any event, he did not), however, is how far Winthrop personally sought to implement elements of the reform program elaborated by Comenius, Plattes, Moraien, and other members of the Hartlib circle once back in America.[52]

The great practical benefit of the Comenian scheme was that it made universal reform a topic for public discussion and planning. Whereas utopian movements were speculative and the Rosicrucian brotherhood secretive, pansophism invited natural philosophers to collaboratively examine how they might implement meaningful, science-based reformation. This invitation not only energized individuals' desire to contribute to the general good; it opened the door to the formation of new alliances for progress among natural philosophers. Winthrop was to cultivate such relationships throughout his European visit, and he attempted to lead a concerted reform program after he returned to America.

Like Comenius, Hartlib had a substantial interest in both America and Christian alchemy. In the late 1620s he and John Dury had joined with a group of Polish-German intellectuals to form a secret alchemical society, modeled after J. V. Andrae's Christianopolis, called Antilia. As plans for this "collegium of learned men" developed, the issue of most concern besides attracting suitable patronage was selecting an appropriate location for Antilia's establishment. Devastation and military uncertainty on the Continent were measured

---

51. Wilkinson, "John Winthrop, Jr. and the Origins of American Chemistry," 79; Comenius, *The Way of Light*, ed. and trans. Campagnac, preface; Young, *Comenius and the Indians of New England*, 6; Young, ed., *Comenius in England*, 5, 16. No correspondence between Winthrop and Comenius survives, but Cromwell Mortimer's dedication of the *Transactions* of the Royal Society for 1741 (to Winthrop's grandson John Winthrop, F.R.S., 1681–1747) mentions Mortimer's seeing correspondence between Winthrop and Comenius. Cromwell Mortimer, dedication, Royal Society, *Philosophical Transactions*, XL (1741), A-r, B-v.

52. Cotton Mather, *Magnalia Christi Americana: or, The Ecclesiastical History of New England* . . . (London, 1702), book 4, 128; [Mather], *Ratio Disciplinae Fratrum Nov-Anglorum* . . . (Boston, 1726), 6; Young, *Comenius and the Indians of New England*, 2; Albert Matthews, "Comenius and Harvard College," Colonial Society of Massachusetts, Publications, *Transactions*, XXI (1919), 146–190.

against the distance and dangers involved in American colonization, leaving planners uncertain whether Prussia, Poland, Liffland (northwestern Latvia), or Virginia (North America's east coast from Florida to Newfoundland in the early seventeenth century) presented the fewest obstacles to successful plantation. Dury rejected a 1629 offer from John Davenport on behalf of the newly formed New England Company to serve as a pastor to a congregation intending American settlement. The company's offer was unrelated to the Antilia project, but Virginia and New England subsequently came under consideration by Dury and others as potential Antilia locations. A plan was put forward in late 1629 for a member of the Antilia society to visit Virginia to ascertain its suitability, but the next summer the German Antilia members rejected the idea of establishing a colony in Virginia because of its remoteness, the hardships their families would endure in journeying there, and its vulnerability to attack by Spain. They later reconsidered, however, and Hartlib continued to seek backers for an American-based Antilia. Despite persistent efforts, the project never materialized, and for Hartlib Antilia itself remained "a certain iland which is cleerly discovered afare of but when one comes neere it is vanished." Hartlib nevertheless maintained an active interest in New England and its possibilities, even to the point of gathering recommendations from Abraham Kuffler, Theodore Haak, and Constantijn Huygens for protecting New Englanders against winter cold.[53]

Hartlib's encounter with Winthrop made a lasting impression on both men. Winthrop sought to convince Hartlib of the mineral potential of New England. He showed him ore samples, probably of black lead, possibly of iron, which Hartlib recalled fifteen years later as "as rich oare as any is in the world." Hartlib replaced one of the books Winthrop had lost to the Dunkirk privateers with a book from his own collection. The two men might have shared discussions about their mutual patrons, Lord Brooke and John Pym, and undoubtedly discussed Plattes's *Macaria* and the Comenian reform schemes. By the end of 1642, hopes that Parliament would provide funding to transform

---

53. The name "Antilia" referred to the mythical Atlantic island supposedly settled in the eighth century by seven Portuguese bishops. They fled to the island after the Moors defeated the Spanish king Don Rodrigo in A.D. 743 at the Battle of Salamanca and overran the Iberian Peninsula. On Hartlib's Antilia scheme, see Turnbull, *Hartlib, Dury, and Comenius*, 69–76. On Andrae and Christianopolis, see Manuel and Manuel, *Utopian Thought*, 289–308; Johann Fridwald to Samuel Hartlib, June 18, 1629, Hartlib Papers [27/34/3–3b], Nov. 28, 1629, [27/34/5a–5b], Nov. 22, 1630, [27/34/7], Oct. 11, 1632, [27/34/8a–8b], Apr. 30, 1633, [27/34/11a–b], John Davenport to John Dury, Oct. 27, 1628, [4/5/1a–b], Nicodemus Farber to Samuel Hartlib, Aug. 30, 1632, [27/30/2a–b], Johann Fridwald to Samuel Hartlib, Apr. 30, 1633, [27/34/11a–11b], Mar. 24, 1639, [37/13a–14b], Nov. 28, 1629, [27/34/5a–5b], June 1630, [27/34/6a–6b]; Samuel Hartlib, "Ephemerides," 1640, part 3, [30/4/53a–60b]; 1635, Hartlib Papers [30/4/56b].

the College of Chelsea or the Savoy Hospital into the Universal College had faded, as the country dealt with the deep divisions and distractions of the Civil War. Hartlib surely welcomed the idea of Winthrop's making an effort to begin implementing at least part of the pansophic program upon his return to New England. As Frank Manuel and Fritzie Manuel have noted: "While his [Hartlib's] eye was on the 'Idea,' the design, and its national and even universal dimensions, he was willing to start with the small model, a private little body. He would implement utopia in easy stages." Certainly Winthrop, with his alchemical enthusiasm and pansophic entrepreneurial goals for New England's improvement, must have done much to modify Hartlib's previous thinking about New Englanders. In a 1640 entry in his commonplace book, the "Ephemerides," Hartlib had been less than optimistic about the refuge of the saints. He noted that not only were the New England Puritans smugly complacent, but they had undermined the religious intent of their removal from England by implementing a more restrictive religious environment abroad than they had left at home. Moreover, Hartlib suggested, New Englanders had not taken the opportunity to institute reforms in their educational system and so had institutionalized old mistakes.[54]

> Those of New England are lesse capable of the meanes of Reformation especially in learning because they bee of that Laodicean thinking, [thinking] themselves as rich godly wise etc. that they cannot bee richer etc. They commit one ordinary fault as the Ancients did in their Plantations which was to put downe that maine grievance and tyranny under which they groaned at home and to substitute a worse. . . . As for other defects they take not notice of them but establish them as they had at home. So wee see in New England the matter of schooling et Universities.[55]

Winthrop presented a different image of the way at least some New England Puritans thought and acted. More tolerant, more collaborative, and more focused on the potential gains from practical deployment of scientific knowledge, Winthrop was an ideal candidate to represent the pansophic quest

---

54. Samuel Hartlib, "Ephemerides," 1656, part 2, May–June 1656, Hartlib Papers [30/4/59b–60a], "Ephemerides," 1640, [30/4/61a–68b; 61a]; John Winthrop, Jr., to Samuel Hartlib, Dec. 16, 1659, [40/3/1a]; Charles Webster, ed., *Samuel Hartlib and the Advancement of Learning* (Cambridge, 1970), 3; Manuel and Manuel, *Utopian Thought*, 327.

55. Samuel Hartlib, "Ephemerides," 1640, Hartlib Papers [30/4/59b–60a]. The Laodicean Church was one of the early Christian churches, located in a town famous for its beauty and wealth in what is now Turkey. In the Book of Revelation the Laodicean Church was criticized for its self-satisfied complacency and self-righteousness (Rev. 3:14–19).

across the Atlantic. Winthrop returned to America with a vision of how New England might fit into the scheme of universal reformation, and over the next decade he would devote most of his energy to bringing that vision to life. The next time he heard from Hartlib, Hartlib would send him "a great number of books and manuscripts" while praising Winthrop for "how you were so nobly useful to your generation in that famous plantation."[56]

By the time John Winthrop returned to America at the beginning of June 1643, the broad outlines of his alchemical worldview had been set for life. At times he would have doubts about the attainment of alchemy's great mysteries, and the intensity of his enthusiasm for universal reformation would wax and wane, but his belief that alchemy provided a set of essential tools with which to improve the lot of New England and its people never changed. While in Europe, the model of Christian service through alchemical implementation that had been adumbrated during his early quest to make contact with the Rosicrucians became fully formed. In the company of so many likeminded natural philosophers of the republic of alchemy, the path to world improvement became clearer, the way of duty less obscure.

Buoyed by the pansophic moment's optimism in the face of a sense of increasing pan-European peril, Winthrop returned to New England intent on implementing a pragmatic, technologically focused pansophic reform agenda in the Zion in the wilderness. Using the whole array of alchemical skills he had obtained—in mining, medicine, industrial technology, raw material processing, and agriculture—Winthrop set out to transform New England's economic base while making it a model for the worldwide reform that seemed imminent. The belief that such a transformation was possible had a long heritage, extending from the rediscovery of Hermes Trismegistus to the works of Paracelsus and the manuscripts of John Dee, the books of Robert Fludd, Plattes's *Kingdom of Macaria,* Comenius's *Way of Light,* and Glauber's furnaces. Winthrop embarked from the Old England in 1643 on an alchemical errand into the wilderness. It was to change the development of New England in ways that would surprise even Winthrop.

56. Samuel Hartlib to John Winthrop, Jr., Mar. 16, 1660, Hartlib Papers [7/7/1a].

(THREE)
Founding a New London

John Winthrop, Jr., returned to New England in 1643 filled with a sense of possibility. Inspired by the alchemical contacts he had made in England and in Europe, Winthrop had formed a pansophic vision of New England's potential to serve as a vanguard in the restoration of knowledge and improvement of the human condition. In cooperation with other alchemical philosophers, he set out to implement a series of projects that would help transform New England and make it an example to the world of a society improved by the application of godly science. Through the creation of the New England ironworks and, more important, the establishment of a regional riverine plantation in what is now central Connecticut, Winthrop hoped that New England might become a magnet for alchemical research and a model for pansophic regeneration. Because the grandiose scientific aspects of this program ultimately failed—and because they were advanced under the conditions of guarded disclosure inherent in alchemical matters—both the scope of and the intentions behind Winthrop's projecting activities in the middle of the 1640s have gone unnoticed. A fresh examination of these projects highlights the importance that alchemical culture played in the second wave of New England colonial expansion and offers an insight into one Puritan approach to colonial settlement and cultural interaction.

THE CORRESPONDENCE of Johann Moraien underscores the power of alchemy and pansophism as tools of colonization. J. T. Young notes that the common thread uniting the two (and the effort to convert American natives as well) was the impulse to mastery. Alchemy was not just about gaining understanding of creation but achieving dominion over it.

> All alchemists were chemists, but not all chemists were alchemists. The distinction is between the mere student and the practitioner or adept, between passive understanding of Nature's forms and active domin-

ion over her spirit. The chemist was, as it were, the cartographer of a newly discovered country; the alchemist colonized it.[1]

Similarly, the educational goals of pansophism were designed to help individuals achieve control over the external world while mapping a socialized control system onto their internal consciousness. Conversion of the natives involved this same set of controls; Christianity depended upon the civilizing process to enforce the behaviors appropriate to Christian life.[2]

While alchemy and pansophism supported colonial agendas, our descriptive terms underscoring colonialist intentions are anachronistic. Alchemical rhetoric did not speak of dominating or colonizing nature but of "helping it, husbanding it, curing it." In the same way, English missionaries did not think of themselves as conquerors, but as liberators, hastening natives along a pre-destined path to perfection. Just as all metals aspired to become gold, Jews and native Americans were destined to mature into Christians. "Colonists, pansophists, and alchemists were only acting as catalysts, helping the rest of Creation along its providentially pre-ordained way." Winthrop's alchemical plan to hasten New England's colonial development was an expression of his desire to serve God's will and mirrored the intentions of many Puritan colonists. But, as Young points out, it was also a projection of Winthrop's and his alchemical partners' own deep-seated impulse to mastery over their environments, over themselves, and, ultimately, over others.[3]

The New England ironworks was the first of Winthrop's projects of transformation. It was one of the largest economic undertakings in New England since the sailing of the Winthrop fleet in 1630. Capitalized at fifteen thousand pounds, it brought a needed economic infusion to New England during its first major financial crisis. Equally important, as Stephen Innes has pointed out, ironworking symbolized the new industrial order. It asserted New England's intention through Winthrop to bring new technologies to bear on the godly transformation of the wilderness. The Walloon indirect process blast furnace that was used at the New England ironworks, which Winthrop presumably studied in his journey into the Brabant, was the most advanced system of iron production then in operation.[4]

---

1. J. T. Young, *Faith, Medical Alchemy, and Natural Philosophy: Johann Moriaen, Reformed Intelligencer, and the Hartlib Circle* (Aldershot, 1998), 165. Young also makes the point that the chemist-alchemist distinction is semantically inaccurate until the late seventeenth century. During almost all of Winthrop's lifetime, the two terms were functionally interchangeable.

2. Norbert Elias, *The History of Manners*, I, *The Civilizing Process*, trans. Edward Jephcott (New York, 1978).

3. Young, *Faith, Medical Alchemy, and Natural Philosophy*, 173, 174.

4. Bernard Bailyn, *The New England Merchants in the Seventeenth Century* (New York, 1964), 62;

While in England, Winthrop had attracted twenty-four investors and secured sufficient working capital to put the ironworks project in motion. With the assistance of his alchemical partner Robert Child, also one of the twenty-four investors, he purchased construction materials to be shipped to New England and contracted with laborers. After a delayed and difficult journey in which he and the ironworkers all became seriously ill, Winthrop arrived back in Boston and soon thereafter launched an extensive survey of the New England coastal area seeking suitable ore deposits and potential mine sites. He and his furnace workers selected a location near Braintree, Massachusetts, that they thought had rich deposits of limonite, or bog iron, the mainstay ore of ironworks in England. He initiated furnace construction, sought local investors, and gained additional concessions from both Massachusetts and the local Boston and Dorchester governments for a short-term, limited-franchise monopoly. The company was granted extensive tracts of land and timber, waterpower rights, and a twenty-year exemption from taxes on all stock and goods employed in the works.[5]

By December 1644 the Braintree furnace was nearly completed, and in May 1645 the General Court reported that there was "sufficient proof that the iron worke is very successfull (both in the richnes of the ore and the goodnes of the iron . . .)." Yet, even as the first blasts of iron were poured from the furnace, Winthrop requested his uncle Emmanuel Downing to take over management of the ironworks or to find someone else to.[6]

This seeming reversal in Winthrop's commitment to the ironworks has been puzzling. Biographer Robert Black has suggested that it was an example of Winthrop's "lack of staying power," a mercurial inability to see projects through to completion that affected Winthrop throughout his life. Bernard

---

Stephen Innes, *Creating the Commonwealth: The Economic Culture of Puritan New England* (New York, 1995), 237; E. N. Hartley, *Ironworks on the Saugus: The Lynn and Braintree Ventures of the Company of Undertakers of the Ironworks in New England* (Norman, Okla., 1957), 11, 12, 56; Robert C. Black III, *The Younger John Winthrop* (New York, 1966), 115; Richard S. Dunn, *Puritans and Yankees: The Winthrop Dynasty of New England, 1630-1717* (Princeton, N.J., 1962), 70-71.

5. Finding skilled ironworkers willing to relocate to New England was a continual problem. Innes notes that England too had a shortage of ironworkers. Irish ironworkers who might have taken such positions had been killed in the Irish rebellion of 1641. Innes, *Creating the Commonwealth*, 252-253; Samuel Eliot Morison, *Builders of the Bay Colony* (Boston, 1930), 276; Emmanuel Downing to John Winthrop, Jr., Feb. 24, 1644, in Samuel Eliot Morison et al., eds., *Winthrop Papers* (Boston, 1929-), V, 5-8 (hereafter cited as *Winthrop Papers*); Boston Records, 1634-1661, in *Second Report of the Record Commissioners of the City of Boston* (Boston, 1881); Innes, *Creating the Commonwealth*, 246; Bailyn, *New England Merchants*, 63; Hartley, *Ironworks on the Saugus*, 92-95.

6. Innes, *Creating the Commonwealth*, 246; Nathaniel B. Shurtleff, ed., *Records of the Governor and Company of the Massachusetts Bay in New England* (Boston, 1853-1854), II, 103-104; Emmanuel Downing to John Winthrop, Jr., Feb. 25, 1645, *Winthrop Papers*, V, 5-8.

Bailyn believed the issue was one of economic loyalty: that the profit orientation of the English undertakers was at cross-purposes with the domestic needs of the colonial authorities, and upon realizing this Winthrop elected to remove himself from the fray. E. N. Hartley has offered a number of possible explanations, ranging from growing disillusion with the venture, to the conflict between company and colony, to the "tug of alternative fields of interest." This last suggestion is the closest to the mark, for, even as Winthrop engaged in the ironworks development, he was pursuing another project that had occupied his attention since before his 1641 journey to England.[7] Winthrop's desire to focus more on this second endeavor was not a case of wanting to abandon the ironworks. Pansophic reformation as envisioned by Gabriel Plattes and Jan Comenius involved progressing simultaneously in a number of areas. Having got the ironworks to the stage where it would begin production, Winthrop was eager to switch his focus to a second project that was altogether more promising. To be sure, he supported Richard Leader, the new manager sent over by the ironworks company, and he continued to help further the development of the ironworks' endeavors. But, by the fall of 1644, Winthrop was turning his concentration away from furnaces for iron to the potential for mines of silver.

The English of the late sixteenth and the seventeenth centuries were preoccupied with mines and mining. Seventy-five percent of all patents granted by the crown between 1561 and 1688 were directly or indirectly concerned with mining. Fully a tenth of William Gilbert's cosmological work *De magnete* was devoted to mining, smelting, and fashioning iron. English North America was thought of as a potential storehouse of minerals. Writers throughout the seventeenth century consistently reported or predicted the presence of important mineral deposits there. John Smith was confident about finding "valuable minerals" in New England; Thomas Morton reported finding as many as seventeen different types of minerals, including gold and silver. In 1639 Plattes had noted, "It is very probable that . . . divers Mynes may be discerned with the eye . . . in many places," and in the 1670s John Josselyn reported finding rubies, diamonds, and emeralds, along with fifteen other minerals during his stay in New England. The Massachusetts Bay Company charter mentioned mining gold and silver on four separate occasions. New Englanders lived in constant hope that wealth could be mined or refined in their stony new world. Winthrop, with his alchemical

---

7. Black, *Younger John Winthrop*, 22; Bailyn, *New England Merchants*, 65; Hartley, *Ironworks on the Saugus*, 105.

knowledge, was in a unique position to help discover these metals and capitalize on their presence.[8]

Along with most of his contemporaries, Winthrop believed that metals, like plants, were animate creations that grew in the bowels of the earth. As a Paracelsian, he believed metals originated from seed. Proof that they mimicked the growth of plants could be found in the way metal ores spread within the earth in branches from a central trunk. "The vein of gold is a living tree," wrote the alchemist Peter Martyr, "that . . . spreadeth and springeth from the root, by the soft pores and passages of the earth, putteth forth branches, even to the uppermost part of the earth, and ceaseth not until it discover it self unto the open air." As living objects, metals were capable of regeneration and, in some cases, natural purification. It had been proven that saltpeter, left alone, could increase in volume, and it was assumed, as late as 1735, that mining sites, if covered and left alone for a lengthy time period, were capable of replenishing their ore. Over time ore could even change states. Paracelsians thought that all metals began from one kind of seed; but, as they were ripened within the earth by what the Puritan alchemist John Webster described as "the motion of the celestial bodies, central sun, or subterraneous fire or heat," they became purified and assumed a higher form. "Natures ultimate labour," Webster wrote, "is in time to bring all Metals to the perfection of Gold." This gradual natural perfection process provided assurance that transmutation was possible. Alchemists, as Johann Glauber maintained, only sought the secret that would hasten this natural process along.[9]

From his earliest days in New England, Winthrop had scoured the countryside for likely mineral prospects. He also engaged Indians to bring him potentially mineral-laden stones from the country's interior. Before his first year in

---

8. Harold Jantz, "America's First Cosmopolitan," Massachusetts Historical Society, *Proceedings*, LXXXIV (1973), 3–25; Charles Webster, *The Great Instauration: Science, Medicine, and Reform, 1626-1660* (New York, 1976), 334; William Gilbert, *De magnete*, trans. P. Fleury Mottelay (New York, 1958); Walter W. Woodward, "Captain John Smith and the Campaign for New England: A Study in Early Modern Identity and Promotion," *New England Quarterly*, LXXI (2008), 94; Raymond Phineas Stearns, *Science in the British Colonies of America* (Urbana, Ill., 1970), 74, 80, 147; Gabriel Plattes, *A Discovery of Subterranean Treasure* . . . (London, 1639), 13; Webster, *The Great Instauration*, 394.

9. Stearns, *Science in the British Colonies of America*, 28, 32, 39; Allen G. Debus, *The Chemical Philosophy: Paracelsian Science and Medicine in the Sixteenth and Seventeenth Centuries* (New York, 1977), I, 95; Pietro Martire, *The Decades of the Newe Worlde; or, West India etc.* (1511), reprinted in Edward Arber, ed., *The First Three English Books on America* (Birmingham, 1885), 173; Carolyn Merchant, *Ecological Revolutions: Nature, Gender, and Science in New England* (Chapel Hill, N.C., 1989), 126; Charles Webster, *Metallographia; or, A History of Metals* . . . (London, 1671), 72, 77; Young, *Faith, Medical Alchemy, and Natural Philosophy*, 168.

New England was over, Winthrop was reminding himself to order sandiver (a chemical flux used in assaying ores), evidence that from earliest settlement he was conducting metallurgical experiments. Winthrop's English connections took an active interest in his mineral explorations. Henry Jacie, a friend and minister who followed Winthrop's New World activities closely, sent him two rocks that had originally come from New England, which Jacie had had assayed by a goldsmith in Norwich. The goldsmith had found them of little value, and Jacie sent them to Winthrop "whereby you may better judge of the same ure if you see the like, and not count it better then it is." In 1633, Plymouth Colony trader John Oldham had sent John Winthrop, Sr., a sample of black lead ore, "whereof the Indians tould him there was a wholl rocke." The term "black lead" was used interchangeably to describe a number of different ores, among them common lead, bismuth, and graphite. The mine from which the ore came was well within Massachusetts's remote interior, which seems to have precluded immediate investigation. By 1641, however, Winthrop had determined the mine's exact location and retrieved additional ore samples.[10]

The black lead ore came from a site that was tantalizingly named—for a Paracelsian believing in the generation of metals from seeds containing both male and female properties—"Tantiusque," the place between two breast-shaped hills—and Winthrop had substantial reason to be interested in its potential. Not only was black lead itself valuable, but it possibly contained considerable wealth in silver.[11]

Lead-bearing mines were a major component of the English mining industry. By 1600 the lead-mining and smelting industry centered in Derbyshire produced Britain's second-most valuable export after textiles. Technological advances had helped to raise the English output of refined lead ore tenfold between 1570 and 1600. A factor in the increased demand for lead ore was the presence of various concentrations of silver in lead deposits. The presence of

10. John Winthrop, Jr., to William Brereton, Nov. 3, 1663, MHS, *Proceedings*, 1878 (1879), 218–219; William Kirby, London, to John Winthrop, Jr., June 22, 1632, *Winthrop Papers*, III, 83. Sandiver was used in assaying all ores, smelting gold concentrates, and separating gold from copper (Georgius Agricola, *De Re Metallica* [1556], trans. Herbert Clark Hoover and Lou Henry Hoover [New York, 1950], 235, 397–398, 464). Among the chemicals Winthrop is known to have brought with him to Massachusetts is aqua fortis (nitric acid), used to separate gold from silver. Henry Jacie to John Winthrop, Jr., June 1633, *Winthrop Papers*, III, 127; Richard S. Dunn, James Savage, and Laetitia Yeandle, eds., *The Journal of John Winthrop* (Cambridge, Mass., 1996), 97; Dunn, *Puritans and Yankees*, 85–87; Ronald Sterne Wilkinson, "John Winthrop, Jr. and the Origins of American Chemistry" (Ph.D. diss., Michigan State University, 1969), 70.

11. George H. Haynes, "'The Tale of Tantiusques,' An Early Mining Venture in Masachusetts," American Antiquarian Society, *Proceedings*, n.s., XIV (1902), 471–497.

more than one metal in the same ore confirmed the belief that metals matured in the earth. Lead was immature silver; the concentration of silver in any given lead ore depended on how long it had "ripened."[12]

Silver-bearing lead mines were considered among the best prospects for successful mining ventures. John Webster, who recommended that an array of English lead ores be collected and assayed for their silver content, described how silver was detected: "Black-lead . . . doth not long indure in the melting pot . . . but is partly changed into that which we call Spume of Silver, partly into that we call the foam of Lead." Since several different kinds of ores were at that time classified as black lead, Winthrop appears to have been unsure of the exact nature of his find. The potential for silver was exciting enough, however, that Winthrop packed samples of the black lead along with the iron ore that he carried to England in June 1641. In a move that coincided with Winthrop's departure, the Massachusetts General Court passed a law just before he departed "for incuragment of such as will adventure for the discovery of mines," which offered a twenty-one-year site monopoly and land usage rights to anyone "at the charge for discovery of any mine within this jurisdiction." Presumably, although Winthrop's European journey ostesibly was devoted to the ironworks project, he traveled to England with a dual mission: to find investors for the ironworks project; and to determine the value of the black lead and, depending on its value, seek investors for the mining venture as well as the iron project.[13]

Before leaving for England, Winthrop discussed the black lead project with Boston merchant Thomas Fowle, who, despite Boston's economic depression and specie crisis, wanted to invest in the black lead mine. Fowle might have worried about losing out to English investors. He wrote Winthrop in England to tell him of his decision rather than waited for his return to the Bay. In England, Winthrop showed the ore to Robert Child and to Samuel Hartlib, who remembered it as "as rich oare as any is in the world." Child also committed to investing in the black lead mine.[14]

At some point after his arrival in England, and perhaps based on positive assessments he was receiving of the ore's potential, Winthrop made a clear

---

12. Innes, *Creating the Commonwealth*, 241.

13. Webster, *The Great Instauration*, 395; Webster, *Metallographia*, 206, 271-272 (in this instance, Webster is describing common lead); Shurtleff, ed., *Records of the Governor and Company of the Massachusetts Bay*, I, 327.

14. "For your blacke leade business," Fowle wrote Winthrop in England, "I am ready to joyne with you th[er]in and I shall attend your further directions." Thomas Fowle to John Winthrop, Jr., Sept. 30, 1642, *Winthrop Papers*, IV, 355; Samuel Hartlib, "Ephemerides," 1640, Hartlib Papers [30/4/59b-60a], University of Sheffield.

distinction between the ironworks he was promoting and the black lead mine. He opened up the iron project to all with capital while reserving the black lead mine project only for himself, members of his family, and others, like the alchemist Child, whom he knew and trusted. Though implementing the black lead project would remain a priority, Winthrop stopped soliciting additional investors until the deed to the mine itself was secured and the excavations were under way.[15]

The black lead project probably influenced Winthrop's Continental journey to Hamburg and the United Provinces. These were the best possible places to learn more about black lead ore and the process of separating silver from black lead. Bismuth, or *plumbum cinereum*—the mineral Winthrop came to believe the black lead was primarily composed of—was found in many places in Germany, and manuals on extracting silver from bismuth and other lead ore had been published in that country since early in the 1500s. Hamburg, an intellectual and cultural center as yet unravaged by the continuing destruction of the Thirty Years' War, offered Winthrop access to people with firsthand knowledge of bismuth mining. The Dutch, on the other hand, were masters at extracting silver from lead ores profitably, with a technical ability unknown to the English. They were the people to whom English lead miners vended their lead exports. "The way of separating Silver from Lead in great quantities, so as to save the greatest part of the Lead, hath been little known or practised in *England*," John Webster wrote, "saving by one experienced person that had been in *Holland,* and seen it done there. . . . The Dutch . . . have for many years last past brought up and transported all the [English] Lead Ore they could possibly buy; and . . . thereby got no small profit." Through his contacts with Augustus Tanckmarus and Johann Glauber, Augustinus Petraeus, the Kuffler brothers, and Johann Moraien, it is probable that Winthrop was able to combine alchemical inquiries with related investigations into the technologies that could help assure the black lead mine's and ironworks' success.[16]

Sometime during his European journey, and certainly by the time he had crossed the English Channel, Winthrop, probably in response to the black lead mine's potential, the crises of the English Civil War and the Thirty Years' War, and the wave of pansophic enthusiasm then cresting in Europe, began to

---

15. That Winthrop intended to keep the black lead project closely held is confirmed by Winthrop's correspondence with his uncle. See Emmanuel Downing to John Winthrop, Jr., May 1, 1645, *Winthrop Papers,* V, 21–22.

16. Robert Child to John Winthrop, Jr., Mar. 1, 1644/5, *Winthrop Papers,* IV, 11; Agricola, *De Re Metallica,* trans. Hoover and Hoover, 612–613; Webster, *Metallographia,* 25, 233.

conceive of a new project for New England. The idea was to establish a plantation where, in Macaria-like fashion, alchemical philosophers could escape the dangers of conflict-ridden Europe and work together to solve the great and lesser mysteries of alchemy. In such a plantation, Comenius's College of Light or some similar design of learning might be implemented, and there also the natives might be educated to the acceptance of civility and Christ. The vision of an alchemical colony that Winthrop was adumbrating in Europe in 1642 and early 1643 might have emulated in its goals, if not in its monastic character, the "Venetian laboratory" for which his friend Plattes petitioned Parliament for funding in March 1643. Plattes hoped to build a "Laboratory, like to that in the City of *Venice,* where they are sure of secrecy, by reason that no man is suffered to enter in, unless he can be contented to remain there, being surely provided for, till he be brought forth to go to the Church to be buried." In this laboratory, Plattes and his fellow alchemists promised to serve England's great needs by producing major improvements in husbandry, medicine, and the transmutation of metals.[17]

How far Winthrop had articulated his American plantation project while he was in Europe is unclear. Given his strict adherence to alchemical codes of secrecy, the absence of details about the plan is predictable. To protect alchemical matters from public disclosure, Winthrop adopted a variety of techniques. He requested correspondents not to include secret information in their letters, but to wait until the information could be communicated orally. In his own letters he frequently alluded to possessing information unsuitable for public correspondence that he hoped to disclose at a future time. In addition, he frequently made use of private codes created for use by the alchemists with whom he was sharing information. (Evidence of his use of at least three such codes is found in the Winthrop Papers.) Winthrop and Edward Howes communicated by a code in which symbols stood for both letters and particular words. With Sir John Clotworthy, Winthrop communicated by means of a casement, a sheet of paper in which rectangular slits had been cut in such a way that, when laid over a letter, the words that showed through the slits revealed another message intended only for the possessor of a similarly configured casement. While in Europe developing the New England alchemical project, Winthrop used an unknown code to communicate with Petraeus and the Kufflers. That the men passed information they considered both secret and important is certain. After Winthrop had left Holland for England, Petraeus

---

17. Gabriel Plattes, "Caveat for Alchemists," in Samuel Hartlib, ed., *Chymical, Medicinal, and Chyurgical Addresses Made to Samuel Hartlib, Esquire* (London, 1655), 87.

wrote him reminding him to "write nothing hereafter to him that is secret without our key."[18]

While his secrecy obscures details of Winthrop's emerging project, certain features can be deduced from his public correspondence. Winthrop was promoting relocation to America among his European alchemical friends, and they were responding enthusiastically. Johann and Abraham Kuffler announced that they were "completely resolved to travel to America." Petraeus expected to come to America also, following a journey he was to make through France and Italy to visit a count in Dalmatia at whose laboratory he expected to attain the secret of the *luna fixa,* an alchemically derived metal with the weight and properties of gold, though lacking its color. Petraeus was ebullient about the prospects of coming to Winthrop's plantation. "I am looking forward to write to the gentleman [Winthrop] and ask him to properly feed the hogges so they will be fat when we arrive." In England, Child also announced his intention to join Winthrop in America and supported his intentions with money. He invested in both the ironworks and the black lead mine. As Winthrop began to conceptualize establishing his alchemical colony in America, he found others equally excited about his vision.[19]

Upon his return to New England, even as he fulfilled his duties as manager of the iron project, Winthrop set in motion steps to implement creating a new London in the New England wilderness. Winthrop's plan, as it was revealed through a series of permissions granted by the Massachusetts General Court in 1644 and 1645, was ambitious. He intended to settle a plantation at the mouth of the Pequot (now Thames) River on land that had become open to English settlement as a consequence of the 1637 Pequot War. Although the land was fairly inhospitable to agriculture, the harbor was one of the finest in New England and was viewed as a prize by both Connecticut and Massachusetts colonies, each of which had claimed title to the region following the

---

18. Edward Howes to John Winthrop, Jr., Nov. 23, 1632, MHS, *Collections,* 4th Ser., VI (1863), 480–482 (reproduces code for use in encrypted communications); Gerhard F. Strasser, "Closed and Open Languages: Samuel Hartlib's Involvement with Cryptology and Universal Languages," in Mark Greengrass, Michael Leslie, and Timothy Raylor, eds., *Samuel Hartlib and Universal Reformation: Studies in Universal Communication* (Cambridge, 1994), 151–161; Sir John Clotworthy to John Winthrop, Jr., Mar. 6, 1635, *Winthrop Papers,* III, 190–191 (a photograph of the casement is reproduced), Augustinus Petraeus to John Winthrop, Jr., Mar. 9, 1643, IV, 368–369 ("Der herr schreibe nichtes hinfuro an mihr das etwa secret ist without our clavis").

19. Augustinus Petraeus to John Winthrop, Jr., Mar. 9, 1643, *Winthrop Papers,* IV, 368–369; Lawrence Principe, *The Aspiring Adept: Robert Boyle and His Alchemical Quest, Including Boyle's "Lost" "Dialogue on the Transmutation Of Metals"* (Princeton, N.J., 1998), 81 n. 58; William R. Newman, *Gehennical Fire: The Lives of George Starkey, an American Alchemist in the Scientific Revolution* (Cambridge, Mass., 1994), 140.

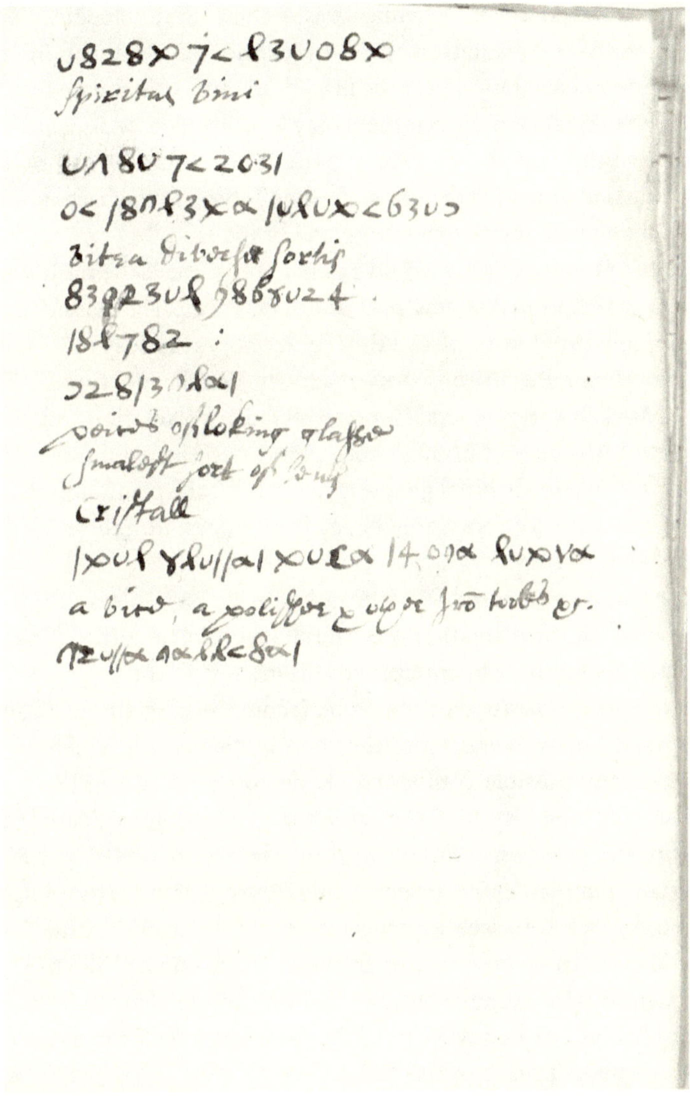

FIGURE 4. Ciphers. Memoranda Book of John Winthrop, Jr., 1631, unnumbered page. Manuscript, Winthrop Family Papers, Massachusetts Historical Society, Boston. *Courtesy of the Massachusetts Historical Society*

war. Winthrop had purchased Fishers Island at the mouth of the harbor in 1641 and was familiar with the area and the Pequot harbor's deepwater potential. This plantation, which Winthrop would insist on naming New London, would become the center of Winthrop's pansophic alchemical undertaking. There he would pursue a range of alchemically related projects and pursue the pansophic vision of generating world reformation through the collaborative work of a group of alchemical researchers. Simultaneously with the establishment of the plantation at the river's mouth, Winthrop intended to establish a second plantation near the black lead mine site, situated at the head of the Pequot watershed, near present-day Southbridge, Massachusetts. This plantation would serve as the support center for the mine. Ore from the mine would be carried down the Pequot watershed's rivers to the harbor plantation for transshipment to processors in England or Holland.

While the black lead mine would be only one of the alchemically oriented projects undertaken in the new plantation, it was clearly the one on which Winthrop, initially at least, placed the highest hopes. In May 1644, with construction of the Braintree ironworks underway, Winthrop secured permission from the Massachusetts General Court to establish a plantation and ironworks at the Pequot River site. Although both grants were issued simultaneously, it is clear that Winthrop's initial goal was to establish the plantation rather than the ironworks, for no effort would be made to process iron there until several years after settlement.[20]

With permission secured, Winthrop began to assemble a group to join him in laying out the plantation site. Simultaneously, he turned his attention to the black lead mine. The intensity of his interest in the project is demonstrated both by the haste with which he secured title to the mine site from the Nipmuck Indian owners and the thoroughness with which he sought to eliminate any possible conflicting claims to the land. Winthrop sent an agent named Steven Day to Tantiusque, charged with investigating the area around the mine site, ascertaining who the rightful Nipmuck owner of the land was, and purchasing the property. Day was to search for more than black lead, though; he was to seek evidence of the richer silver deposits Winthrop believed the site must hold. While Day was prospecting, William Pynchon, a trader whose far-flung enterprises along the Connecticut River valley had given him a substantial knowledge of the region, sent Day a message through an Indian runner. "I spake to this Indian in your behalfe," Pynchon wrote. "I tould him that the Governor sent you to serch for something in the ground, not for Black lead as they suppose but for some other mettell." On the very day he received Pynchon's letter, Day concluded an agreement with the Nipmuck natives Webucksham and Nonmonshot to purchase the Tantiusque site and all land in a ten-mile circumference around the hill where the black lead was. On November 11, Day purchased the mine site again from a second native claimant, Nodawahunt, who sold Winthrop his rights to the land adjacent to

---

20. Order of the Massachusetts General Court on the Petition Of John Winthrop, Jr., Apr. 28, 1644, *Winthrop Papers*, IV, 466.

the mine "for ten miles" and also validated the previous sale by Webucksham and Nonmonshot. Both of these transactions with the Nipmuck owners took place before Winthrop actually had permission to make the purchase. Two days after the second deed was signed, the General Court at Boston gave Winthrop approval to buy the land Day had already acquired, granting him "the hill at Tantousq, about 60 miles westward, in which the black leade is, and liberty to buy some land there of the Indians."[21]

Having secured both the mine site and the harbor plantation, Winthrop notified his uncle Emmanuel Downing that he wished to be released from management of the ironworks project. He also contracted with Thomas King to hire workers and "speedily goe up" and begin mining the ore, for which King would receive forty shillings per ton once he had dug up twenty tons of "good marchantable blacklead" and secured it in a storage house, safe from the Indians. Winthrop composed a discourse for Robert Child, discussing his alchemical analysis of the ore, citing his belief that it was composed of bismuth. This was a positive analysis indeed, for bismuth was frequently found covering rich silver mines. Winthrop called it *mater argenti,* mother of silver.[22]

Samples of the ore were packed for Downing to take with him on his journey to London, to be forwarded on to France and Holland for better assays of the silver content. If the new ore samples continued to show positive signs of silver, Downing was to keep the project closely held; if it proved to be only black lead, he was to seek additional investors. Winthrop's brother Stephen, then in London, was tasked with finding a London merchant to vend the ore. For Winthrop, inclined by both religion and alchemical practice to find the providentialism in all things, the possibility that God would allow him to support a pansophic alchemical colony with funds derived from a Protestant silver mine in the New World must have seemed a specially significant blessing.[23]

---

21. Haynes, "'The Tale of Tantiusque,'" AAS, *Proceedings,* n.s., XIV (1902), 475. Winthrop had received the title of governor of the Connecticut River for his leadership of the Saybrook Plantation project in 1635/6. Pynchon also suggested that Day investigate another black lead deposit, five or six miles south of the Tantiusque site. William Pynchon to Stephen Day, Nov. 8, 1644, *Winthrop Papers,* IV, 495–496, Deed of Webucksham and Nonmonshot to John Winthrop, Jr., Oct. 8, 1644, 496, Deed of Nodawahunt to John Winthrop, Jr., Nov. 11, 1644, 496; Shurtleff, ed., *Records of the Governor and Company of the Massachusetts Bay,* II, 82.

22. Agreement between Thomas King and John Winthrop, Jr., Nov. 27, 1644, *Winthrop Papers,* IV, 497, Robert Child to John Winthrop, Jr., Mar. 1, 1645, V, 10–12; Robert Child, "An Answer to the Animadversor," in *Samuel Hartlib: His Legacy of Husbandry . . . ,* 3d ed. (London, 1655), 133–134.

23. Emmanuel Downing to John Winthrop, Jr., Feb. 25, 1645, *Winthrop Papers,* V, 5–8, Stephen Winthrop to John Winthrop, Jr., Mar. 1, 1645, 13.

When Winthrop journeyed to the Pequot harbor to lay out the plantation the following spring, almost all elements of the alchemical plantation scheme seemed to be pointing toward success. At least five European alchemists had committed to joining Winthrop in America, and a number of influential New Englanders, including the Saybrook minister Thomas Peters, had committed to locating at the Pequot site. Emmanuel Downing wrote with positive news about the ore assays. "Mr. Leader hath tryed your leade oare and fyndes yt to be a silver Myne, therefore I am resolved not to sell any parte thereof." In addition Downing reported that he had "a freind preparing to come over with me who doth resolve to make a plantation by your myne, who hath monie enough, and purposeth to improve some therein." Robert Child, who had received Winthrop's discourse on the black lead as well as additional ore samples, wrote urging optimistic caution. He provided a lengthy alchemical analysis, which argued that the black lead was, not bismuth, but graphite. Nevertheless, he wrote, "I am unwilling to beate you out of your great hopes; nay I hope I shall not discourage you from digging lustily about it, for the commodity as I have tould you, wisely managed, will maintain it self." Child still promised to be "active in the business, to the uttermost of my power" when he arrived in New England later that year, and agreed to pay one-fourth of the start-up costs.[24]

In a subsequent discussion about the black lead ore, Child noted that graphite was itself something of a rare commodity in Europe, coming from only one mine in England, which was opened once every seven years. Graphite had four main commercial applications. It was used in pencils for mathematicians. It was also used by painters and limners. Coppersmiths used it to reduce friction when hammering copper. In addition, when large pieces were mined, combs were made from them "because they discolour gray hairs, and make black hair of a Raven-like, or glittering blacknesse, much desired in *Italy, Spain, etc.*"[25]

Child's analysis that the ore was only graphite was not shared by others. A third assay by Richard Hill not only again confirmed the presence of silver

---

24. In addition to Child, Petraeus, and the Kufflers, Edward Howes had asked Winthrop to procure him a few acres of land in Cambridge and mentioned some friends "left behind" who might join him in settling in the Narragansett country, which was adjacent to the Pequot site. Edward Howes to John Winthrop, Jr., Feb. 25, 1645, ibid., 8–9, Emmanuel Downing to John Winthrop, May 5, 1645, 21–22. Child's belief that the ore was graphite was correct. However, his own subsequent analysis of the ore showed the presence of fifteen pounds of silver per ton of ore. Child, "An Answer to the Animadversor," in *Samuel Hartlib*, 133–134; Robert Child to John Winthrop, Jr., Mar. 1, 1645, *Winthrop Papers*, V, 11–12.

25. Child, "An Answer to the Animadversor," in *Samuel Hartlib*, 133–134; Wilkinson, "John Winthrop, Jr. and the Origins of American Chemistry," 118–120.

in the ore; it brought a compliment from the assayer. "[I] am glad to heare yow have soe well spent your time as I understand yow have, in Finding out that mine of black Lead," wrote Hill. If Winthrop thought it worthwhile, he would be happy to come to New England and "spend some time and paines" on developing the mine.[26]

ALTHOUGH THE BLACK lead mine was to be the most important initial project of the new plantations, Winthrop and Child developed additional plans to introduce pansophic utilitarianism into the laboratory that was New England. Husbandry was an area of special interest. Plattes had promised Parliament that his Venetian laboratory's experiments would lead to a doubling of England's agricultural output. In similar fashion, Winthrop and Child sought to expand on New England's productive capacity. As Karen Kupperman has pointed out, colonists to New England were still working out the implications of farming in a climate that was at once much hotter and much colder than comparable latitudes in Europe. In addition, there were very few experienced farmers among New England's settlers, and, through the first decades of settlement, colonists had had to become pupils of the American natives and rely on native crops. In the 1630s they had begun to succeed at growing traditional European crops, but the results were inconsistent.[27]

Winthrop and Child hoped to expand New England's agricultural base and improve the ratio of success in growing European crops. At Child's urging, Winthrop had sent seed samples from New England to John Tradescant, for planting at the Ark, his garden of rarities in Lambeth Park. This collection of rare plants, gathered from locations throughout the world, had been started by Tradescant's father (also named John), former keeper of his majesty's gardens. The purpose of the Ark was to bring honor to England and to "benefit such ingenious persons as would become further enquirers into the various modes of Natures admirable works." Child received "diverse sorts" of seeds from Tradescant and forwarded them to Winthrop with "5 or 6 sorts of vines in a Caske . . . with some prune grafts, some pyrocanthus trees, and very many sor[ts] of our common plants, and seeds." Child urged that "they may be carefully planted with all Expetition." Child, who had journeyed to France to learn about winemaking, was confident that within three years "wine may be made as good as any in France." He intended upon his arrival in New England

---

26. Richard Hill to John Winthrop, Jr., June 16, 1645, *Winthrop Papers*, V, 28–29.
27. Karen Ordahl Kupperman, "Climate and Mastery of the Wilderness in Seventeenth-Century New England," in David Grayson Allen and David D. Hall, eds., *Seventeenth-Century New England* (Charlottesville, Va., 1984), 3–38.

to settle near the mine site if the place pleased him and to "undertake a vineyard with all Care and industry."[28]

Child's interest in planting "very many sor[ts] of our common plants, and seeds" was not just ecological imperialism; it represented agricultural innovation. Gardening—raising a variety of food crops rather than depending almost exclusively on grain staples such as oats and barley—was, according to Child, a relatively new approach to food production in England, and he hoped to expand upon its potential in New England. In "A Large Letter concerning the Defects and Remedies of English Husbandry Written to Mr. Samuel Hartlib" in 1651, Child noted that the "Art of Gardening" had only begun to "creep into *England*" about a half century before. "Some old men in *Surrey* . . . report, That they knew the first *Gardiners* that came into those parts, to plant *Cabages, Colleflowers,* and to sowe *Turneps, Carrets,* and to sow *Raith* . . . *Pease, Rape,* all which at that time were great rarities, we having few or none in *England,* but what came from *Holland* and *Flanders.*" Despite its utility, Child noted, gardening was still not universally practiced. "Many parts of *England* are as yet wholly ignorant. . . . I could instance divers other places, both in the *North* and *West* of *England,* where the name of *Gardening,* and *Howing* is scarcely known, in which places a few *Gardiners* might have saved the lives of many poor people." Child saw the garden itself as a source of pansophic regeneration, and the plethora of plants and seeds he forwarded to New England as storehouses of improvement. "Dayly new Plants are being discovered, useful for *Husbandry, Mechanichs,* and *Physick,* and therefore let no man be discouraged, from prosecuting new and laudable *ingenuities.*" In his letter, Child also claimed that finding a way to make good English wines was an issue of national pride. "I would scorne to honour *France* so much as men do usually," he noted. Child hoped to acquire from Glauber his alchemical secret for hastening the maturation of grapes by applying a chemical compost to their roots.[29]

Winthrop sought almost immediately to make improvements in pastoral farming and livestock production at the new plantation. In addition to hiring a laborer to tend to the protection and increase of hogs and goats on Fishers

---

28. Karen Ordahl Kupperman, *Indians and English: Facing off in Early America* (Ithaca, N.Y., 2000), 22; Jim Bennett and Scott Mandelbrote, *The Garden, the Ark, the Tower, the Temple: Biblical Metaphors of Knowledge in Early Modern Europe* (Oxford, 1998), 88–89; Robert Child to John Winthrop, Jr., Mar. 1, 1645, *Winthrop Papers,* V, 10–12.

29. Robert Child, "A Large Letter concerning the Defects and Remedies of English Husbandry, Written to Mr. Samuel Hartlib," in *Samuel Hartlib: His Legacie of Husbandry* . . . (London, 1651), 9–12, 26, 38, 81. Ann Leighton's study of early New England narratives suggests that gardening was common among New England's first settlers. Ann Leighton, *Early American Gardens: "For Meate or Medicine"* (Boston, 1970), 32–33.

Island, he initiated a program to replace the native broomstraw, wild rye, and spartinas grasses, which were relatively poor livestock feed, with nutritively rich English grasses such as bluegrass and wild clover. The first planting season after settlement he ordered grass seed from English planters who had developed successful pastoral areas in the Narragansett region. He aggressively pursued a program to plant English grasses, directing the sowing of twenty-four bushels of five different kinds of English hayseeds during the first spring planting season. Although the English at Nameaug (the name the site would have until it was officially named New London) would for a time continue to rely on Indian corn as a primary food source, Winthrop also ordered the planting of rye and winter wheat. He requested and received trees from John Endecott of Massachusetts, whose farm was called Orchard. Endecott also sent Winthrop indigo seeds for the new plantation. Winthrop gathered breeding stock for his Fishers Island goat, sheep, hog, and cattle enterprise, and within two years of settling the plantation he would have sufficient stock to be able to establish relationships with merchants in New Haven Colony and Massachusetts to sell goats and pork and beef.[30]

WHILE WINTHROP WORKED in 1645 and 1646 to lay the foundation for his visionary new settlement, a number of European natural philosophers, including a remarkable group of English natural philosophers, were observing its progress with much more than passing interest. Cromwell Mortimer, in his dedication of the 1741 edition of the *Philosophical Transactions* of the Royal Society to Winthrop's grandson and namesake, mentioned seeing a body of correspondence (now lost) that passed between Winthrop and many of seventeenth-century Europe's leading natural philosophers. Citing letters he had read from "*Mr.* [Robert] *Boyle, Dr.* [John] *Wilkins, Sir* K[enelm] *Digby,* etc," Mortimer noted, "Had not the Civil Wars happily ended as they did, Mr. *Boyle* and Dr. *Wilkins,* with several other learned Men, would have left *England,* and, out of Esteem for the most excellent and valuable Governor, JOHN WINTHROP the younger, would have retir'd to his newborn Colony, and there have establish'd that SOCIETY *for promoting Natural Knowledge,* which these Gentlemen had formed, as it were, in *Embryo* among themselves." The

---

30. William Cronon, *Changes in the Land: Indians, Colonists, and the Ecology of New England* (New York, 1983), 141–143; Kupperman, *Indians and English,* 3, 161–165; Roger Williams to John Winthrop, Jr., May 1647, in Glen W. LaFantasie, ed., *The Correspondence of Roger Williams* (Providence, R.I., 1988), I, 234–235; Robert Williams to John Winthrop, Jr., Apr. 18, 1647, *Winthrop Papers,* V, 149, John Winthrop, Jr., to Thomas Peters, Sept. 3, 1646, 100–101, John Endecott to John Winthrop, Jr., Mar. 19, 1646, 67–68, Nicholas Davison to John Winthrop, Jr., Oct. 2, 1648, 264; Alexander Bryan to John Winthrop, Jr., August 1648, 249, John Clark to John Winthrop, Jr., Oct. 8, 1649, 372–373.

Society for Promoting Natural Knowledge that Mortimer wrote of ultimately became the Royal Society, of which Winthrop became the first colonial member in 1661. Mortimer's claim, like Cotton Mather's statements regarding Winthrop's invitation to Comenius to assume the presidency of Harvard College, is problematic. At the time the Pequot plantation was settled, the Civil War was in full bloom, and the polymath Robert Boyle was a precocious nineteen years old. The cryptographer and mathematical magus John Wilkins, though, was indeed meeting with a group of natural philosophers in London and had hopes of establishing a "Mathematico-Chymical-Mechanical" school there. At the same time, the Catholic alchemist Sir Kenelm Digby, chancellor to Henrietta Maria, was serving as the envoy of the Catholic Royalists to the pope. While it is a bit difficult to think of Digby translating himself from the Vatican to Puritan New England, Mortimer's letter suggests strongly that Winthrop's efforts to establish a New London in America were, at the time, an issue of much more than passing interest to natural philosophers in the old one.[31]

Seventeenth-century Connecticut history (with the exception of the Pequot War) has often been treated as a pastoral and largely insignificant mirror of the history being written to the north and east in Massachusetts Bay. This all-too-common tendency to treat Massachusetts's history as the equivalent of New England history produced a historical vision with significant blind spots. Not only has it ignored the implications that followed from the intended establishment of an ecumenically inclusive scientific community along the Long Island Sound; it significantly understated very real conflicts between the Bay Colony Puritans and their Connecticut neighbors. Several of these conflicts focused on, and produced sharp contention during, the settlement of New London. The errand of John Winthrop, Jr., into the wilderness was not just an effort to establish an outpost for the pansophic pursuit of knowledge. It was a project that pitted colony against colony, Puritan against Puritan, science against religion, Indian against Indian, and, in the case of John Winthrop, Jr., himself, father against son.

---

31. Cromwell Mortimer, dedication, Royal Society, *Philosophical Transactions*, XL (1741), a–b3; Barbara J. Shapiro, *John Wilkins, 1614–1672: An Intellectual Biography* (Berkeley, Calif., 1969), 147–150; Bennett and Mandelbrote, *The Garden, the Ark, the Tower, the Temple*, 61; Principe, *The Aspiring Adept*.

(FOUR)

Which Man's Land?
Conflict and Competition in Pequot Country

Two conflicts that surfaced with explosive force in New England in 1637 reverberated with particular impact on Winthrop's new plantation in the mid-1640s. For more than a decade, the success or failure of the alchemical project hinged on how the issues raised by these earlier events would be resolved. The first of these conflicts, the Pequot War of 1637, had reduced the once powerful Pequot nation to servile status and exacerbated already strained Indian relations in the former Pequot territory. At the same time, the war had created competing claims to the former Pequot lands between the English colonies of Connecticut and Massachusetts as well as among their Indian allies, all of whom had played a part in the Pequots' overthrow. The decision of John Winthrop, Jr., to establish his new plantation in the heart of the former Pequot country heightened the tensions produced by destabilized Indian relations and intracolonial English competition, as various protagonists, including Winthrop, sought to manipulate events to achieve the greatest advantage.

The second conflict of 1637, the antinomian, or free grace, controversy surrounding Anne Hutchinson and her followers, left in its wake still-unresolved questions about the limits of acceptable Puritan practice in New England. These questions would, for a time, come to focus on ecumenically inclusive practitioners of alchemy like Winthrop, whose tolerance for religious diversity raised substantial concerns among some Puritan leaders about the potential ecclesiological consequences of establishing an alchemically focused research center. Immigration laws had been tightened during the Hutchinsonian free grace crisis to exclude from residence in Massachusetts anyone whose religious beliefs were thought in conflict with the emerging Independent orthodoxy. Alchemists such as Winthrop, despite his personal adherence to the New England way, readily associated with and in fact eagerly recruited natural philosophers representing a broad range of confessional beliefs. Their potential presence in New England, even—or perhaps especially—in a remote area such as Winthrop's new plantation, caused some Puritans great concern. This was especially true after Winthrop's associate Robert Child confronted Massachusetts's limited religious tolerance head-on, presenting a remonstrance to the General Court demanding that Bay officials loosen their restrictive church membership and political enfranchisement policies. In the wake of Child's remonstrance in Massachusetts, just as Winthrop's plantation was starting up, Connecticut's wariness of Winthrop's settlement increased, and its willingness to support his plantation, even against incursions from the English settlers' Indian opponents, was limited. Massachusetts, too, cracked down on at least some of the alchemists in its colony, intimidating them to the point that they left the country. While the backlash against alchemy in New England was short-lived, its timing was significant and dampened the ambitions of the New London plantation.

The backlash also negatively affected relations between John Winthrop, Sr., governor of Massachusetts, and his son and namesake. In principle, the elder Winthrop supported his son's alchemical practices and his new plantation scheme. The younger Winthrop had built a chemical furnace at his father's house in Boston, which he presumably made use of during visits to Boston, for he stored both chemicals and alchemical texts there. Furthermore, the grants the younger Winthrop had secured from Massachusetts allowing him to establish the plantation in the contested former Pequot territory were obtained while the elder Winthrop was a Massachusetts magistrate, and presumably with his support. On close inspection, though, it is clear that the father had misgivings about the spiritual rectitude of his son's pansophic undertaking and even greater concerns about the strategy his son followed in his relations with the Pequot Indians. The result was, if not a breach between father and son over the project, a very clear expression of disagreements.

Understanding how the residual tensions from the free grace controversy and the ownership of the contested Pequot lands affected the early days of Winthrop's plantation helps us better contextualize alchemy's status in early New England. Because of the benefits it could provide—in medicine, mining, agriculture, and industrial processing—many New Englanders valued alchemical practitioners and welcomed them into their communities. Yet alchemists who too openly embraced ecumenical religious perspectives or who seemed to flirt with unacceptable magical arts could come under critical and sometimes harsh scrutiny. At the New London plantation, critical scrutiny of the project preceded its later valued acceptance and made the younger Winthrop's difficult task of founding an alchemical settlement in the wilderness even harder.

AT THE END of the Pequot War in 1637, all parties to the victory hungered for the spoils of war. These included not only the former Pequot lands—

an extensive region along the Long Island Sound extending up the western bank of the Pequot (now Thames) River watershed to the palisaded Mohegan settlement at Shantok (see Map 1)—but control over the surviving Pequots as well. While Pequot warriors known to have fought against the English were executed (bounties were paid to Indian allies for bringing English authorities Pequot warriors' heads and hands), noncombatants were distributed among the victors as servants or tributaries. Connecticut and Massachusetts leaders took as many as three hundred Pequots as servants into their households, but what these expansion-minded colonies wanted most was access to and control of the former Pequot lands. The Mohegans and Narragansetts, who had allied with the English, were also interested in the former Pequot lands, particularly in the hunting rights to them, but were even more interested in the human spoils. The Narragansetts had suffered from a severe epidemic of smallpox in 1633 and saw absorbing former Pequot members into their tribe as an opportunity to offset some of their recent population decline. For the Mohegan Uncas, who had been a sachem of relatively limited power before the war, subjugating and exacting tribute from the former Pequots was part of an aggressive plan to increase Mohegan authority within the region. Some surviving Pequots, too—despite a declaration by Connecticut in the 1638 Treaty of Hartford that their existence as a tribe should be eradicated in both name and in fact—sought vigorously to retain a hand in their own destiny. An untold number of former Pequot warriors who had avoided capture were quietly welcomed into the villages of their former Indian opponents (in which they often had relatives); other Pequot survivors sought to stay together in remote places under the protection of their own sachems. Michael Leroy Oberg has described the result of this confused mix of postwar agendas and survival strategies: "'Mohegan' and 'Narragansett' villagers were not always what they seemed. Indians who had been Pequots could be Mohegans, Narragansetts, or Niantics while remaining Pequots."[1]

This fluidity of identity and allegiance made postwar intercultural relations in the region particularly difficult, and always potentially dangerous. It also made forging intercultural alliances essential. "No one of the participants," Oberg notes, "could unilaterally determine the nature of the relationships that developed." Here, I describe as succinctly as possible the situation on the

---

1. Andrew Lipman, "'A Meanes to Knitt Them Togeather': The Exchange of Body Parts in the Pequot War," *William and Mary Quarterly*, 3d Ser., LXV (2008), 3–28; Michael L. Fickes, "'They Could Not Endure That Yoke': The Captivity of Pequot Women and Children after the War of 1637," *New England Quarterly*, LXXIII (2000), 58–81, esp. 61; Michael Leroy Oberg, *Uncas: First of the Mohegans* (Ithaca, N.Y., 2003), 87.

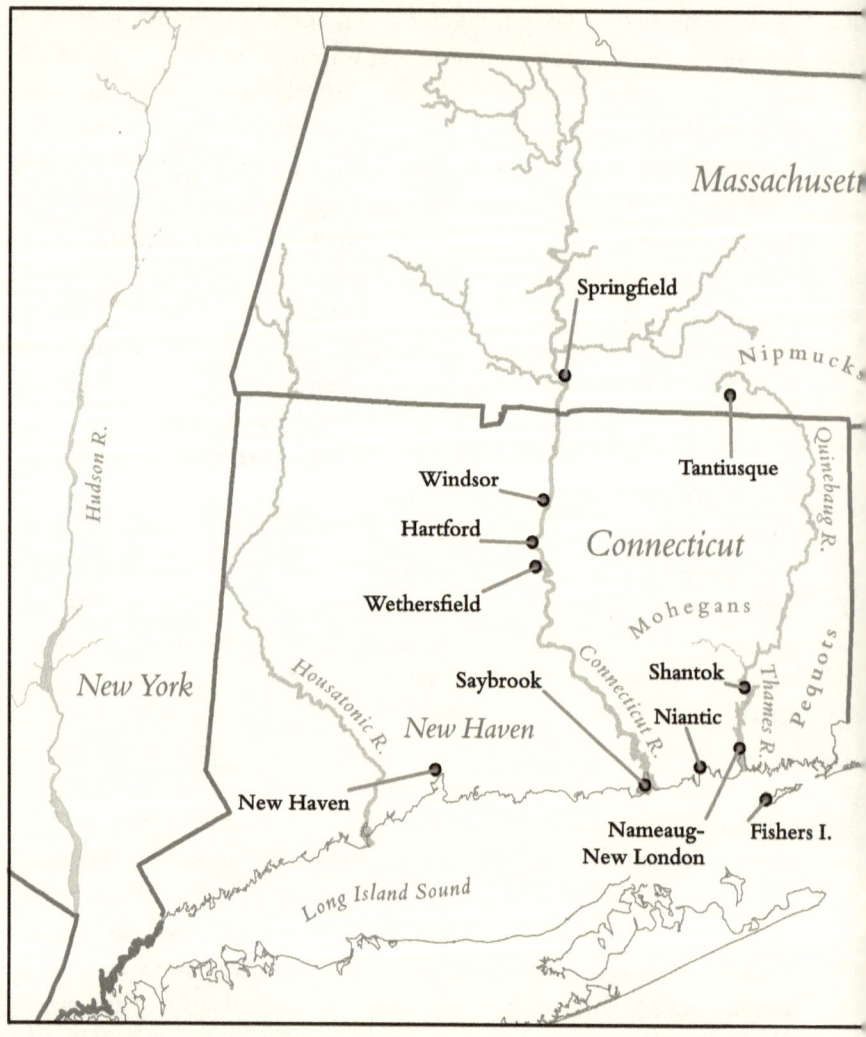

MAP 1. The New England World of John Winthrop, Jr. *Drawn by William F. Keegan*

ground as Winthrop arrived to lay out his proposed colony in 1645 and how it affected both his reception and the events that followed.[2]

---

2. Oberg, *Uncas*, 87. The Niantics were another regional tribe, often, but not always, subject to Narragansett control. In addition to Oberg, see Neal Salisbury, *Manitou and Providence: Indians, Europeans, and the Making of New England, 1500–1643* (New York, 1982); Francis Jennings, *The Invasion of America: Indians, Colonialism, and the Cant of Conquest* (Chapel Hill, N.C., 1975); Alden T. Vaughan, *New England Frontier: Puritans and Indians, 1620–1675* (Boston, 1965); Jenny Hale Pulsipher, *Subjects unto the Same King: Indians, English, and the Contest for Authority in Colonial New England* (Philadelphia, 2005). The notes and editorial comments on the situation in Glen W. LaFantasie, ed., *The Correspondence of Roger Williams* (Providence, R.I., 1988), are particularly useful.

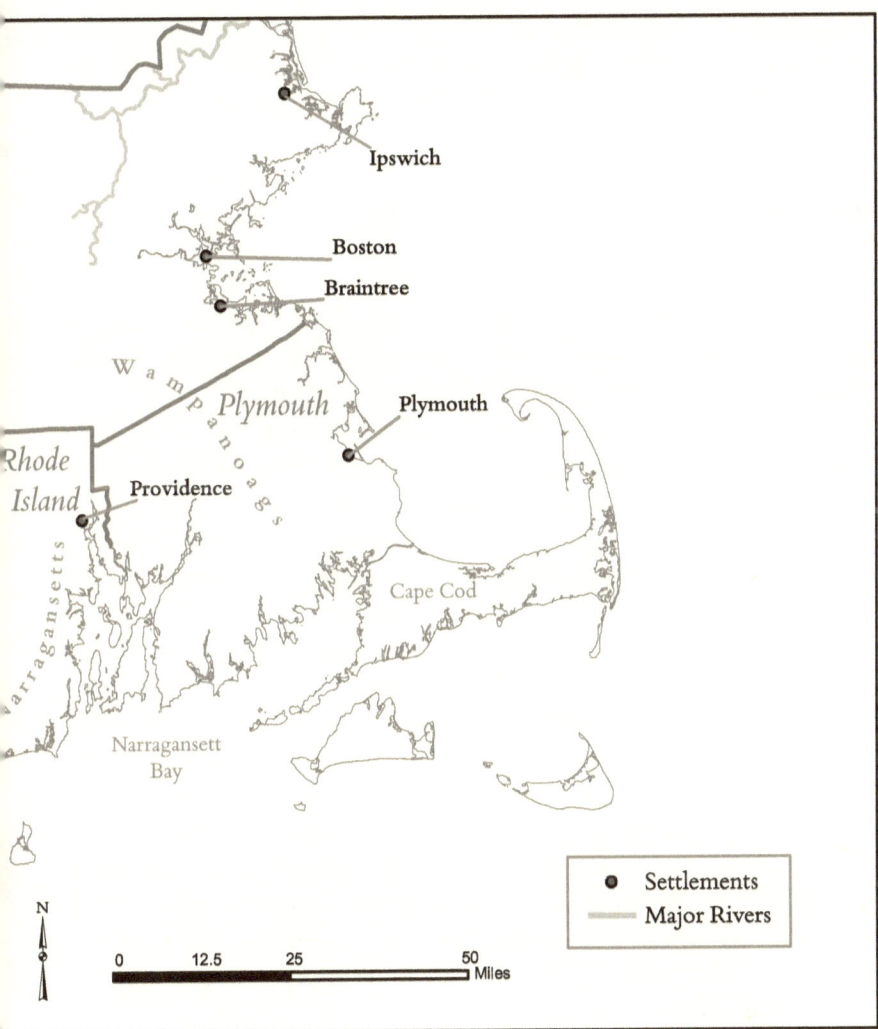

WHEN WINTHROP CAME with the Reverend Thomas Peters of Saybrook to do the initial groundwork for his new harbor plantation in the summer of 1645, whether the former Pequot territory belonged to Massachusetts or to Connecticut was still in dispute among the English. Both colonies had claimed the former Pequot territory by right of conquest, and both had sought recognition of their land claims from their former Indian allies, who were competing among themselves to assume regional dominance. Each colony had established separate postwar treaties with the Mohegans and Narragansetts, the primary native allies of the English in the war. At the risk of oversimplifying, it may be said that Massachusetts was allied with the Narragansetts while Connecticut supported Uncas's efforts to build the Mohegans into a strong regional power. The Narragansetts acknowledged Massachusetts's rights to the former Pequot lands in 1638; Uncas deeded all his tribe's lands—except for those on which his people currently resided or that they had under cultivation—over to Connecticut in 1640, promising not to let any English settle there without Connecticut's prior permission. While each English colony pursued its own particular interests in its relations with native peoples, both demonstrated to the Indians that, despite their intercolonial competition, they would come together as they had in the Pequot War to counter future violence. All actors in the region, however, understood that the two colonies had distinctly separate and competing interests regarding the former Pequot lands.[3]

As the two English colonies vied to solidify their land claims, conflicts among native groups competing to assume the Pequots' former regional hegemony increasingly dictated events. Uncas, under the protection of Connecticut, used strategic marriages, coercion of former Pequot tributaries, and the covert practice of giving former Pequot warriors sanctuary to rapidly build the Mohegans into a substantial regional power.

As Uncas prospered, Narragansett resentment against him and the English who protected him smoldered. The Narragansett sachem Miantonomi, who had always seen Uncas as an upstart and minor player in the region, came to fiercely resent what he believed was the unfair favoritism consistently shown Uncas by the English, especially in any matter involving disputes between the Mohegans and Narragansetts. Despite his early alliance with Massachusetts, he ultimately became convinced that he could no longer get fair treatment in either Boston or Hartford, and by 1641 Miantonomi was trying to foment a pan-Indian regionwide anti-English uprising. The postwar Narragansett-Massachusetts alliance rapidly degenerated while Uncas's stock among the English leaders in both colonies increased.

From the standpoint of his relations with Massachusetts and Connecticut, the one English leader Miantonomi consistently trusted turned out to be an unfortunate choice. Roger Williams had been banished from Massachusetts before the outbreak of the free grace controversy because of, among other

---

3. The September 21, 1638, Treaty of Hartford, often presented as the treaty concluding the war, was only a unilateral treaty between Connecticut and the indigenous participants; Massachusetts Bay authorities were not party to the agreement, and the terms of the treaty favored Connecticut; see LaFantasie, ed., *Correspondence of Roger Williams*, I, 187, 194, 197, 198, 200, 207, 226; Pulsipher, *Subjects unto the Same King*, 22–23. Winthrop was clearly displeased that Connecticut had tried to undermine the Narragansetts' alliance with the Bay Colony in the Treaty of Hartford. Pulsipher, *Subjects unto the Same King*, 23–24; Henry A. Baker, *History of Montville, Connecticut, Formerly the North Parish of New London, from 1640-1896* (Hartford, Conn., 1896), 11; Oberg, *Uncas*, 72–86, 89–90; LaFantasie, ed., *Correspondence of Roger Williams*, I, 177–178 n. 5.

things, his advocacy of complete separation between church and state and his rejection of the king's right to dispose of Indians' lands as he had done in the Massachusetts charter. Williams had established the Providence plantation in Narragansett country, which soon became a refuge for exiles from the free grace controversy. There he had also become an ally to Miantonomi and defender of the rights of the Narragansetts. He had accompanied Miantonomi to Hartford as his trusted translator at the 1638 meetings leading to the Treaty of Hartford, but he was prevented from entering Massachusetts in the same capacity in 1640 during critical negotiations between that colony and the Narragansett sachem. Williams wrote of his service to the tribe: "I was here their Councellour and Secretary in all their Wars with Pequits Monhiggins Long Ilanders Wompanoogse.... I never denied them ought they desired of me." Williams's readiness to come to the aid of Miantonomi's tribe was particularly disturbing to Massachusetts, for it suggested the possibility of growing anarchy in the wilderness through a dangerous and unwelcome alliance between radical English exiles and potentially hostile and powerful native groups.[4]

The English responded to Miantonomi's effort to foment collective Indian resistance by forming an intercolonial confederation for coordinated defense, an extralegal alliance among Connecticut, New Haven, Plymouth, and the Massachusetts Bay colonies called the United Colonies of New England, deliberately excluding Roger Willams and the English of Rhode Island. The formation of the confederation was timely, for by August 1643, when the first meeting of the commissioners for the United Colonies was held, the Mohegans and Narragansetts were at war, and Uncas had taken Miantonomi captive in battle. Uncas, having received a letter he might have believed was from the Massachusetts authorities ordering him to release Miantonomi, had turned both the Narragansett sachem and the decision regarding his fate over to the authorities in Hartford, who in turn passed the baton to the commissioners. Seeing an opportunity to end the increasing threat posed by Miantonomi without having to take direct action against him, the commissioners instructed Uncas to kill the Narragansett sachem, but not at Hartford. Connecticut soldier observers accompanied Uncas's Mohegans as they marched Miantonomi back to Mohegan country and watched as Uncas's brother Wawequa clubbed the hated rival to death.[5]

---

4. Roger Williams to Assembly of Commissioners, Nov. 17, 1777, in LaFantasie, ed., *Correspondence of Roger Williams*, II, 748–755, esp. 752, and I, 75, 182–289, 202–206, II, 620–621 n. 7.

5. Harry M. Ward, *The United Colonies of New England, 1643–90* (New York, 1961). The letter demanding Miantonomi's release actually came from a group of religious radicals in Rhode Island associated with Samuel Gorton, who was then very much at odds with Massachusetts. Gorton might have implied to Uncas that he was acting under Massachusetts's authority. Richard S. Dunn,

If the English hoped eliminating Miantonomi would end intratribal conflict in the region, they were most mistaken. The Narragansetts' outrage over their leader's execution was exacerbated by the fact that, before killing him, Uncas had apparently accepted a sizable ransom payment from them for Miantonomi's release. Pessicus, Miantonomi's youthful successor, desperately wanted revenge. Seeking, perhaps, to exploit the differences he still perceived between Massachusetts and Connecticut as a result of their competing claims to the Pequot lands, Pessicus sent wampum as a peace offering to Boston along with a request that the Bay colonists not interfere with the efforts he intended to make to avenge Miantonomi's death. He made no such request of Connecticut, presumably because he believed they were more closely tied to the Mohegan leader. On this matter, however, Massachusetts and Connecticut were in agreement. Uncas, having executed the United Colonies' first official intercultural act, must be protected.

Although Massachusetts explicitly ordered them not to seek retribution, the Narragansetts dispatched a war party in June 1644 that attacked an outlying Mohegan village, killing six men and five women. Uncas, having been promised at the time he killed Miantonomi that the English would defend him against any Indian effort to seek revenge, immediately appealed to the commissioners for aid. In response, the United Colonies sent agents to warn the Narragansetts that, if they attacked Uncas again, they risked war with both the English and Mohegans. Undeterred, the Narragansetts recruited allies from neighboring tribes during the fall and winter of 1644–1645, many of whom also looked with alarm at the growing power of Uncas and his tribe and prepared to launch a spring offensive against their now-beleaguered Mohegan enemy.

IT IS A SIGN of the intensity of Winthrop's commitment to the alchemical plantation and to the lure of the silver-mining scheme he envisioned that he decided, despite the volatile atmosphere and almost certain threat of intratribal warfare, to come to the heart of the Pequot region in the spring of 1645 to begin laying out the future New London. It is also an indication of Massachusetts's growing concern over its weakening ability to make good its claims to Pequot lands by right of conquest. The ascendance of Uncas, whose loyalties seemed ever more tightly bound to the authorities in Connecticut, stood in sharp contrast to Massachusetts's broken alliance with the Narragansetts. It was clear that Massachusetts no longer could count on them as an Indian ally

---

James Savage, and Laetitia Yeandle, eds., *The Journal of John Winthrop* (Cambridge, Mass., 1996), 471; Oberg, *Uncas*, 91–107.

or on their support for Bay Colony efforts to become the dominant English presence in the Pequot region. Recognizing that having Massachusetts colonists on the ground could prove decisive in asserting territorial rights, the Massachusetts General Court approved on June 28, 1644 — the very same day it was received — Winthrop's petition to "make a plantation in the said Pequott Country." Bay authorities also tried to force a resolution of the disputed land claims by asking the United Colonies commissioners to decide once and for all, at their meeting in September of 1644, which English colony was entitled to claim jurisdiction over the former Pequot lands. Had the vote been taken then, it seems likely it would have gone in Massachusetts's favor. At least that is what events subsequent to Massachusetts's request imply.[6]

When the September meeting arrived, the request to take up the issue of the disputed lands was blocked by United Colonies commissioner George Fenwick of Saybrook, who claimed that Pequot harbor and the adjacent lands were, not spoils of war at all, but part of the original Warwick patent under which the Saybrook plantation had been founded at the mouth of the Connecticut River in 1635.

This was disturbing news for Massachusetts. Should Fenwick's assertion be proved true, the Warwick patent claim would precede and therefore supersede any claim to ownership of the Pequot lands by right of conquest. Fenwick asked the commissioners to defer disposition of the lands in question until the Saybrook proprietors in England had been notified, which the commission agreed to do. As it turned out, however, Fenwick's action was performed not so much on the Saybrook proprietors' behalf as on Connecticut's, for, within three months of the meeting at which action on the Pequot lands was tabled, Fenwick sold Connecticut the fort and lands at Saybrook and all the proprietors' patent rights, a move that ultimately helped assure that jurisdiction over the Pequot lands would go to the river colony.[7]

Fenwick's surprise move might have instigated Winthrop's decision to hurry and lay out his plantation in early 1645 despite the looming threat of hostilities. It is perhaps also significant to note that the companion to help him lay out the plantation was the Reverend Thomas Peters, who was then the minister at Saybrook but apparently ready to cast his lot, not with Fenwick and Connecticut, but with Winthrop and the new plantation. Not surprisingly, Winthrop's sudden arrival in the Pequot region — having given no prior

---

6. Samuel Eliot Morison et al., eds., *Winthrop Papers* (Boston, 1929–), IV, 466 (hereafter cited as *Winthrop Papers*).

7. David Pulsifer, ed., *Acts of the Commissioners of the United Colonies of New England* (Boston, 1859), vols. IX–X of Nathaniel B. Shurtleff and David Pulsifer, eds., *Records of the Colony of New Plymouth in New England,* I, 19.

notice of his intentions to the authorities in Hartford—strongly influenced how the Connecticut authorities subsequently interpreted and reacted to his presence there.

Winthrop claimed publicly to be indifferent whether his new plantation fell under the jurisdictional control of Connecticut or Massachusetts, instead presenting himself as a stabilizing presence come to pour English oil on the troubled waters of regional intratribal conflict. Nevertheless, Connecticut authorities, upon his arrival and for a long time thereafter, viewed him with suspicion, as an interloper from the Bay Colony who intended to secure by occupation land that was not his by right. Even after the United Colonies finally granted Connecticut formal jurisdiction over the Pequot region in 1646, the colony's leaders remained suspicious and continued to view Winthrop's undertaking with thinly veiled hostility for several years.[8]

Winthrop's ties to Massachusetts were not the only source of concern Connecticut authorities had regarding the new plantation. Connecticut leaders, and Uncas, too, were at first surprised and, later, infuriated by the approach Winthrop followed at this new plantation in regard to the settlement's relations with Indians.

WINTHROP BELIEVED from the start that the success of his prospective alchemical plantation scheme would hinge on the silver-bearing lead mine project, whose effective operation would itself be contingent on many factors. For the mining project to succeed, Winthrop had to establish a second plantation at the lead mine site that would support the mining operations. He also had to mine sufficient ore to make commercial shipment possible, and he simultaneously had to establish good relations with a number of indigenous groups whose support of, or at least acquiescence to, the project would be crucial. The mine site was deep in the interior of Massachusetts; transporting the ore to where it could be shipped to England for processing and sale posed a very formidable challenge. Fortunately for Winthrop, and not coincidentally, the mine site was situated along the same riverine watershed that flowed into the Long Island Sound at the site he had chosen for his New London. Although transporting the ore via this water route would require at least one portage, it was by far the best of all possible transportation routes. The rivers flowed, however, through lands settled by several tribes; successful transportation would require the consent of a number of native groups. These included the Nipmucks, who occupied the lands surrounding the mine site and along the northern Pequot river watershed; the Mohegans, who occupied

---

8. Ibid., I, 79.

the central Pequot river lands and who resided just north of Winthrop's new plantation; the Niantics and Narragansetts, who lived east of the new harbor plantation; and a surviving group of perhaps as many as five hundred Pequots, ostensibly tributaries of Uncas, who had, since the end of the 1637 war, regrouped at the mouth of the Pequot River harbor in an effort to retain their tribal integrity (see Map 1).

Given the intertribal conflict that had characterized postwar Indian relations and their exacerbation by the death of Miantonomi, it took an Englishman with supreme confidence in his ability to interact with Indians to believe he could form alliances with and pacify all these competing factions. Winthrop not only had that kind of confidence; he had a reputation for relating to and with Indians that suggested he just might be able to pull it off. Not only was he considered an unusually effective Indian negotiator, but he also arrived in the Pequot territory with a unique intercultural advantage—a special relationship with Robin Cassacinamon, the sachem of the five hundred or so Pequots living at Nameaug, the Indian town along the harbor's edge, who could supply an in-depth understanding of the local intertribal political landscape.

IT HAS BEEN argued that, after the Pequot War, English authorities abandoned their previous policy of winning acceptance from native groups through adopting native patterns of cultural communication and negotiation, insisting instead that native people subordinate themselves to English cultural practices and accept political domination. Winthrop chose otherwise. Long known as an Englishman especially capable of interacting effectively with native people, yet sharing his contemporaries' cultural attitudes regarding the unquestionable superiority of all things English, Winthrop based his strategy for establishing authority among Indians on an assumption that *he* had to understand Indian power dynamics, not vice versa, and represent *his* claims to leadership within the terms of the indigenous culture. The source of this attitude came, at least in part, from his English alchemical partner Edward Howes, who helped shape Winthrop's attitudes toward America's Indians and who was the first to call Winthrop the Sagamore of Agawam, reflecting the approach to the natives that he had recommended Winthrop follow with local Indians when he had founded the town of Agawam (later Ipswich, Massachusetts) in the 1630s. "Sagamore" was an Indian term for the political leader of a tribe.[9]

9. On changing attitudes to cultural interaction with native groups, see Pulsipher, *Subjects unto the Same King*, 25, 35–36.

Howes viewed Indians, not adversarially, but as inferior allies, whose allegiance must be earned rather than commanded. Writing of one Indian leader, he told Winthrop, "I conceive it were very good, to bestowe respect and honor unto such as he . . . it is a rule in warre, to aime to surprise and captivate the greate ones, and the lesse will soone come under, soe winn the hartes of the Sachems and you win all." Howes advised Winthrop to work within existing native social and political hierarchies and to gain allegiance through amity, beneficence, and the display of cross-cultural esteem. Through witnessing Winthrop's demonstration of friendliness and largesse to native leaders, Howes argued, the common people among the Indians would come to naturally yield him respect and deference. "The more love and respect you shewe to the Sagamores and Sachems the more love and feare shall you gaine from the common natives."[10]

Howes also believed in intercultural proximity as a means to acquire information about the environment, gain access to resources, and help accelerate the assimilation of Indians to English ways. He assumed that Winthrop might take Indian servants into his household (which Winthrop did) and sent him a grammar he considered especially useful for teaching the Indians English. The grammar, "of such a rare method that it is admirable to conceive," came into Howes's hands, he noted, by a special providence. "I hope for the good of N: E: and the speedy bringinge of English and Indians to the perfect understandinge of our tongue and writinge truly." Howes had also acquired a "booke of Characters, grounded upon infallible rules of Syntax and Rhetorick," which he promised to send portions of to Winthrop for a trial among the Indians. These books were probably composed by Thomas Harriot, the alchemist and natural philosopher who had participated in the Roanoke settlement and who had "created an unprecedented way of recording unfamiliar languages for use in America."[11]

Howes not only advised Winthrop to bring Indians into his household; he recommended that Winthrop settle among the Indians and rely on their experience for selecting plantation sites. "Learne by reports and your owne observations where (on that River) the natives have lived longest and healthfullest and in greatest aboundance, though it be 50, 60, or 70 or more miles up in the land. . . . Gett theire good will; if possible you can to sitt downe with them or by them." Howes urged Winthrop to "be as neere [to the Indians] as may be, soe it be a place comodious for trade and husbandrie, and not easilie

---

10. Edward Howes to John Winthrop, Jr., Mar. 26, 1632, *Winthrop Papers*, III, 73–75, esp. 74.
11. Ibid., June 22, 1633, 131–133, Apr. 20, 1632, 76–77; Karen Ordahl Kupperman, *Indians and English: Facing off in Early America* (Ithaca, N.Y., 2000), 80–82.

surprized by an enimie." Those measures would not only lead to future prosperity, Howes believed; they would also hasten the Indians' conversion. With God's help Winthrop would "lay a foundation for a Cittie of Peace, to the honor of his great name; in your religious cohabiting together."[12]

Such views about living in proximity to native people and building a peaceful society through "religious cohabiting together," especially when adopted and implemented by New England's leading natural philosopher, complicate arguments that English colonists and natural philosophers were busy during this period recasting Indians as physically and technologically inferior. Joyce Chaplin has argued that, during the same period that Winthrop was settling New London, New England's Puritans developed a science-based aversion to physical proximity to Indians, based on an epidemics-generated belief that Indians' bodies were physically inferior to English bodies. Although "trade and evangelization might require some amicable contact," Chaplin asserts, it definitely did not mean "setting up house with the Indians."[13]

But setting up house with the Indians was exactly what Winthrop did, and nothing in his writings suggests he had, or ever developed, a view of native people as physically inferior or potentially contaminating. Not only did he lay out his new harbor plantation immediately adjacent to the five hundred residents of the Pequot town called Nameaug; he sought from the beginning to incorporate native groups into his pansophic development program, some as active participants, others as passive supporters. Moreover, Winthrop's decision of where to settle his new town, his efforts to secure the mine site from its indigenous owners, and his understanding of how to approach intertribal relations in the region were all arrived at in consultation with and with reliance upon the Pequot sachem Robin Cassacinamon, whom Winthrop not only respected but to whom he developed a relentless loyalty that severely strained his relations with many Englishmen, including his own father.

Like the Mohegan chief Uncas, Robin Cassacinamon was a leader especially skilled in both intertribal and intercultural relations. Like Uncas, Cassacinamon understood the strategic value of developing a special relationship with a well-placed English patron and using that relationship to advance the interests of his people. Uncas had closely aligned himself with the English of

---

12. Edward Howes to John Winthrop, Jr., Aug. 4, 1636, *Winthrop Papers*, III, 290–293.

13. Joyce E. Chaplin argues that Indians' disastrous susceptibility to contact period epidemics led the English to see them as physically inferior. Working outward from Indians' inferior bodies, colonists also recast native technologies, medical practices, and other knowledge-based enterprises as equally inferior, in a process that ultimately led to a kind of early modern scientific protoracism. *Subject Matter: Technology, the Body, and Science on the Anglo-American Frontier, 1500-1676* (Cambridge, Mass., 2001), 180, 187.

Connecticut and the Connecticut military leader John Mason; Cassacinamon aligned himself with the claims of Massachusetts and, more particularly, with Winthrop. The alliance with Winthrop, which endured the fiercest opposition from outside forces both English and Indian, was crucial not only to the success of Winthrop's plantation but also to the survival of the Pequot people as a distinct tribe.[14]

Cassacinamon first appears in colonial records after the Pequot War, where he is identified as a tributary of the sachem Uncas; he had not been identified as a tribal sachem or a warrior during the Pequot War (if he had been, he would have been executed) and was one of the surviving noncombatant Pequots placed in servitude to Uncas as part of the dispersal of the vanquished tribe. Cassacinamon must have come from a leadership lineage, though, as he was the sole Pequot among a group of ten Indian men sent to Boston by Uncas to secure the release of a female Pequot servant from Massachusetts governor John Winthrop, Sr. Uncas wanted to marry this woman, who had presumably been "given" to the Massachusetts governor as a spoil of the recent war, as part of his process of strengthening Mohegan tribal authority through strategic marriages. Uncas instructed Cassacinamon's party to offer Governor Winthrop ten fathoms (sixty feet) of wampum as compensation for the woman. According to the Rhode Island leader Roger Williams, who heard the story from an Indian informant, Uncas also instructed Cassacinamon that, in case Winthrop declined the offer, he was to attach himself to the Winthrop household and stay there until he arranged for the woman to escape. This is apparently exactly what Cassacinamon did; the woman did escape, and he was rewarded by Uncas with the ten fathoms of wampum that Winthrop had turned down.[15]

These would seem to be rather inauspicious circumstances with which to mark the beginning of a supposedly close intercultural relationship, and yet somehow this incident of intercultural deception led to the formation of a partnership between Cassacinamon and the younger Winthrop of significant future consequence. Over the next eight years (the records are silent on how it came about) Cassacinamon managed both to serve simultaneously as a servant in Governor Winthrop's household and to acquire (or retain) sachem status among the five hundred or so Pequots gathered together at Nameaug, a site along the Pequot River harbor. During this same period, he and the younger

---

14. Kevin A. McBride, "The Legacy of Robin Cassacinamon: Mashantucket Pequot Leadership in the Historic Period," in Robert Steven Grumet, ed., *Northeastern Indian Lives, 1632-1816* (Amherst, Mass., 1996), 74-92.

15. Roger Williams to John Winthrop, July 23, 1638, in LaFantasie, ed., *Correspondence of Roger Williams*, I, 168-169.

FIGURE 5. American Indian (formerly identified as Ninigret II). Oil on canvas by Charles Osgood after portrait by unknown artist (circa 1681). 1837–1838. *Courtesy of the Massachusetts Historical Society*

It has been suggested the subject is a Pequot sachem.

Winthrop came both to trust each other and to see each other as indispensable allies in achieving their individual goals. Cassacinamon became an adviser to Winthrop about Indian relations in the Pequot region and was undoubtedly, probably much more than Winthrop's English confidant Edward Howes, instrumental in Winthrop's decision to locate his new plantation virtually among the wigwams of the Nameaug Pequot community. Cassacinamon also became a facilitator for Winthrop in handling critical intercultural relations. In return for Cassacinamon's help in achieving goals among the region's Indians, Winthrop became the Pequots' strongest and sometimes only supporter among the English, with a persistence that ultimately helped them overcome the provisions of the Treaty of Hartford calling for the tribe's cultural and political eradication.[16]

WINTHROP UNDERSTOOD THAT, for his plantation projects to succeed, he needed to establish authority among the numerous competing indigenous groups in the region. He also understood that the often-conflicting policies of Massachusetts and Connecticut had created confusion among many Indians as to where the locus of English power resided. With that in mind, and consistent with his understanding that he should claim leadership in terms readily understandable within Algonquian culture, Winthrop attempted, with Cassacinamon's assistance, to cultivate a perception of himself as an English paramount sachem, more powerful than the other sachems living in the region.

Winthrop's personal circumstances helped provide the foundation for this intercultural persona. As the son of the governor of Massachusetts and as a former governor of Saybrook, Winthrop appeared to natives as a hereditary sachem, first in the line of succession of an English ruling lineage. Patrilineal principles permeated the formal leadership structures of native society in New England. Most native leaders in the region knew, from prior dealings with Massachusetts, of the Winthrop family's preeminence in Massachusetts government, and Cassacinamon could be counted on to communicate young Winthrop's lineage and status among the English to any who didn't. Winthrop sought to further enhance this perception by presenting himself to the Indians using the title "Governour," even though his only service as a colonial governor up to that time had been a one-year appointment as governor of the Connecticut River during the settlement of Saybrook a decade earlier. Winthrop's English supporters helped reinforce this intercultural self-fashioning

---

16. McBride, "Legacy of Robin Cassacinamon," in Grumet, ed., *Northeastern Indian Lives*, 81; Frances Manwaring Caulkins, *History of New London, Connecticut: From the First Survey of the Coast in 1612 to 1860* (New London, Conn., 1895), 45, 51–52.

effort. The Springfield trader William Pynchon's use of the title "Governor" when describing Winthrop to the natives at the Tantiusque mine site helped underscore the impression of a royal lineage between the two Winthrops. The implications of this trumpeted genealogical patrimony were not lost on the region's native leadership (nor, as we shall see, on the English leaders of Connecticut). The Narragansetts and Niantics, for example, finding young Winthrop a direct and formal political link to his father, quickly began to channel their communications to the authorities in Boston through New London.[17]

Promoting himself as someone possessed of an English sachem's lineage was just one element of Winthrop's intercultural self-fashioning. He also took care to assert his claims to authority to Indians through the careful observance of native customs and rituals. One place this is seen clearly is in the extraordinary care with which Winthrop purchased the mine site and surrounding lands from the indigenous owners. Despite the fact that his English agents had made two previous purchases of the land around the Tantiusque site, Winthrop deemed it advisable to conduct yet a third transfer of the land in February 1645. Unlike the first two sales, which had involved simply the exchange of goods and signatures, this final transfer was confirmed with ritualized performances specifically intended to give it cross-cultural authority. The "Deed of Webucksham and Washcomo to John Winthrop, Jr." contained an explicit description of the land being sold and an enumeration of the goods being provided in payment, validated by appropriate signatures and marks, as was the English custom. But it also contained a detailed description, intentionally incorporated into the language of the document, of the native ritual of validation that was conducted by the Nipmuck sachem and his son.

> According to English Custom I have given Possession of all my Lands aforesaid unto Amoss Richason Servant to said Winthrop Governour of the English for said Winthrops Use. To have and to hold to him the said John Winthrop his Heirs and Assignes for Ever. In Everlasting Remembrance and Wittness hereof I lay this Wishkeeg or Writing on Washcomos my Son and Heirs Breast and sett my Mark and Seal and Wascomos my said Son according to Indian Custom freely makes his

---

17. His father's selection as president for the July 1645 meeting of the commissioners for the United Colonies, which dealt forcefully with native unrest in the Pequot territory, highlighted for the Indians Winthrop's elevated lineage among the English. Pulsifer, ed., *Acts of United Colonies*, I, 31; Ann Marie Plane, *Colonial Intimacies: Indian Marriage in Early New England* (Ithaca, N.Y., 2000), 21; Kathleen J. Bragdon, *Native People of Southern New England, 1500-1650* (Norman, Okla., 1996), 180-181; William Pynchon to Stephen Day, Oct. 8, 1644, *Winthrop Papers*, IV, 495-496; Roger Williams to John Winthrop, Jr., Apr. 7, 1649, in LaFantasie, ed., *Correspondence of Roger Williams*, I, 277-279.

Mark and Seal hereunto on my Breast. This don with Consent of all the Indians of Tantiusques.[18]

By conducting the land transfer ceremony according to both English and Indian customs, Winthrop did more than formalize a bicultural transfer of property. He validated native cultural practices and demonstrated his sensitivity to symbolic forms important to indigenous cultures. At the same deed signing, Winthrop, with the help of Cassacinamon, also incorporated an additional demonstration of his authority in terms native people could readily understand. In Indian land transfers, sachems customarily made use of "ahtaskoaog," or "principal men," who represented the sachem at the land exchange and provided assurance of the sachem's consent to the transaction. Winthrop followed this same procedure in purchasing the mine site. Observing the sachems' custom of acting through agents, Winthrop did not attend the deed signing with Webucksham and Washcomos; rather, Cassacinamon did, and the title "Governour and Chief Councilor among the Pequots" adjacent to his mark on the sale agreement confirms he was serving at this transaction as an Indian ahtaskoaog for Winthrop. This was a role Cassacinamon would assume for his English ally many times as part of their symbiotic alliance. It underscores not only the importance of their collusive relationship but also Winthrop's belief in the necessity of maintaining clear cross-cultural communications. While Winthrop unquestionably sought an indisputable documentary record of ownership that conformed to English legal standards, he paid equal attention to observing and recording the land transfer at Tantiusque according to native customs and rituals of transfer.[19]

In other settings, Winthrop's alchemical practices, coupled with his well-known medical skills, enabled him to present himself to Indians not just as an English sachem but also as a magico-religious specialist, similar to an Indian shaman, or *powaw*. This profile, combined with his reputation as an English sachem, provided Winthrop a channel through which he might at-

---

18. Deed of Webucksham and Washcomo to John Winthrop, Jr., *Winthrop Papers*, V, 4–5.

19. Historians are divided in their assessment of Europeans' sensitivity to indigenous people's cultural practices related to land transfer. Patricia Seed found that, in most cases, Europeans' rituals of possession surrounding land transactions primaily targeted fellow Europeans, who would ultimately determine the legitimacy of land transfer. Faren Siminoff's study of early land transactions on eastern Long Island, however, suggests European settlers recognized that native land claims were legitimate until both parties agreed they had been properly extinguished. On "ahtaskoaog," see Bragdon, *Native People of Southern New England*, 142. Deed of Webucksham and Washcomo to John Winthrop, Jr., *Winthrop Papers*, V, 4–5; Patricia Seed, *Ceremonies of Possession in Europe's Conquest of the New World, 1492–1640* (Cambridge, 1995), 11; Faren R. Siminoff, *Crossing the Sound: The Rise of Atlantic American Communities in Seventeenth-Century Eastern Long Island* (New York, 2004), 5–8, 110–129, Plane, *Colonial Intimacies*, 154–159, 172–176.

tain uniquely elevated status in native circles. Powaws were, like alchemists, "intermediaries between the physical and spiritual worlds." Their "access to a range of spirits both within and outside of themselves" gave them great influence in their tribes. Moreover, while it was unusual in Algonquian culture for the roles of shaman and sachem to overlap in one person, the rare individual possessed of both abilities was believed to be exceptionally powerful. None of the sachems Winthrop encountered along the Pequot watershed claimed dual status as shaman and sachem. But Winthrop did; by presenting himself to the Indians as a leader who combined political authority with special healing powers, he could hope to establish among them the perception that he was a superior sagamore.[20]

In this regard, Winthrop's medical reputation and extensive practice served him well. Alchemical medicine was the one means through which almost every New Englander derived personal benefits from alchemy, and Winthrop was the foremost alchemical physician in colonial New England. Wherever he went, his medical skills were in demand. Indian messengers, often hired by the families of diseased colonists to carry requests for medical assistance great distances through the wilderness to Winthrop at New London, were highly aware of the English respect for Winthrop's healing powers. From the beginning of settlement, Winthrop sought to establish this same respect for his healing among the watershed tribes in the Pequot region.[21]

Providence seemed to offer support to this strategy. Just as Winthrop was arriving in the Pequot region in the spring of 1645 to begin laying out the port town, the Mohegan chief Uncas was attacked by Narragansett warriors in their continuing effort to gain retribution for Uncas's role in the murder of their sachem Miantonomi two years before. Winthrop and his aide Thomas Peters hurried to Uncas's fort at Shantok, north of the new plantation site, to offer medical assistance. Their action, by demonstrating the efficacy of the healing powers Winthrop was bringing to the region, advertised to Uncas one of the advantages he could derive from having Winthrop settled in the region. Uncas, appreciative of the aid, and perhaps not yet fully cognizant of the extent of Winthrop's alliance with the Pequot group that Uncas himself claimed as tributaries, welcomed Winthrop to the area with gifts of wampum and great professions of joy.[22]

---

20. On powaws, see William A. Starna, "The Pequots in the Early Seventeenth Century," in Laurence M. Hauptman and James D. Wherry, eds., *The Pequots in Southern New England: The Fall and Rise of an American Indian Nation* (Norman, Okla., 1990), 33–47, esp. 43; William S. Simmons, *Spirit of the New England Tribes: Indian History and Folklore, 1620-1984* (Hanover, N.H., 1986), 42–43.

21. Roger Williams to John Winthrop, Jr., June 22, 1645, *Winthrop Papers*, V, 30.

22. Thomas Peters to John Winthrop, ca. May 1645, ibid., 20–21.

Winthrop's effort to present himself as an English powaw possessed of special medical powers was also enhanced by his alchemical practices. He was conducting experiments and making medicines at New London soon after settlement; his alchemical apparatus and the smoke and smell of his chemical fires made a profound impression. John Arundell, who visited Winthrop at New London after the plantation was well established, was astonished by what he saw. "Upon my first entrance to your home," he wrote, "I believed I was consulting the oracles of the Tripod or the pharmacy of Aesculapius." Native observers, whose rituals placed a sacred significance on smoke as a medium of transition between the natural and spirit worlds, probably shared Arundell's awe, if not his allusions. Among northeast people, smoking tobacco was thought to attract Manitou. The native concept of Manitou, the spirit that pervades all things, was homologous to the *spiritus mundi,* the living spirit that many alchemists believed animated all beings in the cosmos. The ability of Indian shamans to act as intermediaries between the natural and spirit worlds directly correlated with the role of the alchemical magus who believed that through a heightened prayer state during experiments he might achieve divinely granted insights into the secrets of nature. The alembics and furnaces in which Winthrop conducted his experiments and produced his alchemical medicines, present in a town in which Indians and English were thoroughly intermingled, might well have conveyed a cross-cultural message of magical power. Certainly, they helped reinforce the popular respect given to Winthrop's medicines and medical practice by both English and native people. Although no Winthrop medical account books survive for the early years of the plantation, Winthrop undoubtedly rendered medical services to the Pequot people at Nameaug and elsewhere just as he and Peters had to Uncas's injured warriors. Winthrop's reputation for healing Indians spread among natives as well as the English. Soon the English in other towns such as Roger Williams's were requesting him to send medicines "fit for Indian Bodies." Such requests were significant, for, given the fierce opposition of some native powaws to English healing, adoption of Winthrop's medicines by native people represented a substantial vote of confidence in the English leader's powaw-like abilities.[23]

23. John Arundell to John Winthrop, Jr., Jan. 16, 1656, trans. and quoted in Harold Jantz, "America's First Cosmopolitan," Massachusetts Historical Society, *Proceedings,* LXXXIV (1973), 3–25. The oracles of the "Tripod" refer to the tripod upon which Pythia, the prophetess of Apollo at Delphi, sat to issue her oracles. The tripod was placed in a small, belowground chamber filled with gases that issued from underground springs. On smoking, see Bragdon, *Native People of Southern New England,* 204; Roger Williams to John Winthrop, Jr., August 1651, in LaFantasie, ed., *Correspondence of Roger Williams,* I, 335–336. Puritan missionaries came to believe that medicine was an

An unintentional side-effect of Winthrop's medical practice with his English patients might have had major symbolic importance to native observers, inadvertently reinforcing Winthrop's cross-cultural image of superior authority. Winthrop often hosted within his household for extended lengths of time the wives of English leaders from other towns, who came to him suffering from a variety of illnesses and resided in his home during treatment. The presence of such women, among them the wife of John Haynes, governor of Connecticut, Mary Pynchon, wife of the founder of Springfield, and the wife of Thomas Doxey of New London, became a semipermanent aspect of a household that from its inception was occupied by two mistresses, rather than one. The first English woman at New London was, not Winthrop's wife, Elizabeth, but her sister, Margaret Lake, who would live for years as a member of the Winthrop household.[24]

The practice of polygyny, the marriage of one man to several women, was widespread in native societies during this period. Only the elite practiced such unions; most men and women lived monogamously, with a single partner. In some of these elite polygynous marriages, men married two or more closely related women, perhaps even taking sisters as co-wives. After the Pequot War, Uncas had embarked on an aggressive campaign of entering into multiple marriages to solidify Mohegan political and social alliances and strengthen his tribe. On five different occasions during the three years following the conflict, he either married or expressed a desire to marry seven different women. In practicing polygyny for political purposes, Uncas was following a widely accepted practice. The Narragansett leader Miantonomi had unsuccessfully sought to form an alliance with Uncas by taking one of his daughters as a wife. Wequashcook, a Niantic / Pequot sachem, had married the mother of Sassacus, who had led the Pequots in the 1637 war, within a year of the war's conclusion.[25]

---

essential aspect of missionary work. Indians were reluctant to give up the spiritual healing of powaws without confidence that an equally powerful spiritual medicine was available from the English. "These powows are factors for the devil, and great hinderers of the Indians embracing the gospel," noted Daniel Gookin, for they worried: "If we once pray to God, we must abandon our powows; and then, when we are sick and wounded, who shall heal our maladies?" Daniel Gookin, *Historical Collections of the Indians of New England* . . . (Boston, 1792), 14.

24. Caulkins, *History of New London*, 44; John Haynes to John Winthrop, Jr., Sept. 12, 1652, *Winthrop Papers*, VI, 220–221; John Pynchon to John Winthrop, Jr., July 26, 1654, in Carl Bridenbaugh, ed., *The Pynchon Papers*, I, *Letters of John Pynchon, 1654–1700*, Colonial Society of Massachusetts, Publications, LX (Boston, 1982), 7–8.

25. On polygyny, see Plane, *Colonial Intimacies*, 5, 21–23. Eric S. Johnson, "Uncas and the Politics of Contact," in Grumet, ed., *Northeastern Indian Lives*, 39–44, McBride, "The Legacy of Robin Cassacinamon," 76–77.

Given the pervasiveness of elite polygyny in the Pequot region, the presence of so many women in the Winthrop household—his wife and sister-in-law at all times, and women from the households of what could be seen as minor English sachems at others—sent a cross-cultural message about Winthrop's status and authority. Winthrop did not, of course, intentionally set out to create a native perception of himself as a polygynist; it was a natural outgrowth of his sister-in-law's estrangement from her husband and his own popularity as a healer. But, while Winthrop did not purposefully create the situation in which numerous elite women normally occupied his household, he certainly understood the message such a household configuration suggested. It unintentionally reinforced a perception Winthrop did try deliberately to project in other ways: that he had great political authority as an English sachem and remarkable healing powers as a powaw, the unique combination of traits that often produced an exceptional paramount sachem.[26]

Winthrop's special relationship with Cassacinamon, his kinship ties, his political leadership of the new English plantation, his alchemical practices, his medical skills, and the presence of many women in his household, all supported his culturally perceptive effort to establish leadership among the many contesting tribes inhabiting the Pequot region. He also attempted, as would be expected of any paramount sachem, to exercise leadership in a culturally sensitive manner—through conference and consensus. Before settling at the river mouth to lay out his plantation, Winthrop called together Indian representatives from all neighboring bands into one assembly, to obtain clear agreement on the boundaries between the former Pequot lands (which he intended to settle) and its adjacent territories. The meeting was a success; Uncas helped delineate a clear northern boundary between Mohegan territory and the new plantation, and whatever other boundary issues there might have been with adjacent tribes were readily resolved. This was not, however, to be a harbinger of things to come.[27]

Despite his considerable skills at intercultural diplomacy and his carefully crafted intercultural self-representation, Winthrop's settlement of the Pequot region met with serious resistance from both Indians and English. Winthrop must have misjudged how much Uncas's influence in the Pequot region had increased since the end of the Pequot War. The Mohegans are estimated to

---

26. Bert Salwen, "Indians of Southern New England and Long Island: Early Period," in William C. Sturtevant et al., eds., *Handbook of North American Indians,* XV, Bruce Trigger, ed., *Northeast* (Washington, D.C., 1978), 160–176; William Wood, *New Englands Prospect,* ed. Alden T. Vaughan (Amherst, Mass., 1977), 99; Roger Williams, *A Key Into the Language of America* . . . (London, 1643), 230–231.

27. Plane, *Colonial Intimacies,* 23; Caulkins, *History of New London,* 51.

have grown from about 400–600 total members in 1637 to more than 2,500 by 1643, and Winthrop does not seem to have factored in the increased power that growth had provided the Mohegan sachem. Certainly, Winthrop misread the strength of the alliance that had developed between Uncas and Connecticut as well as how much mistrust Connecticut would greet his own arrival in the region with. Yet, even without Winthrop's miscues, the enmity between the Narragansetts and the Mohegans was too strong, the power relations among the multiple regional tribes too unstable, for anyone, especially any Englishman, to forge a regional power structure that would resolve the long-standing conflicts exacerbated by the Pequot War. As a result, almost before his plantation had settlers, Winthrop faced a combination of dangerous Mohegan harassment and harsh English criticism that threatened the very survival of the plantation.[28]

WINTHROP'S SPECIAL RELATIONSHIP with Cassacinamon's Pequots at Nameaug (the Indian name until New London became official) posed a direct challenge to the authority of both the Mohegan sagamore Uncas and the magistrates of Connecticut. By agreeing to settle among Cassacinamon's people and serve as their English advocate while insisting that Uncas define clear boundaries between his territory and theirs, Winthrop rejected the Pequots' status as servile tributaries of Uncas, the role to which they had been assigned by Connecticut in the 1638 Treaty of Hartford. This simultaneous rejection of Connecticut's authority over the tribe's postwar disposition, when coupled with Connecticut's perception of Winthrop as an interloper from Massachusetts, led to serious problems for the fledgling English plantation almost from its inception.

It is a sign of Winthrop's success at presenting a cross-cultural claim to authority that the Mohegan sachem Uncas came to view him as such a formidable rival. Uncas, who had profusely welcomed Winthrop to the area when Winthrop had provided the sachem's wounded warriors medical aid after the 1645 Narragansett attack, quickly realized that the newcomer posed a major threat to his own ambition to assume regional dominance. Once the Englishman's direct challenge to Uncas's authority over Cassacinamon's Pequots as well as the extent of his regional plantation scheme became clear, Uncas's attitude shifted from welcoming to hostile. Even as the first permanent English settlers were arriving at Nameaug in the spring of 1646, Uncas launched a sudden, violent campaign of harassment and intimidation—ostensibly targeted at Winthrop's resident Pequots—that was manifestly intended to send a warning

28. Johnson, "Uncas and the Politics of Contact," in Grumet, ed., *Northeastern Indian Lives*, 35.

to both the incoming English and the restive Pequots. According to testimony presented by "the inhabitants of New London" to the commissioners of the United Colonies at their September 1646 meeting:

> He [Uncas] comes downe from mohegen in A hostile way with 300 men into the english plantation   so soone as he came within sight of mr. winterops wigwome he gave the word whereupon they divided themselfs into squadrans And soe fell on upon the indians presentlye tearing up their wigwoms cutinge And sloshing and beatinge in a sore maner which was A sad sighte to the beholders takinge there wompum there skins there baskets tearing there breaches there hose from there legs there showes from there feet forcinge them in the water and there shootinge at them allsoe forcinge into English mens houses friteinge the women And children takinge indians out of there houses also carying away A great deal of mr. winterops wompum pege caryinge away a hat and coat of mr. Peters also a coat and severall skins of other mens.[29]

If anything, the written account understates what must have been the terrifying impact on the new English settlers of the sudden appearance of three hundred well-organized and battle-ready Mohegans who, after marching deliberately to within sight of Winthrop's Indian-style temporary residence, began a carefully orchestrated but frenzied assault on the Nameaug Pequots. In addition to destroying their wigwams, injuring them physically, and stealing their furs and household goods, the Mohegan warriors forcefully stripped off the Indians' clothes and drove them naked into the river, where they shot weapons at them (whether with arrows or muskets is unclear). The attacking Mohegans left no doubt of their attitudes toward the new settlers. They broke without hesitation into the English people's houses to drag out Pequots who might have sought refuge there, "friteing the women And children" and undoubtedly many of the English men as well. Without question, Uncas wanted Winthrop to understand that this raid was intended as a warning to him personally as well as to Cassacinamon's Pequots, which is why the Mohegan raiders made sure to break into Winthrop's property and steal "A great deal of mr. winterops wompum pege." To further underscore the new settlers' vulnerability, they stole a hat and coat belonging to the second-most important leader in the settlement, the spiritual leader the Reverend Thomas Peters, and a number of other English men's furs, drove away English cattle,

---

29. Petition of the Inhabitants of New London to the Commissioners of the United Colonies, Sept. 15, 1646, *Winthrop Papers*, V, 111.

stole corn from their fields, and, even threatened to shoot one of the Englishmen in the back. Winthrop emphasized this outrageous act in his report of the incident. "One of the Monheges presented his piece [that is, aimed his musket or other firearm] at Cary latham when his back was towards him."[30]

TO WINTHROP, in Boston at the time, Uncas's actions were completely unconscionable. No matter how the Mohegan leader might try to justify himself, this had been nothing less than a thinly veiled attack on the new and vulnerable English settlement, and *could not* be allowed to go unpunished. Permitting such a brazen assault to go unanswered would serve as an open invitation to other disgruntled native groups to act similarly. Winthrop, therefore, assumed with some confidence that the commissioners of the United Colonies would impose severe correction on Uncas for his provocation of the new settlement, and he decided to try to capitalize on Uncas's blunder by recommending that the commissioners formally release Cassacinamon's Pequots from their assignment as tributaries to the Mohegan leader. "If these Indians that we must live neere be still under Uncas command," Winthrop wrote Thomas Peters, "there wilbe noe living for English there." As long as Uncas could claim the Pequots as a subject tribe, he would have an open excuse to assault the English town in a similar fashion again. Winthrop urged Peters to detail for the commissioners the full extent of Uncas's aggression and to stress to them the jeopardy that allowing the Pequots to continue as Mohegan tributaries would pose for the new English settlement.[31]

Winthrop's unannounced arrival in the Pequot region from Massachusetts, coupled with his self-proclaimed assumption of the role of English protector of the Pequots, had offended more than Uncas. It also generated negative attitudes among Connecticut's magistrates. For that reason, Winthrop informed Peters that he himself was not willing to take the lead in presenting the Pequot matter to the commissioners, "because it may be conceived my intentions are other then they are." Worried that Uncas's supporters in Connecticut might portray his opposition to Uncas's control of the Pequots as just a self-serving power grab, Winthrop elected to remain in the background and let Peters, who had come to the plantation from nearby Saybrook, be the spokesman for the complaint against Uncas. Winthrop explained to Peters—perhaps coaching him on what to tell the commissioners if Winthrop's self-interest came up—his only reason for wanting the Pequots released from Uncas's control: "I looke at the quiet of our plantation principally, and conceive [it] a greater

---

30. Ibid., 111–112.
31. John Winthrop, Jr., to Thomas Peters, Sept. 3, 1646, ibid., V, 100–101.

security to have a party of the Indians there, to have their cheife dependance upon the English. They will easily discover any Indian plotts, etc." By allowing the Pequots at Nameaug to break with Uncas and ally with Winthrop, the region would be far more stable, and the English would also have a trustworthy and continuous access to native intelligence about plots against the English.[32]

WINTHROP OBVIOUSLY was concerned about how Connecticut authorities might portray his motives to the commissioners, but he could not have anticipated just how fully the commissioners would end up siding with Uncas against Winthrop and the settlers of the new town. In what seems an almost counterintuitive response to the Mohegan effort to terrify and intimidate new English settlers, the commissioners for the United Colonies came down squarely in support of Uncas. In part, this was due to Uncas's masterful understanding of the realpolitik involved in intercultural relations and to the loyalty he had acquired as an English ally both during and after the Pequot War. By executing the Narragansett Miantonomi at the explicit command of the United Colonies commissioners in 1643, he had cemented his reputation as a trustworthy Indian ally and had secured a commitment from the commissioners that they would aid him should Miantonomi's followers seek vengeance. While Winthrop and Peters might have seen Uncas's actions at the new settlement as an unmitigated act of aggression meant to intimidate the incoming English settlers, Uncas presented it—and the commissioners saw it—as a warranted defense of his regional rights.

Called before the commissioners, Uncas acknowledged that he had been wrong in "vindicatinge his owne right soe neare the English plantations" but asserted that, otherwise, his actions were fully justified. He claimed that many of the Pequots at the English plantation were people who had formerly lived under his authority at the Mohegan town of Shantok but who had, since Winthrop's arrival, fled his service under color of submitting to the English plantation. He further asserted that the primary object of his harassment had not been the English at all, but a Niantic leader named Wequashcook, who had been given permission by the incoming English to hunt on Uncas's lands without obtaining Uncas's permission. That had been a particularly reprehensible act on the part of the new settlers, Uncas reminded the commissioners, because the Niantics Wequashcook led were, like the Narragansetts, sworn enemies of both Uncas and the English.[33]

32. Ibid.
33. Pulsifer, ed., *Acts of United Colonies*, I, 72.

The Niantics and Narragansetts were indeed currently perceived as enemies by the United Colonies. After the Narragansetts had attacked Uncas the year before (the attack that had brought Winthrop and Peters to Uncas's aid), the English commissioners had decided that the two tribes posed an immediate war threat and, further, that the English willingness to defend Uncas against those tribes would be viewed by other Indians as a test case of the English willingness to support native allies. They had authorized raising an army of three hundred to subdue the two tribes. To avoid war, the Niantic and Narragansett sachems had agreed to pay the English a huge surety in wampum (two thousand fathoms) and provide sons of sachems as hostages to authorities in Boston to assure their good behavior in the future. The promised ransom payments had been pitifully slow in coming, the promised hostages had not been readily delivered, and the commissioners were angry. Uncas's strategy of claiming that his assault on Nameaug had actually been in response to the unauthorized privileges granted to the Niantic sachem Wequashcook was a brilliant defense. Not only had the upstart English settlers under Winthrop given support to a declared enemy of both the English and the Mohegans, he argued, but they had done so by abrogating rights Uncas had earned through his former service in defense of the English colonies.[34]

The commissioners, as it turned out, could not have agreed more. In their decision regarding the complaint against Uncas, they totally turned the tables on the incoming settlers: "It [was] disorderly and unwarrantable for any English plantation to entertaine Neckwash Cooke or any of the Narraganset or Neanticke Sagamors or their companies into a league, protection, or submission untill they [the Narragansetts and Niantics] have fully performed all their covenants with the Colonies." In other words, the new settlers of Nameaug, by contracting with the Niantic sachem to hunt for them before he had fully complied with his prior agreement to provide surety payments and hostages to the commissioners, had greatly overreached, to the detriment not only of Uncas's land rights but of the authority of the commissioners as well. The commission assured Uncas that no Narragansetts or Niantics, including Neckwash Cooke, were authorized to hunt in his lands, "nor will they [the commissioners] allow any English plantation to countenance such disorderly huntinge." They further noted, in a direct and explicit rejection of Winthrop's proposal that the Pequots be freed from Uncas's custody, that they would

---

34. Ibid., 55; Wendy B. St. Jean, "Inventing Guardianship: The Mohegan Indians and Their 'Protectors,'" *New England Quarterly*, LXXII (1999), 367–372; Oberg, *Uncas*, 110–138; McBride, "The Legacy of Robin Cassacinamon," in Grumet, ed., *Northeastern Indian Lives*, 74–85, Johnson, "Uncas and the Politics of Contact," 29–47; LaFantasie, ed., *Correspondence of Roger Williams*, I, 220–223, 244–248.

not allow any of Uncas's tributary Pequots to be withdrawn by any English plantation. Finally, as if to forestall another attempt by Winthrop to act unilaterally in the future, they awarded jurisdiction over the new plantation to Connecticut. On every issue, the commissioners had found in favor of Uncas and in the interests of Connecticut, interpreting Winthrop's plantation intentions in the worst possible light. The newcomers had aided English enemies; they had usurped Uncas's authority; furthermore, they were attempting unlawfully to wrest away Uncas's control over his human capital—the Nameaug Pequots—as well as his land.[35]

Uncas, on the other hand, was given the mildest of reprimands. In a decision completely rejecting the settlers' assertion that Uncas's assault was consciously designed to intimidate them and not just the Indians among them, the commissioners told Uncas to acknowledge to the settlers that he had erred in beginning his "quarrel with Neckwash Cooke" so near to their plantation.[36]

Examining the commissioners' findings side-by-side with the settlers' report of what Uncas actually did during the assault on their town, it is hard not to conclude that the commissioners of the United Colonies chose to all but ignore Mohegan aggression, deciding instead to issue a harsh and virtually one-sided condemnation of the incoming English settlers.

It is perhaps relatively easy to see why Connecticut—a colony with a strong historic alliance to Uncas and a competing claim to the lands Winthrop was settling—might have endorsed such a critical view of Winthrop's and his fellow planters' intentions. It is somewhat more difficult to understand why the commissioners as a body came to that conclusion. The commission consisted of two representatives each from Massachusetts, Connecticut, Plymouth, and New Haven colonies, and their decisions required approval of at least six of the eight commissioners. Though we do not have roll call votes on any of the commissioners' decisions, one might assume that Massachusetts's representatives would have been reluctant to censure a project undertaken with their own colony's authorization, especially when headed by one of its leading citizens, the son of the Bay Colony's governor. One might also surmise that at least one of the representatives from Plymouth and New Haven would have had some reluctance about appearing to sanction an Indian assault on a new English frontier plantation.[37]

Perhaps the commissioners from the smaller colonies voted as a block to

---

35. Pulsifer, ed., *Acts of United Colonies*, I, 72, 73, 79.
36. Ibid., 72–73.
37. St. Jean, "Inventing Guardianship," *New England Quarterly*, LXXII (1999), 368–369.

curb what they saw as an unsolicited incursion by Massachusetts into the Pequot region. Massachusetts's efforts a few years before to expand into the Narragansett country had raised howls of protest among both the Narragansetts and a group of Rhode Island's radical religionists headed by Samuel Gorton. The commissioners had backed the Bay Colony in its dispute with Gorton, but Winthrop's entrance into the disputed Pequot country without prior consultation with Connecticut possibly was seen as a similar expansionist move, though against, not radicals, but coreligionists. Some evidence supports the theory that the commissioners from the small plantations voted as a block to restrain the ambitions of their larger partner. In 1648, the Massachusetts General Court expressed concern about acts of the United Colonies commissioners that "may in a short time prove not onely prejuditiall, but exceedingly uncomfortable" to the Bay Colony, and sought amendments to the articles of confederation that would allow Massachusetts to act independently in those cases where it disagreed with the other three colonies. Whether Massachusetts was reacting to the assignment of Winthrop's plantation to the jurisdiction of Connecticut two years earlier is unclear.[38]

The fact remains, however, that in the fall of 1646 the English commissioners collectively exhibited more mistrust of Winthrop's new plantation than they did of Uncas. This mistrust would hang over the project for most of its first few years, and it consistently and negatively influenced the United Colonies' policies toward the town and the resident Pequots during much of that time. While concerns about Massachusetts's imperialist tendencies might have been a factor in this mistrust, it is equally likely that the concerns about the New London project were based less on Winthrop's ties to Massachusetts than on the actions against Massachusetts launched in 1646 by Winthrop's friend and fellow investor in the alchemical plantation Robert Child.

Child, who had helped fund Winthrop's ironworks and had promised to be "active in the [black lead mining] business, to the utermost of my power" upon his return to New England, had arrived back in New England in September 1645, full of enthusiasm for his and Winthrop's new endeavors. He had talked to Samuel Hartlib of the country's promising mineral potential.

> The Country abounds with minerals especially Iron stone. We have discovered about 10 or 12 severall sorts which I have sent to Mr Buckner an Apothecary in Bucklerberry, and to Dr. Mericke dwelling

---

38. The complex issues surrounding this effort are detailed in Pulsipher, *Subjects unto the Same King*, 25–32; Philip F. Gura, *A Glimpse of Sion's Glory: Puritan Radicalism in New England, 1620–1660* (Middletown, Conn., 1984), 276–303; Ward, *United Colonies*, 62; Nathaniel B. Shurtleff, ed., *Records of the Governor and Company of the Massachusetts Bay in New England* (Boston, 1853–1854), II, 245.

there, where you may see them, if you please, and other stones, which promise better thinges, and I hope will not deceive us.[39]

Child was equally enthusiastic about the developing ironworks. "I doubt not (by the Grace of God) but wee shall prosper in Iron works and make plenty of Iron Spedily." He found that horticulture, too, especially the gardens that were so important to Child, did very well in New England, despite the bitter winters. "Summer is hot enough for [Vines?] I suppose," he noted. "Apples and cheries Peares and Apricoks and all kind of Garden ware flourish incredibly."[40]

But, for all the rich potential he found in the prospective development of New England's natural possibilities, Child's interactions with Bay officials quickly led him to a much darker conclusion: that the pansophic program simply could not flourish in the face of what he saw as Massachusetts's astonishingly restrictive intolerance. Massachusetts's doubly choking policies of limiting the political franchise only to church members, and then restricting church membership only to those who had run the spiritual gantlet of making an acceptable public profession of faith, had created, Child believed, a society with an intolerably constrained capacity for the exchange of ideas upon which scientific and economic progress depended. This was not an environment in which mutual collaboration, economic development, or a pansophic program of godly world improvement could proceed, much less progress. Almost before it had begun, Child saw the rigid and constricted orthodoxy embedded in the New England way as choking the designs he had nurtured of making New England a laboratory for the utilitarian deployment of natural philosophy.

In an effort to force the Puritan magistrates to accept a more open society, Child had joined with a diverse group of Massachusetts's dissidents in 1646 in issuing a remonstrance to the General Court forcefully demanding that the Bay Colony provide its residents the same political franchise and the same church membership privileges to which they had been entitled back in England. As spokesman for the group, Child threatened to seek redress from Parliament if the Massachusetts authorities did not respond favorably and immediately to the remonstrance. The Bay magistrates, wanting no part of a parliamentary inquiry—or of greater tolerance in church and political admissions—had levied a huge fine of two hundred pounds against Child, placed him under house arrest, and refused to let him leave the colony until the fine was paid. Child remained in detention until the following year.[41]

---

39. Robert Child to Samuel Hartlib, Dec. 24(?), 1645, Hartlib Papers [15/5/1A–2B], University of Sheffield.
40. Ibid.
41. Robert Emmet Wall, Jr., *Massachusetts Bay: The Crucial Decade, 1640–1650* (New Haven, Conn.,

Although many in New England might have sympathized with Child's and his associates' desire to get Massachusetts's godly elites to embrace the type of Neoplatonic toleration then common among many of England's Puritan parliamentarians, the threat to force Massachusetts's hand by taking the complaint to England backfired. The Bay colonists closed ranks around their leaders, soundly rejected Child's remonstrance, and assiduously defended their New England way. Having been rocked by the Hutchinson controversy over free grace in 1637 and forced more recently to defend their actions against Samuel Gorton before critics in England, they wanted no part of this upstart outsider's programs or policies. Massachusetts's agent in England, the Plymouth colonist Edward Winslow, wrote a pamphlet for consumption at home and abroad branding Child as a probable Papist and implying that his search for and interest in New England's natural resources provided a cover for his activities as a Jesuit spy. This was hardly an apt label for a Presbyterian alchemist with Neoplatonic confessional attitudes, but it effectively cast a pall of evil intentions on Child's insistence on religious toleration. By the time Winslow published his second tract in defense of New England in 1648, Child had become the leader of a "blacke Regiment" out to restore "all the old Ceremonies and odde Holidayes." For a time, it seems, other New England alchemists, or at least alchemists associated with Child, also came under suspicion. George Starkey, the New England–trained alchemist who emigrated to England in 1650, where he influenced both Robert Boyle and Isaac Newton, also reported being confined in 1647 under suspicion of being a "Spie or a Jesuit," a detention that lasted two years.[42]

Such concerns about the hidden agendas being pursued by the alchemists in their midst might easily have been transferred to Winthrop's new alchemical plantation project, especially given Child's close association with it. In fact, evidence suggests that Winthrop's own father might well have had concerns about the project, at least during its formative stages. In 1643, the year young Winthrop was traveling Europe recruiting alchemists to come to New England, the elder Winthrop had sent him a letter in which he expressed seri-

---

1972), 157–211; George Lyman Kittredge, "Dr. Robert Child the Remonstrant," Colonial Society of Massachusetts, XXI, *Transactions,* 1919 (Boston, 1920), 1–145; Margaret E. Newell, "Robert Child and the Entrepreneurial Vision: Economy and Ideology in Early New England," *New England Quarterly,* LXVIII (1995), 223–256.

42. Edw[ard] Winslow, *New-Englands Salamander, Discovered by an Irreligious and Scornefull Pamphlet, Called New-Englands Jonas Cast up at London . . .* (London, 1647), 7–8; [Edward Winslow], *Good News from New-England . . .* (London, 1648), 14–15; G. H. Turnbull, "George Stirk, Philosopher by Fire (1628?–1665)," Colonial Society of Massachusetts, Publications, XXXVIII, *Transactions,* 1947–1951 (Boston, 1959), **222**; William R. Newman, *Gehennical Fire: The Lives of George Starkey, an American Alchemist in the Scientific Revolution* (Cambridge, Mass., 1994), 50.

ous reservations about his son's pansophic pursuits. "Study well, my son," the father warned, "the saying of the Apostle, *Knowledge puffeth up*. It is a *good Gift of God*, but when it lifts up the mind above the *Cross of Christ*, it is the *Pride of Life*, and the High-way to *Apostacy*, wherein many Men of great Learning and Hopes have perished." The younger Winthrop seems to have been able to allay his father's fears that the pansophic pursuit of alchemical knowledge might threaten his commitment to New England's Puritan practice. Massachusetts had, after all, authorized the younger Winthrop to establish both the plantations on which the alchemical project hinged. After Child's remonstrance, however, such concerns seemed warranted and, for a time at least, appear to have become widespread.[43]

The United Colonies commissioners certainly seem also to have worried that Winthrop's new alchemical plantation might become a haven for the kind of dissent Child was promoting in Boston. Even as they assigned the jurisdiction of Winthrop's settlement to Connecticut, they advised all colonial governments, expressly because of the "petitions [that] have beene lately putt up," to keep a due watch "at the doores of Gods house, that none be admitted as members of the body of Christ, but such as hold foorth effectuall callinge ... and enter by an expresse coven[an]t." Warning that "under a deceitfull colloure of liberty of conscience" the godly could be subjected to Anabaptism, Familism, antinomianism, and other heresies, they urged colonial governments to make sure that efforts to liberalize church admission were duly suppressed. The commissioners' reference to Child's remonstrance was explicit, their wholesale rejection of its demand for religious toleration even more so.[44]

WHETHER BECAUSE of Winthrop's associations with Robert Child, with Massachusetts, or with the Pequots under Cassacinamon, his New London plantation received critical scrutiny from Connecticut authorities and the United Colonies commissioners throughout the first five years of settlement. At times, Winthrop was the object of what appears to have been thinly veiled hostility. English efforts to curtail his activities did not, however, alter his approach to settlement. Although he adapted his tactics and even modified his goals when prodded by necessity, he pursued his original development scheme with a per-

---

43. Cotton Mather, *Magnalia Christi Americana; or, The Ecclesiastical History of New-England* ... (London, 1702), book II, 33.

44. Pulsifer, ed., *Acts of United Colonies*, I, 81; Wall, *Massachusetts Bay*; Kittredge, "Dr. Robert Child the Remonstrant," Colonial Society of Massachusetts, Publications, XXI, *Transactions*, 1919, 1–146, esp. 16–93.

sistent single-mindedness that those seeking to regulate the new plantation found off-putting. Because the success of the riverine silver-mining scheme ultimately depended on the ability to safely transport ore through several different tribes' territories, and in an effort to reduce the region's rampant intertribal contention, Winthrop continued to pursue cooperative relationships with Indians, despite the commissioners' efforts to curb his intercultural alliance building. Except for his increasingly conflicted relations with the Mohegan sachem Uncas, Winthrop achieved a good measure of success. The Narragansetts and Niantics came to consider him as a powerful English sachem and a cultural broker to the English in Massachusetts. They asked Winthrop to serve as spokesman on their behalf with the authorities in Boston and before the United Colonies commissioners on various issues surrounding payments of the wampum surety the English had demanded in 1645. In a move suggestive of native social divisions and understandings of gendered power, Niantic women also presented gifts of corn to Elizabeth Reade Winthrop, and on one occasion Niantic or Narragansett sachems asked Elizabeth Winthrop to intercede with her husband on their behalf. In a similar sign of Winthrop's acceptance as an intercultural mediator, the Nipmuck people who lived adjacent to the black lead mine site turned to him to help broker their relations with the English commissioners. The Pequots under Cassacinamon also remained tightly aligned with Winthrop, despite increasing hostility from Uncas.[45]

Buoyed by the commissioners' 1646 rejection of the settlers' complaints against him, the Mohegan sachem expanded the range of his harassments in a way that sent a clear cross-cultural message of intimidation. Much as Uncas had done to the Pequots at Nameaug, his brother Nowequa attacked the Nipmucks adjacent to Winthrop's mine. He confiscated their wampum, copper kettles, hemp baskets, and many bear- and deerskins. After assaulting the Nipmuck settlement, Nowequa traveled with fifty Indian warriors to Winthrop's property of Fishers Island, where they destroyed a canoe and so badly frightened one of Winthrop's servants that he refused to remain in service without protection. Lest Winthrop miss the Mohegan's intentions, on the way back from Fishers Island Nowequa and his men, many armed with guns,

---

45. Pulsifer, ed., *Acts of United Colonies*, I, 87; Roger Williams to John Winthrop, Jr., Aug. 20, 1647, in LaFantasie, ed., *Correspondence of Roger Williams*, I, 235, Dec. 3, 1648, I, 260, Apr. 13, 1649, I, 281. "For a token of his fidelitie to Mrs Wintrop Nenekunat he prayes me to write that all women of his towne shall present Mrs Wintrop with a present of Corne at Pwacatuck if she please to send in any conveyance to Pwacatuck for it." Roger Williams to John Winthrop, Jr., Oct. 10, 1648, I, 252, Roger Williams to John Winthrop, Jr., mid-November 1648, I, 258–260, Roger Williams to Elizabeth Reade Winthrop, Apr. 3, 1649, I, 276–277.

"hovered against" Winthrop's settlement at Nameaug in a manner that once again deeply frightened both the English and the Pequots.[46]

This continuing harassment—meant to show Winthrop that the Mohegans had the ability to undermine every aspect of his plantation project—only reinforced Winthrop's belief that the best hope for stability in the region and security for his plantation resided in his being given guardianship of Cassacinamon's Pequots. As long as Uncas could use the excuse of the Pequots' tributary status to him as a justification for intimidation or new assaults, the viability of the Nameaug settlement would remain in doubt. Who could expect settlers to come or to stay in a place that was regularly under attack, or the threat of attack, by Indians?

In a repeat effort to gain guardianship over the Pequots, Winthrop presented, at the July 1647 meeting of the United Colonies, a petition subscribed by sixty-two Pequots and Niantics living at Nameaug asking that they be made English wards. The petitioners charged Uncas with continuing tyranny and provided the commission with an extended list of the Mohegan's injustices, including wholesale intimidation, repeated extortion of excessive tribute, refusal to pay gambling losses, cutting the Pequots' fishing nets, stealing corn, even threatening to subject members to arbitrary execution. Winthrop and Cassacinamon argued, "Though Uncus [had] seemed glad that Mr John Winthrop came to settle an english plantation at Pequat, and prsented him with wampan, yet without cause . . . he quickly tooke offence, [and] fell to outrages." The real intention behind Uncas's continued assaults was unmistakable. Uncas's "carriage hath bene such," they told the commissioners, "as if he intended by alarums and affrightments to disturb and breake that plantation." Uncas's counselor Foxon, present at the hearings, answered the petitioners' charges with a variety of defenses, implying that Uncas's change in attitude toward Winthrop had been caused by the Englishman's efforts to undermine his authority over his vassal tribe.[47]

The long listing of provocations presented to the commissioners leaves little doubt that Uncas was indeed engaged in an aggressive campaign of psychological intimidation against the infant English settlement and threatened its future viability. But, just as they had done the year before, the commissioners chose to downplay the consequences of Uncas's actions while drawing a series of negative conclusions about the intentions of Winthrop and the Indian petitioners.

---

46. Pulsifer, ed., *Acts of United Colonies*, I, 101–104.

47. Ibid., 99, 102. To further support the guardianship petition, Winthrop informed the commissioners that Uncas himself had already previously assigned custody of the Pequots over to him.

They found the Pequots' petition "false and deceitful" and saw no reason to consider altering their condition as tributaries to Uncas. Among the commissioners, contempt for the Pequots as a conquered enemy remained as strong as ever. Recalling the terms of the Treaty of Hartford, the commissioners stated:

> Remembring the proud wars some yeares since made by the Pequatts, and the just resolutions of the English that . . . the remnant of that nation should not be suffered (if the English could help it) either to be a distinct people, or to retayne the name of Pequatts, or to settle in the Pequatt country, but that they should all be devided betwixt the Narragansetts and Mohegens Indians, and that under a tribute to the English, they. . . . ordered those Pequats foorthwith returne a due subjection to Uncus.[48]

The commissioners did agree that Uncas's "insolency and outrage" had been greater than they previously suspected, but their efforts to prevent its recurrence were nominal. They ordered Uncas to behave in the future and fined him one hundred fathoms of wampum, to be paid to Winthrop only *after* the Pequots had left Nameaug and returned as subjects to Uncas's jurisdiction. Neither Winthrop, Uncas, nor the Pequots would have missed the point that Uncas could have extracted the one-hundred-fathom fine from the Pequots themselves (the very form of extortion the Pequots had just charged him with repeatedly doing). As to Uncas's brother's harassment of the English servant on Fishers Island, the commissioners ordered Nowequa to repair the broken canoe but simultaneously ordered Winthrop's man to return a gun he had taken away from one of the Mohegans during the incident.[49]

Although Uncas received relative forbearance, the commissioners repeatedly acted to put Winthrop in his place. Not only did they flatly dismiss the guardianship petition, but they rendered unfavorable decisions on several other matters affecting Winthrop and his settlement. Someone, probably Winthrop himself, reopened the question of which colony should have jurisdiction over the Pequot plantation. Although Winthrop professed personal indifference regarding the outcome, he did note that some of the town's new settlers "should be much disappointed if that plantation fall and be setled under any other jurisdiction" than Massachusetts's. The commissioners, unmoved by the settlers' desires, once again confirmed, and this time for good, "that the Jurisdiction of that plantation doth and ought to belonge to Connecticut." On another land matter, Winthrop sought confirmation of his

48. Ibid., 97, 100–101.
49. Ibid., 100–102.

claim to "a greate quantity of land" between the new plantation and the fort at Saybrook that he purchased from the Niantics before the Pequot War of 1637. Although Winthrop had no document recording the sale and the sachem who had conducted the transaction was apparently dead, Winthrop presented affidavits from Indians and the English interpreter Thomas Stanton verifying that Winthrop had indeed made the purchase. Connecticut, the colony in which the lands in question were located (just confirmed in its authority over Winthrop's new town), challenged the United Colonies' jurisdiction over the matter as well as Winthrop's claim itself on several substantive grounds. Among their objections was that, at the time Winthrop had supposedly made the purchase, he was in the employ of the English grandees who funded and founded the Saybrook plantation. "It seems somewhat uncomely (at least)," Connecticut's commissioners noted, for Winthrop to have been buying land on his own behalf so close to the plantation he was then starting as the agent of others. Even if Winthrop had bought the land as he claimed, they implied, he had done so in a highly unethical manner.[50]

The commissioners found Connecticut's argument questioning the United Colonies' jurisdiction persuasive and refused to decide the case, effectively killing Winthrop's effort to validate his claim. They also issued an unsolicited caution to Winthrop about a prospective title to land on Long Island he was also rumored to be trying to purchase. Should he complete that purchase, the commissioners cautioned, his deed "will be fownde weake," since several persons from Connecticut and New Haven colonies had already invested considerable sums of money to secure a patent there.[51]

Winthrop's 1647 appearance before the commissioners of the United Colonies was a resounding failure. On every matter in which Winthrop had an interest, the opposing viewpoint carried the day. It is difficult to know how issues like Winthrop's links to Robert Child or his ties to Massachusetts might have influenced the harsh treatment he received. The backlash against Child's remonstrance was then at its peak, and Winthrop's renewed effort to secure the new plantation for the Bay Colony certainly would not have been appreciated by Connecticut. It is clearer, however, that Winthrop's alliance with the Pequots—which he presented as the key to establishing regional stability—had made him no friends.

Despite the commissioners' unequivocal demand that Winthrop return

50. Ibid., 96, 97, 103, 104.
51. Ibid. On the competing claims to land on Long Island and how different claimant "communities of interest" attempted to counter them, see Siminoff, *Crossing the Sound*, 5–8, 91–96, 110–129, 153.

Cassacinamon and the other Indians residing at Nameaug to the jurisdiction and subjection of Uncas, Winthrop dragged his feet. He simply was not prepared, apparently under any conditions, to surrender authority over the Nameaug Pequots to Uncas.

Winthrop's recalcitrance led, more than a year later, to a crisis that would ultimately resolve the Pequot issue. At the September 1648 meeting of the United Colonies, the commissioners noted that, even though they had previously sent Edward Hopkins, one of the commissioners from Connecticut (as well as its governor), to Winthrop's plantation to formally order in the presence of both Winthrop and the Indians that the Pequots return to Uncas's subjection, "noe Conformety hath hithrto been yealded Theareunto by them." The commissioners therefore commanded Winthrop to follow through with their order: if he didn't comply this time, they had authorized Uncas to come and take the Pequots by force.

> In case a Reedy atendance bee not forthwith yealded hereunto, Unquas shall have order, and Liberty by Constrainte to Inforce them; and it is desiered that the Gorment of Conitacott wil pvide hee be not therein opposed by any English Nor the Peaquats or any of them habored or shiltered in any of their howses: whiles noe Just offence is given by them by him or any of his.

A month later, after Winthrop still had not complied, John Mason, the English military commander during the Pequot War and now Uncas's closest English ally, showed up at Winthrop's plantation with the Mohegan sachem. Mason presented Winthrop a letter stating his intentions:

> SIR, whereas it is ordered by the Comissioners that Onkos shall have libertie to fetch his Indians to theire former place who are now residing at Nameag in prosecution of which order he is now come with Comission from the Comissioners of the United Colonies there being severall Eng: [b]oth to witnes to the Carriag of the desig[n] and that there be noe wronge done to the English of Na[meag] the English of Nameage are required by the Comissioners order that they doe not Enterteyne any of Nameag Indians or there goodes unto their uses nor any way hinder Onkos in the prosecucion of this service: thus much is desired that you should be made acquainted with as alsoe the rest of your neighbours.[52]

---

52. Pulsifer, ed., *Acts of United Colonies,* I, 111–112; John Mason to John Winthrop, Jr., October 1648, *Winthrop Papers,* V, 263.

After reading the order, Mason and armed English observers from Connecticut stood watch over an angry town of English settlers while a group of Mohegan warriors roamed through the settlement dragging the town's Indian settlers into captivity. Uncas, under Mason's watchful eye, took custody of at least some, if not all, of the Pequots; thus, the commissioners' decree regarding the Pequots was finally put into effect.

Forcing the Pequots into Uncas's custody was one thing, however; keeping them there, quite another. During the following months, an unknown but substantial number of Pequots slipped away from Uncas's custody and made their way back to Winthrop's plantation. By January, there were enough Pequots back in residence at Nameaug to bring Uncas's warriors and the United Colonies representatives again, and this time they intended to make a lasting impression. Probably between January 17 and January 29, 1649 (based on internal dating in Winthrop's surviving correspondence), Uncas marched into Nameaug to reclaim his recalcitrant tributaries and teach them a lesson as well. He destroyed or stole their possessions, injured men and women, stripped them naked, and carried away their food supplies, all in the middle of winter and all under English supervision. The terrified English settlers watched in astonishment as the armed English observers allowed the Mohegans to maraud the town. In an effort to stop the violence the town's constable tried to intercede, but his efforts were ineffective. The violence proceeded unchecked. The "Protest of the Inhabitants of New London against Uncas" (incorrectly ascribed to 1647) describes what took place:

> We the Inhabitants of Nameag do solemnely protest that the late inrode by Uncas and his Crue upon the indians in this place in Robbing all their wigwams and depriving them of their necessaries for their very life in this very depth of winter as their coats shoes stockins corne beanes hatchets whereby they should provide firing also their kettles and breaking their pots they should boyle their meate in beside all their traies dishes matts baggs sacks baskets they lye upon, and all their wampam, wounding divers men and women, and stripping old, and yong: as it was an unexpected disturbance to our families frighting and amasing our wives, and children and diverting us from the necessary labour of our callings so it is still continually an unsupportable burthen to us, and the serious consideration therof doth give us just cause to feare that we may be as suddainly deprived of all our necessaries and our selves wives and children exposed to the bitterness of a winter flight without succor or shelter especially English appearing in so inhuman an action to our astonishment And doe professe that we

are most barbarously iniuriously and unchristianly dealth withal to have such a people whose tents are yet amonst us, to be unneccessarily provoked and forced upon a condition of absolute despaire."[53]

Because the violent surprise raid had been authorized by the United Colonies commissioners, it provoked a strong backlash against both them and Uncas. Many English considered it an act of excessive violence against both the English and Indians. Roger Williams was appalled "that such a monstrous Hurrie and Affrightmt should be offerd to an English Town either by Indians or English, unpunished." He had no doubt about where the blame for the incident should be placed. "Sir," he wrote Winthrop, "heap Coales of Fire on Cap. Masons head." Uncas's Indian enemies were also enraged, fearing similar English-sanctioned attacks against them. They saw the raid as an opportunity to gain support for taking revenge on the Mohegan sachem. Williams told Winthrop that the Indians near him (the Narragansetts and Niantics) were "ready to fall upon the Monhiggins at your word."[54]

Winthrop's father also believed the commissioners had gone too far. Writing from Boston on March 3, 1649, he commiserated with his son over "Uncas his outrage (though I wish your Constable had forborne to meddle with them)." He expressed hope that the outrageous assault would "give the Commissioners occasion to take some stricter Course with him [Uncas]." Surprisingly, though, while the Massachusetts governor agreed with his son that the commissioners needed to rein in Uncas, he was not at all supportive of his son's continuing to resist turning the Pequots over to Uncas's custody. At the time, the elder Winthrop was extremely ill and approaching death. That, despite the gravity of his illness, he was agonizing over affairs at his son's new settlement is seen in the deathbed request he made eleven days later. No longer well enough to write and conscious that his demise was near, he had his son Adam write his brother John to inform him that his father "wold requst you as if it wear his last requst, that you wold strive no more about the pequod Indians but leave theme to the commissioners order." Twelve days later Winthrop, Senior, was dead.[55]

The tone of the elder Winthrop's final request — one senses that Pequot custody had been a recurring subject of debate between the two men — coupled

---

53. Protest of the Inhabitants of New London against Uncas, ca. January 1649, *Winthrop Papers*, V, 124.
54. Roger Williams to John Winthrop, Jr., Jan. 29, 1649, in LaFantasie, ed., *Correspondence of Roger Williams*, I, 269; Roger Williams to John Winthrop, Jr., Jan. 29, 1649, *Winthrop Papers*, V, 308–309.
55. John Winthrop to John Winthrop, Jr., Mar. 3, 1649, *Winthrop Papers*, V, 311, Adam Winthrop to John Winthrop, Jr., Mar. 14, 1649, V, 319.

with his earlier criticism of his son's New London project during its formative stages ("Study well my son, the saying of the Apostle, *Knowledge puffeth up*") raises questions about both the nature of the relationship between the two Winthrops, father and firstborn son, and the father's support of his son's alchemical errand into the wilderness.

ACCURATE ASSESSMENT of this familial relationship is complicated owing to the Winthrop family's centrality to the history of early New England. The elder Winthrop's character has often been depicted in a way that makes it an exemplar of whatever the historian believes characterizes "colonial life in general and Puritanism in particular." As Francis Bremer further notes, we have the harsh and intolerant Winthrop, the tolerant and humane Winthrop, Winthrop the peacemaker, Winthrop the misogynist, Winthrop the loving husband, Winthrop the Indian hater, and so on. In a similar manner, his son and namesake has often been treated as a foil to his father, a less capable set piece figure demonstrating the perils and reality of Puritan declension. Thus young Winthrop was "totally unlike his father," and the two men "were not genuinely compatible; their respective viewpoints lay an interplanetary distance apart." Such depictions, however, do justice to neither man, nor are they attuned to the more complex and nuanced relationship between them.[56]

A good place to begin when seeking to assess the actual interpersonal dynamic between Winthrop father and son is to examine different styles of child rearing. Philip Greven posits three characteristic "temperaments" shaping colonial-era child rearing: a repressive evangelical style that emphasized breaking the child's will and enforcing obedience to parental authority, a moderate approach focused on bending a child's will to voluntarily act in dutiful ways through the loving exercise of authority, and a genteel, indulgent approach that encouraged willfulness and individualism. According to Greven, the Winthrop family provides an archetypical example of moderate child rearing.[57]

As a child in a moderate household, young Winthrop would be taught through affectionate means that fulfilling one's duty to parents and other family members was a crucial personal responsibility. Nurturing children to

---

56. Francis Bremer, *John Winthrop: America's Forgotten Founding Father* (New York, 2003), xv-xvi; Richard S. Dunn, *Puritans and Yankees: The Winthrop Dynasty of New England, 1630-1717* (Princeton, N.J., 1962), 63; Robert C. Black III, *The Younger John Winthrop* (New York, 1966), 133.

57. Philip Greven, *The Protestant Temperament: Patterns of Child-Rearing, Religious Experience, and the Self in Early America* (Chicago, 1988), 161-162, 178-179. On paternal interactions with children in early New England, see Lisa Wilson, *Ye Heart of a Man: The Domestic Life of Men in Colonial New England* (New Haven, Conn., 1999), 115-139.

be remarkably obedient and to have great affection for their parents was vital in moderate families. Nevertheless, such behaviors could never be imposed by parental fiat or enforced against the child's will. In moderate families, parental authority and power were limited; obedience was achieved through reasoning and parental example. Children were always encouraged to recognize their obligations voluntarily and perform them out of love and domestic affection. It was understood, however, that the choice whether to do so was always subject to their free will. The obligations to be dutiful were reinforced among the godly by the fifth of the Ten Commandments in Exodus. "Honour thy father and thy mother, that thy days may be long on the land the Lord thy God giveth thee" (Exodus 20:12). The filial call to duty was further reinforced by the model of Christ's obedience to the will of the Heavenly Father. A crucial element of that obedience was its voluntary quality.

Attention to duty was an essential attitude for children to bring to a whole range of family interactions. John Winthrop, Jr., like other children in moderate households with extended kinship connections, was raised to see himself as a member of a large and important family network in which every member down to the most distant relative had a particular status relation with, duties toward, and obligations he or she could expect from the others. "Duty," Greven notes, "symbolized the maintenance of position and place within the familial order, by asserting both the obligations of obedience and the limitations upon the exercise of authority and power."[58]

The voluminous Winthrop family correspondence confirms that ties of love and duty provided the lifelong foundation of the relationship between John Winthrop, Jr., and Sr. From his early years as a student at Trinity College in Dublin, the letters between father and son are inscribed with signs of affection (from both men), recognition of duty (on the part of the younger), and appreciation for duty fulfilled (from the elder). Young Winthrop voluntarily made large sacrifices on his parents' behalf, as when he relinquished his entailed interest in the family estate at Groton, England, to help fund his father's emigration to Massachusetts and to furnish a settlement for his stepmother. His father remembered this act with deep appreciation throughout his life. (Bremer notes that he received the strongest expression of parental love the elder Winthrop ever recorded.)[59]

Far from being men whose worldviews lay "interplanetary distances apart," the two John Winthrops were remarkably similar in their fundamental goals and principles. Both worked toward the creation of a radically better world,

---

58. Greven, *Protestant Temperament*, 179.
59. Ibid., 180–181; Bremer, *John Winthrop*, 319.

pursuing that goal with zeal without being zealots. Both approached their relations with others with toleration and forbearance, recognizing that people of differing opinions could seek similar goals. Where the two men differed and where those differences produced tensions were in their respective views of how world improvement was best achieved and where the limits of toleration should rest.

The radically better world sought by the elder Winthrop was first and foremost a product of Puritan piety and would be realized through the creation of a godly society whose members were intensely and harmoniously focused on acting in accordance with God's will. For the younger Winthrop, the radically better world was simultaneously a product of piety and godly scientific improvement, where new knowledge gained through prayerful research would produce world improvement and hasten the return of Christ. The pursuit of knowledge that the son believed was essential was in his father's eyes, despite its foundations in godly science, a potential distraction from the main goal of godliness, as seen in his early expression of serious concern about the New London project. Yet, in the end, this was a difference over approaches, not goals, and young Winthrop ultimately seems to have pursued the project with his father's support, and certainly without his active resistance.

A more significant source of conflict between the two men arose over their views of the appropriate limits of tolerance. In matters of religion, political authority, and even the prosecution of witchcraft, the elder Winthrop's toleration for deviance was significantly more restricted his son's. The elder Winthrop was quick to take umbrage at anything that he felt posed a threat to the leadership of the Bay Colony. Although his customary response to those who strongly differed with him over religious or political matters was to patiently work to bring them within the fold, once he became convinced that an opponent was not just differing in opinion but challenging the authority of the government, he could be harshly repressive, as his ultimate treatment of the Hutchinsonians demonstrates. His son, while ultimately sharing his father's belief that dissent should not be allowed to threaten the state, had a much higher tolerance for diversity of opinion. It is likely that the younger Winthrop's Neoplatonic perspective on religion led him to oppose the crackdown on Anne Hutchinson and her followers during the free grace controversy of the 1630s. But, rather than publicly speaking out against the prosecution in which his father participated, he remained silent, simply absenting himself from the trial. He pursued the same strategy in the case of his alchemical partner, Robert Child. While young John undoubtedly favored the more open polity and church admission principles advocated by Child, he once again absented himself from Child's trial. Although his abstention did not protect him or his new plantation from the fallout of suspicion and scrutiny that followed, it did keep him from publicly challenging his father, which would have been a clear violation of his filial duty under the fifth commandment.

The efforts the younger Winthrop made to be respectful of his father even when he opposed his positions and the father's willingness to support his son's New London project even while having reservations about it suggest the measure of the affection and sense of duty that undergirded the relationship between father and son. The same mutual respect helps us interpret the dying father's last request to his son concerning the custody of the Pequots. For the elder Winthrop, his son's continued failure to yield to the decision of the United Colonies commissioners represented the same kind of challenge to authority that he had always opposed and acted vigorously to end. Continuing to harbor the Pequots in direct contravention of the commissioners' directives was not just ill advised; it was a completely unacceptable action that could no longer be justified. Yet, just as the son was not willing to publicly speak out against his father in the Hutchinson and Child matters, the elder Winthrop was not willing to publicly oppose his son. Despite what were clearly his extremely strong feelings about what young John should do, he, as he had always done, asked his son for obedience while leaving the decision of what to do up to him.

In the end, young John only half complied with his father's request, but in so doing a way was established whereby the competition between him and Uncas over the Pequots would draw to an end. Winthrop petitioned the commissioners in July 1649 to set aside a place for the Pequots apart from his English town where they would not be under complete subjection to the Mohegans or Narragansetts or any other group that could make them "afraid to comply cordially and solely with the English eyther in discovery of any [trecherous plotts, and so forth] . . . or affording their labours and helpe for hire, or principally in attending to any dispensations of such light of the glorious Gospell, which it may please the Lord in his good time to send amongst them."[60]

The commissioners, while still protesting against the "Pequots Resolute withdrawing from their subjection to Uncas," concluded that finding some place "by the Concent of Conectacot" where the Pequots could settle and plant independently of Uncas was a good idea. In a companion decision that suggests their new attitude might have been a response to the negative feedback produced by the previous winter's raid, they also stipulated that, once

60. John Winthrop, Jr., to the Commissioners of the United Colonies, July 1649, *Winthrop Papers*, V, 354.

the Pequots had settled in the new location and owned Uncas as their sachem, "the things taken from them last winter are to bee restored." Neither Winthrop nor Uncas had got all of what he wanted, but as a result of Winthrop's unyielding intransigence the Pequots would henceforward remain together as a discrete group, and Uncas, despite the commissioners' orders, would never achieve full subjection of them again. Four years later, Winthrop was able to help the Pequots secure full release from their obligations to the Mohegan sachem.[61]

Winthrop's struggle with the commissioners over who should control the Pequots was at the heart a conflict about how English and Indian relations should be conducted. Most of the English, and certainly the majority of the commissioners, wanted clear-cut divisions in Indian relations. They wanted to reward allies and punish enemies. They were not easily prepared to accept either dynamism or ambiguity in their relations with people of other cultures. The Pequots had once been enemies, and they must always be enemies. In similar form, Uncas as a tested and loyal ally deserved preeminence among other Indian groups. Winthrop saw Indian and English relationships as much more fluid and dynamic. He had sought to develop stable relationships with the Indians within his alchemical plantation, and he was willing to accept both dynamism and ambiguity in the quest for meaningful communication. His steadfast commitment to the Pequot group, especially in the face of such clear English resistance to the relationship, is harder to understand. Perhaps he considered the Pequots essential parts of the alchemical development scheme, as a potential labor force on the one hand, and as a test case for pansophic conversion on the other. Perhaps it was a matter of loyalty to Cassacinamon and a commitment made before the plantation scheme had begun. Perhaps it was, as Winthrop himself said, simply a result of his conviction that he needed an Indian group close by who could not be intimidated into silence by other Indian groups, as an intelligence source. Perhaps also it was fear of the power that might accrue to Uncas if he were to obtain full authority over all Indians in the Pequot lands.

In many respects, Uncas was the person most in control of affairs in the Pequot region throughout Winthrop's early years of settlement. He understood the threat Winthrop represented to his personal authority, and he could read in remarkable ways the politics of Winthrop's relations with the commissioners. Until the midwinter assault on the Indians at Nameaug, Uncas had masterfully conducted a policy of intimidation of the new plantation in

---

61. Pulsifer, ed., *Acts of United Colonies*, I, 146; "Order for Resettling the Pequots, with Enclosure," Oct. 23, 1654, *Winthrop Papers*, VI, 465–466.

a way that actually garnered English support among the commissioners. Although Winthrop had considered most of the challenges his new alchemical plantation might present, having to contend with a masterful native politician who could effectively conduct aggressive Indian foreign policy within the framework of English cultural expectations and boundaries had probably not been one of them. Though, in the end, Winthrop was able to keep Uncas from enforcing subjugation of the Pequots and undermining his new plantation, his effort was costly. The various resistance measures implemented by Uncas, coupled with unanticipated problems encountered with the project start-up, forced a fairly rapid reassessment of the New London project and its potential. The result was an endeavor with more limited expectations but informed by a continuing vision of creating a godly center for science at a New London in a New England.

(FIVE)
Alchemical Vision Refined

Uncas's sustained harassment during the initial years of settlement had a chilling effect on the launch of Winthrop's alchemical plantation. Continuous unrest discouraged relocation to the new plantation. It also precluded the possibility of shipping ore from the mine at Tantiusque to the harbor town, because doing so would have required transporting the ore through Uncas's territory. Equally problematic, extracting black lead from the hills of Tantiusque had proved to be a major challenge. Attracting laborers to the remote site was difficult, and the ore proved extremely hard to mine. The exceptionally cold winter, not to mention Uncas's harassment of the local Nipmuck people, might have prevented Thomas King from fulfilling even his initial mining contract.

Robert Child's confrontation with Massachusetts, which ultimately led to his leaving New England, and Uncas's resistance to Winthrop's protection of the Pequots limited and ultimately forced a redefinition of Winthrop's original conception for the alchemical plantation. The difficulties these crises posed undermined much of the original vision with which Winthrop had started his project. Establishing the riverine plantation, Winthrop came to realize, would take decades, not days. The black lead mine, too, would take years to adequately test. To be sure, Child, who had retested the mine's ore samples and this time found silver content, remained enthusiastic about its potential, even while under house arrest in Boston ("I should have bin willing to have ventured an 100£ or two upon your Mine of lead"), and he remained so once back in England. ("I am sorry you have not as yet attempted your Blacke lead mine, that we might know Certaynely what it Conteyneth.") But Child did return to England, and the other European alchemists who had committed to come, for reasons unrecorded, never arrived. If there was truly going to be an alchemical New London in the Connecticut wilderness, it would have to be of New England's own generation, and John Winthrop, Jr., would have to create it.[1]

---

1. Robert Child to John Winthrop, Jr., May 13, 1648, in Samuel Eliot Morison et al., eds, *Win-*

WINTHROP PROCEEDED in a fairly methodical manner—given the impediments directed his way by Uncas—to implement a modified version of the plantation scheme. With black lead ore extraction and transshipment precluded, Winthrop sought other natural resources that might be shipped from New London or brought to the settlement for processing and export. He sent Child, who had elected to serve the Bay Colony's house arrest in the home of his friend Richard Leader, the new manager of the ironworks, a metalline sand to determine its usefulness in iron production. Child asked Winthrop for more of the sand and also "a little of the Clay of long Iland," for use in a glassmaking operation. Winthrop had purchased land on Long Island in November 1645. Others came to Winthrop to take advantage of his alchemical metallurgical knowledge. Jonathan Brewster of Plymouth sent Winthrop several bags of metal from Aquidneck Island, requesting Winthrop's "advise about the silver ower." Brewster noted that, subject to Winthrop's assay, "I shall at Spring of the yeare bring a tounne or 2 which I can doe lying neare the Sea." Roger Williams wrote Winthrop elaborating on the ore's probabilities. "Concerning the Bags of oare it is of Rode Iland where is certainly affirmed to be both Gold and Silver Oare upon Triall."[2]

Other reports from around New England highlighted the continuous expectation of the discovery of mineral wealth that pervaded early New England culture. John Winthrop, Sr., reported on a mine that had been opened near the ironworks. "After 24 hours [they] had a some of about 500 [pounds of ore] which when they brake they conceived to be a 5th par[t] silver." In another letter he reported that John Endecott, the owner of "Orchard," "hath founde a Copper mine in his own grounde." William White, an experienced alchemist and ironworker who had been hired away to Bermuda to help an alchemist there perfect transmutation, reported as he was leaving New England that there were "great Riches concer[n]inge whit glass and 2 other things not to be spoken of . . . within 4 myles of boston . . . it may please god I may se you next springe for there is greate things for me to doe."[3]

New England's alchemists both fed and fed upon this continuing excite-

---

th255 Papers (Boston, 1929–), V, 221–223, Robert Child to John Winthrop, Jr., Aug. 26, 1650, VI, 57–59 (hereafter cited as *Winthrop Papers*).

2. Ibid., Mar. 15, 1647, V, 140–141, May 14, 1647, 160, Deed of Theophilus Bailey to John Winthrop, Jr., November 1645, 46–47, Jonathan Brewster to John Winthrop, Jr., Dec. 1, Dec. 2, 1648, 286, Roger Williams to John Winthrop, Jr., Jan. 29, 1649, 308–309; Roger Williams to John Winthrop, Jr., Dec. 15, 1648, in Glen W. LaFantasie, ed., *The Correspondence of Roger Williams* (Providence, R.I., 1988), I, 263–265.

3. John Winthrop to John Winthrop, Jr., Aug. 14, 1648, *Winthrop Papers*, V, 246, Sept. 30, 1648, 262, William White to John Winthrop, July 24, 1648, 239–240.

ment about the region's mineral potential. In the world of the saints, providentialism and prospecting were subtly fused. In a letter to Winthrop, Robert Williams, Roger Williams's brother, metaphorically captured the connection between the quest for mineral wealth and the quest for grace. "The Lord makes us his true and faythfull Minuralls, and Lapidaries, to enjoye the unvalluable riches thereof, unto that hopefull fruition I commend you." Despite few successes, a current of mineral expectation pervaded New England well into the eighteenth century.[4]

Even as Winthrop was attempting to identify and export vendible mineral products at his new plantation, he established it on a productive agricultural footing and was able within a few years to produce sufficient livestock to vend in multiple markets. As the animal husbandry efforts became self-sustaining, Winthrop turned his attention to alchemically related processing enterprises. In 1648, Robert Child sent him instructions on French processes for making resin and turpentine, an activity whose potential would interest Winthrop for decades, especially in regard to introducing Indians to English industrial labor practices. He expressed renewed interest in setting up another saltworks, and in May 1649 the Massachusetts General Court granted him three thousand acres of land near the Pawcatuck River, provided that he establish within three years a saltworks capable of producing one hundred tons of salt annually somewhere between Cape Ann and Cape Cod. Winthrop rejected an offer from a group of merchants to establish another saltworks in Salem. His uncle Emmanuel Downing then urged, "I hope you will not loose tyme in erecting a salte work there [near new London]. you neede not feare vent here for it."[5]

Winthrop helped Downing establish another processing business—the distillation of strong waters, for which he not only furnished the alembic in which the liquor was distilled but also the German recipe used to convert the rye into alcohol without malting the grain. This use of alchemical knowledge proved to be an immediate success. Lucy Downing, using a playful reference to Winthrop's saltmaking operation, wrote of her regret that there was a shortage of available rye:

4. Robert Williams to John Winthrop, Jr., Sept. 24, 1649, ibid., 368–370.
5. John Coggesholl to John Winthrop, Jr., May 24, 1647, ibid., 165–166; Nathaniel B. Shurtleff, ed., *Records of the Governor and Company of the Massachusetts Bay in New England* (Boston, 1853–1854), II, 241. See also Roger Williams's directions to Winthrop on sowing the hayseed: Roger Williams to John Winthrop, Jr., May 28, 1647, *Winthrop Papers*, V, 168. On livestock in colonial New England, see Virginia DeJohn Anderson, *Creatures of Empire: How Domestic Animals Transformed Early America* (New York, 2004), 141–174; Robert Child to John Winthrop, Jr., May 13, 1649, *Winthrop Papers*, V, 221–222, Emmanuel Downing to John Winthrop, Jr., June 20, 1648, 230–231.

Could you but teach us to kern rye out of the sea watter, that invention I question not would quicklye make the still vapor as far as pecoite [Winthrop's town at Nameaug], and the Indians I beleev would like that smoake very well for the english here have but 2 obiections against it, one its too dear 2 not enough of it. cure those, and wee might all have implyment enough at Salem to make lickquors, and as it is wee could have custome ten times more then pay.[6]

Winthrop had continued, during trips to Massachusetts, to support the efforts of the New England ironworks, which, as all early New England industry would be, was plagued with labor problems. In March 1648 he received a letter of thanks from the English undertakers for his help "for the regulacion of some of our unruly men." Winthrop was also making use of the ironworks' production at New London. Given his metallurgical interests and his intention for New London to be an ore-processing center, Winthrop had attracted a number of blacksmiths to his new plantation. He encouraged them to focus their skills on providing the tools that would civilize the landscape through deforestation and cultivation, which Winthrop and his contemporaries believed would help ameliorate the climate extremes of New England. By March 1649, Winthrop was furnishing hatchets and hoes to merchants in Hartford.[7]

Winthrop was also working to accomplish the pansophic goal of bringing the Indians to Christ. How early Winthrop had begun efforts to convert the Pequots to the Christian faith and what role he might have personally played in such efforts is unclear. In 1648, Winthrop had, through his father, attempted to get Thomas Mayhew, the minister missionary on Martha's Vineyard, to relocate to the new plantation. Mayhew had made a special trip to Boston to discuss the offer. Winthrop elaborated the advantages of "preaching to many more Indians then are at martins viney[ar]d." He also invited not just Mayhew and his fellow English settlers but as many Indians as Mayhew liked to move from the island and settle at the new plantation. Mayhew, who surely was

---

6. Emmanuel Downing to John Winthrop, Jr., June 13, 20, 1648, *Winthrop Papers*, V, 230–231, Lucy Downing to John Winthrop, Jr., Dec. 17, 1648, 291.

7. Stephen Innes, *Creating the Commonwealth: The Economic Culture of Puritan New England* (New York, 1995), 252–253; the Promoters of the Ironworks to John Winthrop, Jr., Mar. 13, 1648, *Winthrop Papers*, V, 209, Thomas Olcott to John Winthrop, Jr., Mar. 2, 1649, 314–315; Frances Manwaring Caulkins, *History of New London, Connecticut: From the First Survey of the Coast in 1612 to 1680* (New London, Conn., 1895), 83; Karen Ordahl Kupperman, "Climate and Mastery of the Wilderness in Seventeenth-Century New England," in David D. Hall and David Grayson Allen, eds., *Seventeenth-Century New England* (Charlottesville, Va., 1984), 3–38.

aware of the state of intercultural unrest at Nameaug, told Winthrop that "all had advised him not to remove as yet," and he declined.[8]

WINTHROP TURNED to John Eliot and, using a tactic that had originally been recommended to him by Edward Howes, asked him to send two of his converted Indians with a coat as a gift for the Niantic sachem Ninigret, so that they might "stirre them up to call on God." Ninigret rejected the mission, though he kept the coat, and he requested a meeting with Winthrop to discuss "mr. Eliots letter and coate." Ninigret and Uncas continued staunchly to reject conversion efforts. Three years after Ninigret received the coat, Eliot would privately complain, "There be two great Sachems in the Countrey that are open and professed enemies against praying to God, namely Unkas and Nenecrot and when ever the Lord removeth them, there will be a dore open for the preaching."[9]

Successes were probably achieved, however, among the Pequots. Winthrop might have aided this process, through the administration of medicines. One of the great barriers in achieving native conversions was the natives' dependence on the healing powers of their powaws. The powaws, through their magico-spiritual healing rituals, helped maintain the vitality of the Indians' religions. Many missionaries commonly reported that Indian resistance to conversion centered on medical issues. "There is another great question that hath been severall times propounded, and much sticks with such as begin to pray," Thomas Shepard reported. *"Namely, if they leave off* Powwawing, *and pray to God, what shall they do when they are sick?"* Winthrop, who frequently found indigenous people among his patients, provided a clear answer to that question. The nature of Winthrop's medical practice at Nameaug must be inferred, just as we must guess at who were the missionaries who gained conversions. But, in a 1651 letter to Augustus Tanckmarus in Hamburg, Winthrop noted with enthusiasm that there had been successes.

> Even though I now write only to greet you and to let you know that I am well (by the grace of God). . . . I am desirous you should know

---

8. John Winthrop to John Winthrop, Jr., July 26, 1648, *Winthrop Papers*, V, 236. On Thomas Mayhew's missionary work among the Wampanoag Indians of Martha's Vineyard, see David J. Silverman, "Indians, Missionaries, and Religious Translation: Creating Wampanoag Christianity in Seventeenth-Century Martha's Vineyard," *William and Mary Quarterly*, 3d Ser., LXII (2005), 141–174.

9. Henry Whitfield et al., *Strength out of Weaknesse; or, A Glorious Manifestation of the Further Progresse of the Gospel among the Indians in New-England* (London, 1652), 8; Roger Williams to John Winthrop, Apr. 13, 1649, in LaFantasie, ed., *Correspondence of Roger Williams*, I, 281–282; John Eliot to William Steele, Dec. 8, 1652, *New England Historical and Genealogical Register*, XXXVI (1882), 294.

that the savages here have accepted the Gospel of Christ, and among them are many who fervently embrace the gospel. May the Lord so order it that from small beginnings a great increment of the Kingdom of Christ may develop to the glory of God's name.[10]

Winthrop's practice of alchemical medicine was an aspect of Christian service that became vital to his fellow New Englanders (the subject of the next chapter). During the years he was establishing the plantation, Winthrop became arguably the most sought-after physician in New England. New London itself became a medical center, to which suffering people from all over New England came to receive aid. Through his contacts with alchemical physicians such as Tanckmarus, Johann Glauber, Johann Moraien, and others, Winthrop had developed into a superb physician whose alchemical healing abilities were respected by people, both English and Indian, throughout the region.

Winthrop sought to develop, in the absence of a colony of European alchemical émigrés to New London, a cadre of New England alchemical philosophers, trained in and dedicated to the alchemical quest. He helped advance the alchemical knowledge of George Starkey. Starkey, an orphan from Bermuda, had come to New England to be educated at Harvard College. In 1648 Starkey, who had already established a comfortable alchemical relationship with Winthrop, wrote asking him to provide some antimony and mercury for experiments as well as the use of one or two of Winthrop's "greater glasses." Starkey also asked to borrow from Winthrop alchemical medical texts such as Jan Baptista van Helmont's *De febribus* and *De lithiasi* as well as books more strictly related to transmutation, such as the *Encheiridion physicae restitutae* and *Arcanum hermeticae philosophiae* of Jean d'Espagnet and the *Theatrum chemicum* of Lazarus Zetzner.[11]

Winthrop shared alchemical interests, and probably books and experiments, with Richard Leader, manager of the New England ironworks. Robert Child, who was a primary source of alchemical texts for Winthrop, recommended Leader to Winthrop as someone who "hath more Curious booke than I: Especially about Divinity businesses." Leader gave Winthrop mercury from Child's stock after Child had returned to Europe and later, in the midst of an experimental quest, asked Winthrop to return four or five pounds of it

---

10. Thomas Shepard, *The Clear Sun-shine of the Gospel Breaking Forth upon the Indians in New-England* (London, 1648), 25; John Winthrop, Jr., to Augustus Tanckmarus, Dec. 27, 1651, *Winthrop Papers*, VI, 157.

11. George Stirk to John Winthrop, Jr., Aug. 2, 1648, *Winthrop Papers*, V, 241–242; William R. Newman, *Gehennical Fire: The Lives of George Starkey, an American Alchemist in the Scientific Revolution* (Cambridge, Mass., 1994), 40.

"with what speede you canne," so he could assess some mineral finds. Winthrop also became acquainted with William Berkeley, a Bermudian alchemist on an extended stay in New England. From Boston, Berkeley sent a message to Winthrop requesting him "to send mee those Glasses which you promised mee with all expedition for I doe intend soe sone as I can receive them, to sett some of them one Worke for your and my one farther satisfaction." Berkeley's sense of urgency was great. When twelve days had passed and Winthrop still had not sent the chemical equipment, he dispatched someone to go get them. "I have hired an Indian to bring the Glasses by whome I would request you to send them. . . . I pray write your mind, and put your father in mind of the three pounds of Quicke silver, that I bee not disapoynted." Winthrop apparently stored chemicals at his father's house in Boston as well as at Nameaug.[12]

The correspondence cited above not only suggests the presence of an active communication network among alchemical practitioners in New England; it shows some of the limitations under which they worked. Glassware as well as chemicals seem to have been, at least at times, in short supply. The sharing of apparatus as well as information underscores the existence of and the collaboration occurring within a New England alchemical fraternity. That Winthrop became the hub of an extended alchemical communications network is clear, despite the limited information the alchemical culture of secrecy has left behind. Winthrop continued seeking knowledge of cryptography and observing the code of secret communication throughout his life. At moments, however, the veil lifts enough to provide a sense of how the colonial republic of alchemy looked from the inside. When the soldier, legislator, and land speculator Humphrey Atherton sought to introduce himself to Winthrop as an alchemical practitioner, he did so in the most guarded manner. Inquiring about Winthrop's reaction to William Berkeley's alchemical mercury, Atherton made clear that he was observing the rule of private communication. He wanted to know "how you like the mercarry you had of mr. bartlat: what you thinke or know of it: if I could see you I could tall you more: Sar if you will lat mee know I shall Requit your keindnes. . . . But I pray how ever, do not lat it be knowne to any but my selfe that I did aske such a question: so knowing a word is enugh for a wise man I rest for thet."[13]

---

12. Robert Child to John Winthrop, Jr., Aug. 26, 1650, *Winthrop Papers*, VI, 57. The meaning of "curious" is ambiguous; Child might have meant it to mean skillful, or clever, or to pertain to more occult arts. Richard Leader to John Winthrop, Jr., Aug. 21, 1648, *Winthrop Papers*, V, 248; William Berkeley to John Winthrop, Jr., June 12, 1648, 229, June 25, 1648, 232.

13. Atherton's orthography was egregious, even for his age, and his Lancashire pronunciation

WHEN JONATHAN BREWSTER, the alchemist who had sent Winthrop the prospective silver ore from Rhode Island, believed he had made an important alchemical discovery, he sent it to Winthrop with a message that he was "sending to you [Winthrop] a gift in writing, which I have had by me sometyme, but could not send it till now. You are the first man that sees it; if you thinke it worthy of acceptance, imparte it to any of your frinds, or any other, whom in your wisdom you thinke mette and fitt for the knowledg of such a great mistery." Brewster clearly meant for Winthrop to control distribution of the information. Winthrop selectively distributed texts and other knowledge to those he thought could best make use of them. Brewster's experiment, for example, had come from a text Winthrop had originally lent him. On another occasion, John Alcock, a Roxbury alchemist who was settling on Block Island, wrote to thank Winthrop for the "little booke I received." He was "so taken with the rarity that I am willing to imploy both mony and friends to gaine the skill and knowledge of it."[14]

Winthrop did more than maintain a New England alchemical communications network. He sought to make New London the alchemical center he had originally intended it to become. His presence attracted Jonathan Brewster to the region, who established a trading house adjacent to Uncas's village of Fort Shantok some miles north of New London. Humphrey Atherton also sought permission to settle a new plantation near Winthrop and became a partner with him in a land company attempting to acquire the Narragansett lands adjacent to New London. James Noyes, who was recommended to Winthrop by the Reverend Thomas Parker as someone "very fitt as I conceave for chemical works," became a minister and alchemical physician at the adjacent town of Stonington. Gershom Bulkeley, the alchemist who, along with Winthrop, was to play a major role in ending witchcraft executions in Connecticut permanently, was New London's minister. William White, the alchemist and ironworks expert who had been lured away to help perfect transmutation on the island of Bermuda, returned to New England to conduct experiments with Winthrop on Fishers Island.[15]

---

often is reflected in his writing. It seems clear in context that he meant Berkeley. Humphrey Atherton to John Winthrop, Jr., Nov. 30, 1648, *Winthrop Papers*, V, 273.

14. Jonathan Brewster to John Winthrop, Jr., Jan. 31, 1657, Massachusetts Historical Society, *Collections*, 4th Ser., VII (1865), 77–78, John Alcock to John Winthrop, Jr., Nov. 8, 1660, 5th Ser., I (Boston, 1871), 390.

15. Caulkins, *History of New London*, 65–66; Humphrey Atherton to John Winthrop, Jr., Oct. 30, 1648, *Winthrop Papers*, V, 273; Rev. Thomas Parker to John Winthrop, Jr., Oct. 27, 1656, Winthrop Family Papers, Mss.W.16.7, Massachusetts Historical Society, Boston; John Langdon Sibley, *Biographical Sketches of Graduates of Harvard University*, I (Cambridge, Mass., 1873), 389–402; *Winthrop*

Was Winthrop's very public expression of interest in leaving New London in 1649 and 1650 a sign of his dissatisfaction with the colony, or a continued personal inability to follow through on plans? At various times, he was in discussion, or rumored to be in discussion, with officials from Dutch Long Island, a group intending to settle the Delaware, and agents of the town of Guilford in New Haven Colony, regarding abandoning the town and moving elsewhere. He even let it be quietly known he was considering returning to England. Certainly Winthrop had one very big reason for considering leaving the area. The decision of the commissioners of the United Colonies to side with Uncas and the Pequots against his desire to become the Pequots' guardian, which had culminated in the winter raid, had been a personal affront and outrage. Winthrop might have believed considering other options for settlement was prudent. He might also have been trying to leverage the threat of his removal into political gains.[16]

He had reasons to be satisfied with the progress his plantation was making. Although the cattle market was depressed, livestock production had gone well. Each year, in spite of the intercultural instability, new settlers had joined the community. Alchemists like Brewster had come to settle in the new town, and Humphrey Atherton had begun his efforts to acquire the Rhode Island lands. Effective trade relationships had been established, and larger alchemical projects were on the drawing boards. Whether Winthrop's public dissatisfaction was strategic or real, it does seem to have produced benefits for the plantation. The General Court of Connecticut, which had officially been given jurisdiction over the Nameaug settlement in 1647, granted tax abatement to the town for three years and extended its boundaries an additional two miles northward from the sea. In addition to these concessions, the town received a sizable increase in population, and its first minister, when Richard Blinman brought a group from Cape Ann to settle at the new town. In the fall of 1650 construction of a mill was authorized, and Winthrop and his heirs were given a monopoly to it in perpetuity. By 1653, a sawmill was operating as well. If Winthrop had given serious consideration to leaving, by the end of 1650 such considerations were put aside.[17]

Papers, V, 239 n. 1; Ronald Sterne Wilkinson, "John Winthrop, Jr. and the Origins of American Chemistry" (Ph.D. diss., Michigan State University, 1969), 146–149.

16. Robert C. Black III, *The Younger John Winthrop* (New York, 1966), 160; George Baxter to John Winthrop, Jr., July 15, 1649, *Winthrop Papers*, V, 355, Edward Elmer to John Winthrop, Jr., Aug. 29, 1649, V, 361, Marmaduke Matthews to John Winthrop, Jr., Apr. 1, 1650, VI, 29–30, Thomas Stanton to John Winthrop, Jr., Jan. 29, 1650, VI, 14–16.

17. "The Inhabitants of New London to Edward Hopkins," Oct. 12, 1649, *Winthrop Papers*, V, 374; Caulkins, *History of New London*, 63–70. The presence of the sawmill prompted Iumse, the sachem

The scale of Winthrop's proposed alchemical projects increased dramatically. He drafted a proposal to "The Merchants at Boston" that they raise a minimum of £3,000 capital and as much as £10,000–£20,000 to fund a saltpeter manufactory. Saltpeter, Winthrop argued, would be a readily vendible commodity overseas and would produce net returns to the country. Winthrop's concept probably originated within the Hartlib circle, where Benjamin Worsley had presented a similar petition to Parliament in 1646. Worsley, with Hartlib's backing, had sought a monopoly on English saltpeter production to reduce that country's dependence on imported saltpeter.[18]

Saltpeter, used for gunpowder manufacture, fertilizer, and medicine, had been in great demand in England throughout the 1640s and had outstripped domestic production capacity. Moreover, traditional methods of saltpeter production were intrusive and onerous. They involved the collection by saltpeter men—who dug out cellars, chicken runs, and outhouses—of black nitrous earth, which was mixed with lime and ashes and then leached for its salts. The intrusive approaches used by England's saltpeter men to collect the nitrous earth from people's homes, barns, and other sites had made its production a provocative national issue. Worsley, in a tract called *De nitro theses quaedam*, applied Paracelsian principles of generation to develop a theory that saltpeter could be generated on an industrial scale in mounds of earth containing suitable mixtures of material. Using this method, he argued, would not only reduce England's dependence on saltpeter imported from other countries; it would eliminate domestic intrusion by the saltpeter men.[19]

Worsley had advanced the saltpeter project as the first in a series of economic reforms that were intended to bring about the pansophic regeneration of England, in the midst of the generation of saltpeter. He was praised by Hartlib circle member Cheney Culpeper for his ultimate aim of the glorification of God "throughout the whole worlde . . . not this family, Cownty, Nation, but whole mankinde." His diversion into promoting a greater economic program prevented the saltpeter scheme from reaching fruition. But the desire for a saltpeter manufactory and the sense that it was a potential economic panacea remained strong.[20]

---

of Quinabaug, to sell his land and all his trees to Winthrop. "Deed of Iumse to John Winthrop, Jr.," Nov. 2, 1653, *Winthrop Papers*, VI, 343.

18. John Winthrop, Jr., to the Merchants of Boston, ca. 1650, *Winthrop Papers*, VI, 8–9; Worsley's Proposal for Saltpeter (1646), Hartlib Papers [71/11/12A–13A].

19. Charles Webster, *The Great Instauration: Science, Medicine, and Reform, 1626-1660* (New York, 1976), 378–384. William Newman notes the probable composition date for *De nitro theses quaedam* is 1654 (personal communication to author).

20. Ibid.; Cheney Culpeper to Samuel Hartlib 1644(?), Hartlib Papers [13/294a–295B]; M. J.

Winthrop's European acquaintance Johann Glauber helped solidify Winthrop's belief in the vivifying potential of saltpeter. In *Miraculum mundi*, a text Winthrop owned, he posited "that Salt-petre truly merits the Name of an Universal Menstruum," a term used to describe a particularly effective generative medium for effecting chemical transformations. In Glauber's view, saltpeter offered marvelous benefits in four areas: husbandry, industry, medicine, and personal fitness. Saltpeter offered husbandmen "a new Method of fattening and enriching their Fields and Gardens, without the usual and customary way of dunging, and thence [they can] yearly acquire a greater profit." Seeds soaked in saltpeter grew better, and soil enriched with saltpeter produced greater yields. Saltpeter offered "Merchants and others who have time and leasure" a way to increase their gold and silver in a "much better and honester manner than putting it to Usury." By investing in saltpeter production and funding its incorporation into farming and metallurgical processing, yields and profits for investors would both be dramatically increased. Glauber, who himself ran a voluntary-fee medical practice, saw saltpeter as a way to help physicians fulfill their duties to Christian service. He described a way that "all Conscientio[u]s Physicians . . . may learn to prepare Salutiferous and Efficacious Medicines with small charge, little labour, and in a short time; that (as becomes Christians) they may help and succour the Miseries of the sick, and acquire to themselves an honest livelihood." Glauber also believed that saltpeter had sanative qualities and could be used to preserve health and prevent disease.[21]

Paracelsian matter theory, economic interest, and pansophic enthusiasm underpinned the saltpeter proposal Winthrop presented to the merchants of Boston, but economic realities kept it from being acted upon. Undaunted, Winthrop shifted venues to Connecticut, presenting Governor Edward Haynes a proposal that demonstrated his renewed interest in mining and metals. Noting that he had received "earnest motions . . . f[ro]m some well willers to the Common good, to make some search and tryall for metalls in this Country," he asked Connecticut's General Court to agree that, "if the said J W shall discover any mine of lead tinne or copper or any other mineralls of vitriall, antimony, Bismuth blaclead allom ston salt, or any other salt spring, and shall sett up any workes or furnaces for the foundeing, melting, or working of any such mettalls, or mineralls, within thre yeares after the discovery

---

Braddick and Mark Greengrass, eds., "The Letters of Sir Cheney Culpeper, 1641–1657," *Camden Miscellany*, XXXIII, 5th Ser., VII (Cambridge, 1996), 105–393.

21. Johann Rudolf Glauber, *The Works of the Highly Experienced and Famous Chymist John Rudolph Glauber: Containing Great Variety of Choice Secrets in Medicine and Alchymy* . . . , trans. Christopher Packe (London, 1689), 186.

thereof," Winthrop should enjoy the mine site and necessary surrounding lands, unless they were situated within a town. The court, perhaps grateful that Winthrop was starting to work with, rather than against, the Connecticut leadership, not only agreed to Winthrop's proposal; it omitted the stipulation that Winthrop must set up operations on found sites within three years, and it assigned the rights to the found mine "forever."[22]

Winthrop sought help from his German correspondents in advancing this new prospecting venture. He contacted Augustus Tanckmarus in Hamburg, the German mining center, and a new acquaintance to whom Tanckmarus had introduced him via letter. Paul Marquart Schlegel was an alchemical physician and surgeon whose museum of natural wonders was widely admired. Winthrop provided him a collection of rarities from New England for display there and asked Schlegel, who apparently had an extensive knowledge of mining and minerals, to provide him with a variety of German ore samples. Winthrop had acquired extensive knowledge of German minerals, either from contacts such as Glauber, mineral texts, or prior experience. In his letter, Winthrop asks

> that you acquire for me some minerals from Germany: especially the [ore?] of Silver, namely translucid red Silver, in German Rodgulden Ertz and Glasertz: and if they can be had [ores?] of white lead or tin and the stone from which lead is refined at Goslar which, that is, is the poorest [ore?] of black lead, but [also] bismith from Sueburg and if your Mastership seeks for yourself other kinds of [ores?] from Germany or from elsewhere, if it please you let there be added some for me, as also the [ore?] of copper from Swabia.[23]

Schlegel, unfortunately, died before he could fulfill Winthrop's request, but Winthrop conducted his prospecting survey of Connecticut, which netted at the least a number of prospective locations for ironworks. Winthrop appears to have put forward his project with the intent of seeking English investment for a new ironworks or mining enterprise. William Osborne, a supervisor at the Hammersmith forge of the New England ironworks said to have built "the hearth that made both the most Iron and best yield that ever was made yet in New-England," wrote Winthrop the month after his

---

22. John Winthrop, Jr., to Edward Hopkins, May 13, 1651, *Winthrop Papers*, VI, 104–105, John Winthrop, Jr., to Connecticut General Court, May 13, 1651, *Winthrop Papers*, 105–106; J. Hammond Trumbull, ed., *The Public Records of the Colony of Connecticut*, I (Hartford, Conn., 1850), 223.

23. Paul Marquart Schlegel to John Winthrop, Jr., Sept. 10, 1649, *Winthrop Papers*, V, 365–366, Johannes Tanckmarus to John Winthrop, Jr., Sept. 10, 1649, VI, 367–368, John Winthrop, Jr., to Paul Marquart Schlegel (translation), Nov. 10, 1650, VI, 78–82.

prospecting petition was granted. He offered his services in the forwarding of a new ironworks. He had heard Winthrop was to go to England late in the year and that he intended to "take severall sortes of your Iron stone with you." Osborne himself intended to spend the winter in England and would be glad to help Winthrop any way he could. Richard Post, another worker at the Hammersmith forge, was also planning on leaving the Hammersmith operation and was eager to hear from Winthrop.[24]

Correspondence from Robert Child continued to keep the vision of general alchemical reformation vivid. Child sent Winthrop news of newly published books, including John French's English translation of Henry Cornelius Agrippa's *Occult Philosophy,* a book of magical arts that French dedicated to Child. Winthrop asked Child to buy books for him, including Van Helmont's works and a text on ciphers by Blaise de Vigenère. Child wanted Winthrop to inform him if he met with "any rare thing, vegetable, minerall, etc., any strange newes," and he pressed him to make a serious effort at the black lead mine.[25]

In one letter, Child announced to Winthrop, "Helia Arista is born." This was news of great significance to those hoping to participate in the alchemical regeneration of the world. Elias Arista was the name of a mystical alchemical adept whose appearance had been prophesied by Paracelsus. Paracelsus had said that Elias, like a latter-day John the Baptist, would appear just before the return of Christ and would reveal all the hidden secrets of nature. The prospect of such an appearance was a great source of excitement for millennial-minded alchemists. The Elias legend helped give alchemy a great surge in interest at the middle of the seventeenth century. Johann Glauber became obsessed with the Elias story, and came to connect Elias with Glauber's own theories regarding the primary importance of salt. Elias Arista was, Glauber believed, actually an anagram for Artis Salia, "the salts of the Art." This, Glauber thought, not only confirmed salt's primacy in matter; it again confirmed the intricate pattern of interconnections found within the microcosm. Glauber, Hartlib, Child, and presumably Winthrop, too, found rumors of the appearance of Elias energizing. Undoubtedly they breathed new life into the pansophic quest.[26]

---

24. Richard Leader to John Winthrop, Jr., February, 1655, MHS, *Proceedings,* 2d Ser., III (1887), 192; William Osborne to John Winthrop, Jr., June 27, 1651, *Winthrop Papers,* VI, 111.

25. Robert Child to John Winthrop, Jr., May 13, 1648, *Winthrop Papers,* V, 221-222, John Winthrop, Jr., to Robert Child, Mar. 23, 1649, V, 324-325, Robert Child to John Winthrop, Jr., Aug. 26, 1650, VI, 57-59.

26. Robert Child to John Winthrop, Jr., May 13, 1648, ibid., V, 221-222. On Elias Arista, see Newman, *Gehennical Fire,* 3, 4, 8-9, 11-12, 14, 42, 57, 68; on Glauber, see, especially, J. T. Young,

In the fall of 1650, Child informed Winthrop that he was going to leave England for Kilkenny, Ireland. He expected either to be part of a new academy there, or "I shall retreate to a more solitary life, as I can Commaund myselfe, with 6 or 7 gentlemen and scollars, who have resolved to live retyredly and follow their studyes and Experiences." Child was dusting off Plattes's Venetian laboratory, the model, or at least metaphor, that informed Winthrop's alchemical vision for America, and he was shifting its venue to Ireland. "These gentlemen for Curiositys and Learning scarcely have their equals in England. Next weeke we are to meet and Conclude." Child did remove to Ireland, where he continued to seek alchemical enlightenment until his death three years later.[27]

Winthrop had helped to create an alchemical network in New England; as in medicine, he was primus inter pares in New England's alchemical brotherhood. In 1654, however, perhaps with Child's Macaria-like retreat into Ireland in mind, he sought to further advance the role of New London as an alchemical center by turning at least part of the informal alchemical network he had helped create into a formal one. A letter from the Boston merchant Thomas Broughton reveals that Winthrop was promoting a project to use Fishers Island, in conjunction with a proposed potash-manufacturing program, as a center for pansophic and, presumably, alchemical education. "I hartyly desire the promotion of that designe of learning," Broughton wrote, "judge Fisheres Iland a meete place to promote it on [and] alsoe esteeme that [project] for potash of greate advantage to the countrey if fully and frugally improved." Broughton was constrained by his business affairs from joining in the project immediately but was handing over his duties to others as quickly as he could. He looked forward to wholly applying himself to the design of learning "as the worke where in I hope God may have much honour and the world greated [sic] advantaged."[28]

Faith, Medical Alchemy, and Natural Philosophy: Johann Moriaen, Reformed Intelligencer, and the Hartlib Circle (Aldershot, 1998), 236.

27. Robert Child to John Winthrop, Jr., Aug. 26, 1650, *Winthrop Papers*, VI, 57–59; Webster, *The Great Instauration*, 67.

28. Thomas Broughton to John Winthrop, Jr., Aug. 12, 1654, *Winthrop Papers*, VI, 414–415. George Starkey, who emigrated from New England to England in 1649, was, however, a far more significant contributor to natural philosophy than Winthrop. A celebrated addition to the Hartlib circle, he conducted alchemical experiments with Robert Boyle, and his tracts written under the pseudonym Eirenaeus Philalethes (peaceful lover of truth) strongly influenced Isaac Newton and were later noted by John Locke and Gottfried von Leibniz (Newman, *Gehennical Fire*; William R. Newman and Lawrence M. Principe, *Alchemy Tried in the Fire: Starkey, Boyle, and the Fate of Helmontian Chymistry* [Chicago, 2002]). Potash, named because it was derived from pot-ashes, was an alkaline substance (potassium carbonate) useful in soapmaking and other operations. It was derived from leaching ashes of vegetable matter and evaporating the resulting contents.

The "designe of learning" seems so consistent with Winthrop's original pansophic intentions that we would love to know more about it. Who was involved? Did it ever formally meet? Was there some kind of Invisible College—similar to the one established by Benjamin Worsley and Robert Boyle as a precursor to England's Royal Society? The dedication of Cromwell Mortimer in the *Philosophicl Transactions* comes to mind here, with his assertion that Boyle, Digby, Wilkins, et al. intended emigration to America. Or was the "designe of learning" just a proposal whose time had not yet come? Nothing in the record suggests that it was ever more than just a thought. With alchemists, however, one often has the sense that the most important information was intentionally never recorded.

What we do know is that New London continued to be a center for experimentation with and the elaboration of potential economic development projects. On Fishers Island itself, Winthrop did experiment with potash manufacture and might have attempted indigo processing. With William White, back from Bermuda and on Fishers Island, Winthrop set up a saltworks similar to the one he had established earlier at Salem. Furthermore, two years after Thomas Broughton claimed he was putting away his business endeavors as quickly as possible to pursue the design of learning, we find him party to a business deal with Winthrop and Richard Leader of the New England ironworks. Broughton and Leader entered into an agreement whereby they promised not to divulge anything about Winthrop's "better, shorter, and cheaper way then hath bin formerly used by any" to make salt by means of evaporation in Barbados. Broughton and Leader further agreed that, if they did disclose anything about the process to anyone in any way, they would forfeit twenty thousand pounds to Winthrop. This is a contrast to the idea of alchemical brotherhood implied in a "designe of learning" where "God may have much honour." The code of virtue and honor on which alchemical disclosure was founded has in this instance been replaced by a legal document with onerous financial penalties. Certainly this confirms William Newman's assertion that one of the benefits of alchemical secrecy was the protection of trade secrets; this agreement is explicitly designed to protect Winthrop's intellectual property. To be sure, there was an inextricable linkage between the millenarian idealism of the pansophic moment and the rising capitalist ethic. The fact that spiritual honor was deemed insufficient to protect the value of an idea is a marker indicating that, even in godly New England, the alchemical moral code was falling behind a rising capitalist impulse.[29]

---

29. Emmanuel Downing to John Winthrop, Jr., Mar. 13, 1654, *Winthrop Papers*, VI, 370; William White to John Winthrop, Jr., July 26, 1656, Winthrop Family Papers, Mss.W.19.150; Wilkinson,

Leader traveled to Barbados and spent six hundred pounds building a windmill-driven salt-evaporation system that did in fact appear to be a significant improvement on any previous salt evaporation process. Winthrop obtained a twenty-one-year monopoly on the same process from the Massachusetts General Court. In Barbados, however, rains came and washed the project out, the windmill broke, and Leader, physically ill, returned home. This particular project ended unsuccessfully, but the idea of saltmaking, and of successfully implementing this scheme, would remain important to Winthrop throughout his life.[30]

Despite frequent setbacks, the range of development options Winthrop could provide to towns eager to establish a more solid economic base, coupled with his remarkable abilities as a physician and his status as a member of the colonial elite, made him a much-sought-after person. Additional projects called to and for Winthrop, and by the mid-1650s towns vied with one another to attract him away from New London in much the way that cities today attempt to lure high-technology businesses to their locations. Providence, New Amsterdam, Hartford, and New Haven all sought the alchemical genius of this New World visionary and offered economic incentives to attract him.[31]

For a time in the mid-1650s, it seemed that New Haven Colony might actually succeed in captivating him. There, a commitment to building a local ironworks, coupled with an intense desire to implement Comenian educational reform, was enough to get Winthrop to consider seriously a change of venue. In many ways, the story of the New Haven ironworks project mirrors the story of the ironworks at Braintree and Lynn. It was beset by labor problems, construction delays, an insufficient supply of quality ore, and limited capital investment. Promoted by Winthrop and Stephen Goodyear, one of New Haven's principal citizens, it was first advanced in March 1655, and a company was organized in February 1656. William Osborne of the England ironworks' Hammersmith forge at Lynn was consulted, and Goodman Post of Hammer-

---

"John Winthrop, Jr. and the Origins of American Chemistry," 251; Bruce D. White and Walter W. Woodward, "'A Most Exquisite Fellow'—William White and an Atlantic World Perspective on the Seventeenth-Century Alchemical Furnace," *Ambix*, LIV (2007), 285–297; "Agreement between Richard Leader, Thomas Broughton, and John Winthrop, Junr.," Apr. 1, 1656, MHS, *Proceedings*, 2d Ser., III (1887), 194; William Newman, "George Starkey and the Selling of Secrets," in Mark Greengrass, Michael Leslie, and Timothy Raylor, eds., *Samuel Hartlib and Universal Reformation: Studies in Intellectual Communication* (Cambridge, 1994), 192–310; Max Weber, *The Protestant Ethic and the Spirit of Capitalism*, trans. Talcott Parsons (New York, 1958).

30. Richard Leader to John Winthrop, Jr., Aug. 14, 1660, MHS, *Proceedings*, 2d Ser., III, 196–197.

31. Black, *Younger John Winthrop*, 173–174. On the competition between Hartford and New Haven to secure Winthrop as a resident, see chap. 3.

smith was hired to help build the new furnace, which was undergoing trials there by the spring of 1657. But it was not until 1663 that real output was achieved from the works; Winthrop, who had received a 25 percent share of the mine for agreeing to become involved in the project, leased his interests to William Paine and Thomas Clarke of Boston in the fall of 1657. Winthrop purchased a house in New Haven—he declined the gift of a house from the town—and in 1656 he was active there providing medical services and overseeing the mine construction. For a time, it must have seemed that New Haven had even greater potential than New London as a place to implement pansophic reformation, because there Winthrop would find a person whose connections to the Hartlib circle were older and better than his own.[32]

New Haven has usually been considered the most restrictive of all the Puritan colonies, and much of its conservatism is attributed to its founding minister, John Davenport. Another side of Davenport's theology has gone largely unnoticed: his unbridled support for the union of Protestant churches proposed by John Dury and his desire to implement Comenian educational reforms in New England.[33]

Davenport's connection with the Hartlib circle had preceded the sailing of the Winthrop fleet in 1630. In 1628, on behalf of the New England Company and on the recommendation of Hartlib, Davenport had invited Dury to lead a congregation of emigrating Puritans to New England. Dury had declined, but Davenport's letter initiated a relationship between the two men. Davenport fully subscribed to Dury's notion that all Protestant sects should be united into one. He was, Dury reported to Hartlib, "forward earnest and judicious in the work." In 1631 Davenport was one of the signers of Dury's "Instrumentum theologorum anglorum," a text advocating church unity, and Dury always cited him as one of the early proponents of reunification.[34]

Davenport had also become an enthusiastic supporter of Comenius and the pansophic agenda advanced by the Hartlib circle. "For myself," he wrote Dury from New Haven, "I looke at Yourselfe, and Mr Hartlib, and Mr Comenius as three witnesses against this unthankfull Age, by whom God hath offered singular advantages for Religion, Learning, and Universall Welfare."[35]

32. Ibid., 172–178; E. N. Hartley, *Ironworks on the Saugus: The Lynn and Braintree Ventures of the Company of Undertakers of the Ironworks in New England* (Norman, Okla., 1957), 280–288.

33. Edmund S. Morgan, *Visible Saints: The History of a Puritan Idea* (New York, 1963), 108–109, 135–136; Charles M. Andrews, *The Colonial Period in American History*, II (New Haven, Conn., 1936), 156–159; Isabel MacBeath Calder, ed., *Letters of John Davenport, Puritan Divine* (New Haven, Conn., 1937), 8–11; Black, *Younger John Winthrop*, 141–142 n. 4.

34. John Davenport to John Dury, Oct. 27, 1628, Hartlib Papers [4/5/1A–B], Samuel Hartlib to John Dury, Aug. 31, 1630, [7/11/1A–3B]; Calder, ed., *Letters of John Davenport*, 2–11.

35. John Davenport to John Dury, June 25, 1660, Hartlib Papers [6/5/1A–2B].

Like Winthrop, Davenport hoped to see Comenian educational reforms implemented in America. In 1643, during the time Winthrop was in England meeting with Hartlib, Hartlib wrote Davenport a letter describing his own interest in helping the New England plantations. Davenport, or perhaps Winthrop on his behalf, had asked for an update on Comenius's program. Hartlib, already aware of Davenport's interest in Comenius, expressed surprise that he was not already fully informed. "I did persuade myself that you had full information concerning Mr. Comenius and his endeavors," he noted. He then went on to describe his efforts to interest patrons in promoting a Comenian education scheme in New England. "I have spoken to Mr. Durham, Mr. Hawkins, and others with a great deal of earnestness to promote the good of these Christian Plantations, to Mr Hawkins and others to Intimate my readiness to promote by ways of reformed Learning and some other Meanes the good of those most Christian and Heroic plantations. . . . But my tenders were never entertained with any due respect and by a courtesy of excuses till you had gotten . . . firmer footing in those parts."[36]

If Winthrop contacted Comenius during this period to offer him a college presidency—as Cotton Mather reported he did—it seems likely that the intended college would have been one in New Haven, not Cambridge. However, as Hartlib's patrons confirmed, such an offer would have been premature, for Davenport's new plantation was not yet stable enough to support a college. By the time Winthrop came to New Haven to establish the ironworks, however, it seemed that that firmer footing had been achieved.

The New Haven General Court had acted to make the establishment of a college possible, and, perhaps through Winthrop, a professor had been identified and solicited for the work. William Leverich, a Cambridge graduate, minister, and Indian missionary, had contacted Winthrop from Oyster Bay, Long Island, in 1653 about bringing a group of people with him to relocate to New London. Whether Winthrop was Leverich's conduit to New Haven is unclear, but he subsequently was offered the college position at New Haven. Hartlib was informed about the college formation, and members of the Hartlib circle proposed ideas for the new school. John Beale envisioned it as a model of post-Aristotelian empiricism, with advanced optical equipment and a laboratory:

> I mervayle our Academyes doe not provide themselves of the best Tubes. What were it for a colledge to lay out £100. And what is a

---

36. Samuel Hartlib to John Davenport, Jan. 30, 1643, Hartlib Papers [7/35/1A–2B]. The recipient of this letter is identified in the Hartlib Papers tentatively but incorrectly as Winthrop.

library without Mathematicall Instruments. And what instruments more instructive then Tubes and Optic glasses of all sorts—And hence artificers would bee encouraged to excell each other in perfecting thiese Inventions—With thiese and Thermometors of all dimensions, and a Labouratory, They may learne more truth, and valuable truth in one houre, Then by reading a Cart-loade of Monckish philosophy, and as soone posesse heaven and the stars by a Telescope, as by Schoole-divinity. This advertisement does also belong to Mr Davenports Illustrious Coll.[37]

Comenius's idea of a Universal College, "a living laboratory supplying sap, vitality, and strength to all," seemed for a time to be unfolding somewhere near the New Haven green. For Winthrop, the college must have been as powerful an attraction as the ironworks. But, like the "designe of learning" on Fishers Island, the dream far exceeded the reality. Leverich's wife was adamantly opposed to his accepting the college position, so he remained at Oyster Bay. A major bequest to fund the college by Edward Hopkins, former governor of Connecticut, was tied up and substantially reduced by feuding executors, and the hoped-for college dwindled down to a grammar school. Davenport wrote Dury in 1660, "I am now discouraged from the Colledge worke in these parts, upon which mine heart hathe been so many yeares." By then, Winthrop had gone from New Haven. On May 21, 1657, the Colony of Connecticut had elected him, in absentia, as its governor. After six months of careful consideration, Winthrop assumed the office of governor at Hartford on December 3. Henceforward Hartford would be his primary residence. On March 24, 1658, however, at the second General Court of the new governor, the legislators passed an act that they had resisted for almost a decade. When, in 1648, the inhabitants at Nameaug had requested Connecticut to grant their town the name "New London," the court had declined, recommending instead that the inhabitants accept the name "Fair Harbor." Nameaug residents, undoubtedly led by their eminent first citizen, had refused. Now, as governor, Winthrop put the verbal seal to the vision that he had carried back with him from England in 1643. Hartford would be his residence, but for the rest of his life New London would be his home.[38]

---

37. William Leverich to John Winthrop, Jr., April 1653, *Winthrop Papers*, VI, 274–275; John Beale to Samuel Hartlib, Dec. 21, 1658, Hartlib Papers [51/52A–54B].

38. Robert Fitzgibbon Young, ed. and trans., *Comenius in England: The Visit of Jan Amos Komenský (Comenius)* . . . (London, 1971), 82; John Davenport to the General Court of New Haven Colony, in Charles J. Hoadly, ed., *Records of the Colony or Jurisdiction of New Haven* . . . (Hartford, Conn., 1858), 369–374; John Davenport to John Dury, June 25, 1660, Hartlib Papers [6/5/1–2B].

THE STORY of the founding of New London as the fulfillment of a pansophic vision of alchemical regeneration should, and does, end back near Winthrop's new plantation, in the trading house of Jonathan Brewster at Mohegan, as he elaborated on the "gift in writing" that he had invited Winthrop to share with those he deemed worthy. What Brewster believed he had found was the great secret of alchemy, "how to worke the Elixer, fitt for Medicine, and healing of all maladyes. . . . Also, a light given, how to desolve any hard substance into the Elixer." Brewster had no doubt that he was on the path to alchemical perfection. "This knowe, that my worke being trew thus farr, by all their writinges, it cannot faylle."[39]

Brewster had begun the experiment he was writing Winthrop about almost four years previously. It would not be completed for another year and a half, and Brewster was worried. "I ffeare I shall not live to see it finished, in regard partly of the Indianes, who I feare will raise warres; as also I have a conceite that God sees me not worthy of such a blessing, by reason of my manifold miscariadges." He could not move the experiment from where it was, for, if it were moved, it would be "as if the world which God mad[e] at first should be moved (as I sayd) in my new Creation, and the station of the heavens and earth should be displaced."

To ensure that the secret would live, even though Brewster died, he decided to entrust it to Winthrop "before I would betrust it with any of my children; for it is such a secrett, that is not fitt for every one, either for secrecy, or for partes, to use it, as Gods secrett for his Glory." He would record "the whole worke in a few wordes, plainly, which may be done in 20 lines, from the first to the last, and sealle it up in a littel box, and subscribe it to your selfe, and leave it at my house at Pequett, in one of my chests, that if it please God I should suddainly be taken away, you may call for it." Brewster hoped, "If it please God that you receave and worke it, and bring it to perfection, remember my poore wyfe and children as your owne."

I have cited Brewster's letter in such detail because it so fully elaborates why the alchemical quest was such a powerful motivating force for godly people in the early modern period and how it transformed the world into a theater of cosmic occurrences, even in the most remote places. Beyond the plethora of utilitarian improvements that alchemists sought, there was a deeper meaning to the alchemical quest that can be read in Brewster's excited, anxious words.

The Elixer, fitt for Medicine, and healing of all malydies. The secret Brewster

---

39. Jonathan Brewster to John Winthrop, Jr., Jan. 31, 1657, MHS, *Collections*, 4th Ser., VII (1865), 77–81. All quotations following are from this letter.

thought he had found, the elixir to cure all diseases, spoke as powerfully to human longings and compassion in his age as the magic words of our own time—"the cure for cancer" or "the vaccine for AIDS"—speak to ours. To be an alchemist was, for many, to dedicate one's life to Christian service to others. It was a quest that could be performed anywhere in the world and whose benefits would reach throughout the world.

That Brewster's experiment was several years old and still had years to come to fruition reveals the procedurual nature of alchemy and reminds us that most of the time one spent as an alchemist was time spent in a state of liminal uncertainty, with experiments neither successful nor failed, but in process. Alchemy had few quick fixes: those who chose to seek perfection at the hearth were committed to lives of anticipation and uncertainty. Brewster's assertion, "By all their writinges, it cannot faylle," has a false bravado to it. He also wrote that to move it even the least amount risked destroying it.

I ffeare I shall not live to see it finished, in regard partly of the Indianes . . . as also I have a conceite that God sees me not worthy of such a blessing, by reason of my manifold miscariadges. The juxtaposition of the risks of frontier life with the quest for assurance of grace that lay at the heart of Christian alchemical practice again emphasizes the intensity of the alchemical experience for the practitioner. Alchemy really was as much about the transformation of self as it was about the transmutation of metals. Brewster, in connecting his fear of Indian war with God's judgment, had melded alchemy to the Puritan world of portents and prodigies.

If it were moved, it would be *as if the world which God mad[e] at first should be moved (as I sayd) in my new Creation, and the station of the heavens and earth should be displaced.* Alchemy was about the impulse to dominate, and the metaphors of alchemical experimentation made the magus, even as he questioned his own worthiness, a cosmic colossus. Yet it is not the power to be God that Brewster wants in "my new creation"; it is the power to make that creation what God wants it to be.

Brewster would entrust the secret with Winthrop *before I would betrust it with any of my children; for it is such a secrett, that is not fitt for every one, either for secrecy, or for partes, to use it, as Gods secrett for his Glory.* The brotherhood of the alchemical quest created powerful personal bonds between people in ways that have yet to be explored. Shared alchemical practice was, as the letters of Howes and Brewster reveal, an occasion of intimacy. In the spiritual communion of souls there could be, as Brewster suggests, a bond greater than the ties of family.

Rendering Christian service. Measuring access to grace. Possessing cosmic power. Achieving human intimacy. Each of these was a compelling motivating factor drawing men, and some women, to the alchemical quest in the seventeenth century. To these, I might add one more: uniformity. The alchemical quest had the same ultimate goal wherever it was performed. Transmutation on the periphery was as important and as powerful as it was in the metropolis. And should one be successful at attaining the alkahest or the philosopher's stone, he or she would become the epicenter of a new world. As Brewster reminded Winthrop when he implored him to maintain the code of secrecy, revelation of his discovery would change everything:

I should never be at quiett, neither at home, nor abroad, for one or other that would be enquiring and seking after knowledg thereof, that I should be tyard out, and forced to leave the place; naye it would be blased abroad into Europ.

From such excitement could come a New London in fact as well as name.

(SIX)

"God's Secret"

John Winthrop, Jr., Alchemical Healing, and the Medical Culture of Early New England

I should never be at quiett, neither at home, nor abroad, for one or other that would be enquiring and seking after knowledg thereof, that I should be tyared out, and forced to leave the place; naye, it would be blased abroad into Europ. —Jonathan Brewster, 1656

Jonathan Brewster's concern that news of his discovery of the "Elixer, fitt for Medicine, and healing of all maladyes," would bring a throng of people to his remote woodland plantation was more than just a projection of imaginative desire. It reflected the reality he had seen in the demand for the alchemical medical services of John Winthrop, Jr. Demand for his advice and medicines came from all over New England and as far away as Barbados and across the Atlantic in England. Suffering people arrived at New London in numbers that strained the capacity of the town and of Winthrop himself to provide for them. Cotton Mather said of Winthrop, "Where-ever he came, still the Diseased flocked about him, as if the Healing Angel of Bethesda had appeared in the place." Mather praised his "noble Medicines" and called him "a true adept." It has been estimated that his patients included half of the population of colonial Connecticut.[1]

Such widespread acclaim accorded by seventeenth-century patients to a medical approach that modern observers will probably consider primarily ineffective or even counterproductive underscores the importance of setting aside contemporary attitudes about medicine when attempting to understand

---

1. Jonathan Brewster to John Winthrop, Jr., Jan. 31, 1656, Massachusetts Historical Society, *Collections*, 4th Ser., VII (1865), 78, 80; Cotton Mather, *Magnalia Christi Americana; or, The Ecclesiastical History of New-England* . . . (London, 1702), II, 31–33. The Angel of Bethesda is a reference to the story in John 5:1–9 describing a pool of water named Bethesda in Jerusalem. The sick, blind, lame, and withered would wait around this pool for the appearance of an angel of the Lord, who came at certain seasons and stirred up the water of the pool. The first person to step into the pool after the waters were stirred was healed. Richard S. Dunn, *Puritans and Yankees: The Winthrop Dynasty of New England, 1630–1717* (Princeton, N.J., 1962), 83.

Winthrop's, or any early medical practitioner's, practice. The stark reality is that it has been only within the last three-quarters of a century that medicine has moved out of the realms of art and mystery into the category of serious science. Prior to that time, despite the array of voluminously theorized medical systems fiercely contested over by the followers of writer-physicians such as Hippocrates, Galen, or Paracelsus, the actual treatment of disease was largely a hit-or-miss affair, with the misses outnumbering the hits. The fact that medical treatment was largely ineffective did little to reduce patients' devotion to their favorite physicians or their belief in the effectiveness of the medications and other treatment regimes those physicians prescribed. As the physician Lewis Thomas noted after pointing out that his father—a highly respected physician in Brooklyn in the 1930s—could rarely actually heal any of the diseased people who relied on him:

> Patients do get better, some of them anyway, from even the worst diseases; there are very few illnesses, like rabies, that kill all comers. Most of them tend to kill some patients and spare others, and if you are one of the lucky ones and have also had at hand a steady, knowledgeable doctor, you become convinced that the doctor saved you.

Such appreciation was based in part on the doctor's ability to provide an authoritative explanation for both the illness and its potential outcome, and in part on the medicines he or she prescribed to treat it. Whether these medicines' ingredients provided physical relief or not (some, but certainly not most, did indeed have therapeutic value), the patient's and physician's belief that they *were* effective *made* them effective. As Thomas, again writing of his father's practice, noted, "Placebos . . . had been the principal mainstay of medicine, the sole technology, for so long a time—millennia—that they had the incantory power of religious ritual." Keeping this in mind while considering the role of alchemical medicine in early New England will help modern readers better appreciate its medical importance despite what we would see as limited therapeutic effectiveness.[2]

The respect accorded Winthrop as a physician underscores the widespread acceptance of alchemical medicine in New England. The alchemical medicines that constituted the majority of his medicinal prescriptions produced physical effects in the recipients that were often demonstrably stronger than the herbal medicines he also dispensed. Their power provided a physical confirmation to the patient of what Christian alchemists had always asserted—that their special medicines manifested a divinely derived improvement on

---

2. Lewis Thomas, *The Youngest Science: Notes of a Medicine-Watcher* (New York, 1983), 14.

traditional medical practice and that God was the direct source of all medical knowledge and all healing.

Through most of the seventeenth century, the terms "chemical medicine" and "alchemical medicine" were often used without distinction, just as alchemy and chemistry were interchangeable terms. To the degree a distinction could be made between alchemy and chemistry, it was that alchemy was the part of chemistry that penetrated to its core, "the spiritual understanding of created matter." Alchemical medicine in New England relied upon and strongly benefited from this explicit linkage with the spiritual. For, while the alchemical medicines New Englanders ingested frequently produced strong physical effects, the perceived value of those effects probably resided as much in their spiritual implications as in their physical benefits. The power of alchemical medicines, frequently administered as purges and diaphoretics, demonstrated to users the divine "blessing on the means" of healing provided by the alchemical physician.[3]

New Englanders were perhaps uniquely receptive to assertions that healing and providential intent were inextricably intertwined. The medical experiences of the early colonists intensified the medical providentialism that was part of English Puritan culture. The simplified medical culture established in New England, which escaped the divisions and contentiousness found in England's pluralistic medical culture, further reinforced a highly providential understanding of medical matters. In New England alchemical medicine could and did thrive, because it provided Puritan patients medical treatment whose underlying explanatory framework blended seamlessly with their cultural expectations.

The examination here of the source of New England's intensified medical providentialism will explore the cultural context in which alchemical medicine became part of a distinctive variant of seventeenth-century English medical culture. Then an overview of the medical practice of Winthrop will illustrate how the practice of the leading exemplar of alchemical medicine in New England worked and how it met a range of social, cultural, and personal needs for both Winthrop and his patients. Finally, to situate Winthrop among the many alchemical healers practicing in early New England, it will survey the alchemical medical landscape of seventeenth-century New England and those practitioners whose practices had alchemical components.

---

3. J. T. Young, *Faith, Medical Alchemy, and Natural Philosophy: Johann Moriaen, Reformed Intelligencer, and the Hartlib Circle* (Aldershot, 1998), 164–165.

MEDICAL PROVIDENTIALISM

Throughout the seventeenth century, New Englanders had virtually no access to the healers who represented the upper echelons of English medical practice. No licensed medical practitioners emigrated to New England during the Great Migration, and no university medical graduates worked there before 1671. Only three medical doctors—Leonard Hoar, Edmund Davie, and John Crowninshield—practiced in New England during the entire seventeenth century. Fellows of London's Royal College of Physicians, who had studied the required seven years beyond the master's degree to obtain the degree of medical doctor and passed the college's scrutiny of knowledge and personal character, simply did not emigrate. Yet, despite this dearth of representatives of England's medical elite and the absence of formal medical institutions to review the qualifications of colonial medical practitioners, New England's Puritans created a medical culture in which they had a great deal of confidence.[4]

Relying on healers with varied degrees of education and experience—only a small minority had received any university medical training—colonists countered disease with a range of treatment regimes and appear to have been, by the standards of the age, quite satisfied with the results. Discussions of illness and treatment pervade their correspondence, but complaints about the rudimentary nature of colonial health care or the qualifications of medical

---

4. Leonard Hoar, president of Harvard College, received the degree of Doctor of Medicine at Cambridge in 1671 and died in 1675; Edmund Davie was a Harvard College graduate who became a Doctor of Medicine at Padua in 1681 and practiced in Boston; John Crowninshield, Doctor of Medicine of Leipzig, arrived in Boston in 1688 and died in 1711. Eight Doctors of Medicine in total are known to have had New England connections in the seventeenth century. Robert Child (M.D., Padua, 1638) took two trips to New England between 1638 and 1641 and again from 1644 to 1647 but did not apparently practice medicine in New England. John Glover, Ichabod Chauncey, Samuel Bellingham, and Henry Saltonstall received medical degrees in Europe after graduating from Harvard but did not return to New England. See C. H. Brock and Eric H. Christianson, "A Biographical Register of Men and Women from and Immigrants to Massachusetts between 1620 and 1800 Who Received Some Medical Training in Europe," in *Medicine in Colonial Massachusetts, 1620-1820,* Colonial Society of Massachusetts, Publications, LVII (Boston, 1980), 117-144; the list of minister-physicians of colonial New England in Patricia A. Watson, *The Angelical Conjunction: The Preacher-Physicians of Colonial New England* (Knoxville, Tenn., 1991), appendix 1, 147-152; and the alchemical physicians identified in William R. Newman, *Gehennical Fire: The Lives of George Starkey, an American Alchemist in the Scientific Revolution* (Cambridge, Mass., 1994), 39-53. On the fellowship requirements of the Royal College of Physicians as well as the traditional Galenic medical focus of the admission examinations, see Harold J. Cook, *The Decline of the Old Medical Regime in Stuart London* (Ithaca, N.Y., 1986), 72-73. Margaret Pelling and Charles Webster clarify the complex components of English medical hierarchy in "Medical Practitioners," in Webster, ed., *Health, Medicine, and Mortality in the Sixteenth Century* (Cambridge, 1979), 165-235.

practitioners are few. New England's medical practitioners themselves seem to have shared a substantial degree of mutual respect, also surprising for the period in which they lived. While surviving correspondence and medical account books suggest there was a lively exchange of ideas and techniques among practitioners and quite a bit of medical, especially pharmacological, experimentation in New England throughout the early colonial period, the region was (until the furor over Cotton Mather's promotion of smallpox vaccination in 1724 Boston) generally free from the conflicts over medical theory and practice that were constant features of early modern English medicine. If New Englanders suffered from inferior medical care compared to their English counterparts—as at least one English physician who practiced in Rhode Island in the late seventeenth century believed—New Englanders themselves did not appear to notice. Admittedly, on a therapeutic level, virtually all early modern medicine was rudimentary in comparison to present-day medical treatments, but this did not constrain the intense debates over the relative effectiveness of various medical modalities among European medical practitioners.[5]

The key to New Englanders' confidence in the medical treatment they received lay in their intense medical providentialism—the unwavering conviction among the godly that God played an active role in both inflicting and healing diseases. Faith that "God worked out His eternal decrees dramatically in the lives and psyches of His Massachusetts Saints" applied with particular force in New England to medical matters. John Cotton's reminder, "There is not any sickenesse befalls us or ours . . . but it is a knocke of Gods hand to turne to him," was a foundational belief among New England's Puritans. For them, illness was a direct, immediate, and sometimes life-threatening reminder that God monitored personal and collective behavior and intervened medically to call saints' attention to their spiritual estates. Disease was never a mere physical malady; sickness always implied, first and foremost, divine censure or admonition. "All our *Sicknesses* are but the Execution of that primitive Threatening in Gen. II 17," Cotton Mather noted. "'In the Day that Thou Sinnest thou shalt Surely dy.'" For the Puritans of Massachusetts, the providential implications of illness were always as important as the physical malaise itself.[6]

---

5. Samuel Lee, a physician newly arrived from London, was highly critical of the informally licensed and officially unregulated nature of New England's medical practice, where, from his perspective, "Practitioners are laureated gratis with a title feather of Doctor." Samuel Lee to Nehemiah Grew, 1688, reprinted in George Lyman Kittredge, "Letters of Samuel Lee and Samuel Sewall Relating to the New England Indians," Colonial Society of Massachusetts, *Publications*, XIV, *Transactions*, 1911–1913 (Boston, 1913), 145.

6. Michael P. Winship, *Seers of God: Puritan Providentialism in the Restoration and Early Enlighten-*

Recognition of the providential foundations of disease did not conflict with medical theories of disease etiology, though it superimposed on such theories a layer of meaning that had important implications for the understanding of and approaches to medical treatment. In this context, Puritan alchemical practitioners worked within a broad providential framework that shaped all aspects of colonial New England life. Cotton Mather, like his Puritan co-religionists, recognized that illness stemmed from natural causes, even as he emphasized that spiritual failure was the trigger that set pathology in action. "If *Crudities,* and *Obstructions,* and *Malignities,* are the *Parents* of our *Sicknesses,*" Mather claimed, "tis very sure, that *Sin* is the *Grand Parent* of them." New England's saints held that God's intervention in disease was all-encompassing: God not only inflicted illnesses, but he regulated their duration and intensity as well. Sometimes he sent disease as mortal affliction, a condemning and fatal judgment. More often, sickness was a sharp corrective meant to refocus the sufferer's attention on spiritual living. The minister-physician Michael Wigglesworth's *Meat out of the Eater,* a long poem that was one of the most popular religious texts among seventeenth-century New Englanders, noted the spiritually therapeutic value of illness.

> Our Bodies Sicknesses,
> Are Physick for the Soul,
> Corrected by a skillful hand,
> That can its force controll.

Divine correction by disease, Wigglesworth asserted, was carefully measured, and, though the Lord administered illness, he took no pleasure in it. Sharing the sufferer's pain, he would ultimately provide the source of relief as well.

> He knows thy Sicknesses;
> He feeleth all thy smart:
> Thy sufferings are his sufferings
> And reach his tender Heart.
>
> And if he know and feel
> All that doth thee agrieve:

---

ment (Baltimore, 1996), 9–10, 13–20, 36–41, esp. 16; David D. Hall, *Worlds of Wonder, Days of Judgment: Popular Religious Belief in Early New England* (New York, 1989), 77–78, 91–94, 116, 122–123, 241; Robert Middlekauff, *The Mathers: Three Generations of Puritan Intellectuals, 1596-1728* (New York, 1971), 143–148; Perry Miller, *The New England Mind: The Seventeenth Century* (New York, 1939), 14–17; John Cotton, *Gods Mercie Mixed with His Justice; or, His Peoples Deliverance in Time of Danger, Laid Open in Severall Sermons* (London, 1641), 10; Cotton Mather, *The Angel of Bethesda,* ed. Gordon Jones (Barre, Mass., 1972), 5.

He will with choicest Cordials
Thy fainting Soul relieve.⁷

The foundation of New Englanders' intense medical providentialism was to be found in the English Puritan movement. Such an expansive conception of God's influence in medical matters did not, however, reflect the prevailing beliefs among England's traditional medical elite, who held to a much more limited conception of God's involvement in the healing arts. English medical providentialism was contested by powerful segments of a polyglot medical community, and attacks on it limited its overall cultural influence in England and restricted its influence outside Puritan circles. Across the ocean, however, where providential medical theories were uncontested throughout the seventeenth century, acceptance of a close and direct integration between the spiritual and physical aspects of healing flourished. English critics of medical providentialism did not extend their interest, influence, or their critiques to the colonies, so providential interpretations of pathology flourished in New England largely unchallenged by counterinterpretation or competing medical perspectives. New England's medical culture came to reflect a more literal view of God's involvement in medical matters than was generally found in England's professional medical communities. Medical care in early New England shared many features with rural England, to which it has often been compared, but it was a religiously focused and intensified variant of English medical culture, not merely a creolized transplant of that culture or an extension of rural English medical practice.⁸

---

7. Mather, *Angel of Bethesda*, ed. Jones, 5; Michael Wigglesworth, *Meat out of the Eater; or, Meditations concerning the Necessity, End, and Usefulness of Afflictions unto God's Children, All Tending to Prepare Them for, and Comfort Them under the Cross,* 4th ed. (Boston, 1689), 101, 176.

8. The view of English medical authorities regarding God's agency in healing is discussed in further detail later in the chapter. See also Peter Elmer, "Medicine, Religion, and the Puritan Revolution," in Roger French and Andrew Wear, eds., *The Medical Revolution of the Seventeenth Century* (Cambridge, 1989), 10–45, esp. 13; Ronald Charles Sawyer, "Patients, Healers, and Disease in the Southeast Midlands, 1597–1634" (Ph.D. diss., University of Wisconsin, Madison, 1986), 242–243; Paul H. Kocher, *Science and Religion in Elizabethan England* (San Marino, Calif., 1953), 263–264; Harold J. Cook, *The Decline of the Old Medical Regime in Stuart London* (Ithaca, N.Y., 1986), 72–73, for a discussion of the polyglot, contested medical culture of seventeenth-century England, from which New Englanders were largely isolated. On the intensification of providentialism in general that occurred in Massachusetts as a consequence of forming a "new Israel," see Winship, *Seers of God,* 9–16. Patricia Watson discusses New Englanders' highly providential view of illness and healing, though her interpretation is, to some degree, anachronistically negative (Watson, *Angelical Conjunction,* 20–22). Richard D. Brown, "The Healing Arts in Colonial and Revolutionary Massachusetts: The Context for Scientific Medicine," in *Medicine in Colonial Massachusetts,* 43–44, C. Helen Brock, "The Influence of Europe on Colonial Massachusetts Medicine," 101–116; Watson,

The medical experiences of New England's founders provided convincing evidence of God's positive intervention in health matters, and it persuaded the early Puritan settlers that they enjoyed particularly favored providential medical status. Especially during the early decades of settlement, New England's godly settlers were spared many of the diseases endemic in Europe; moreover, God seemed to have blessed the Puritans' colony with a genuinely restorative climate. Although the members of the Winthrop fleet experienced a period of serious seasoning diseases similar to those that had initially devastated both Jamestown and Plymouth colonies, once the initial sickness period passed, the physical health of many Puritan immigrants actually improved over what it had been in England. Given the denser population of English cities and the "'norm' of endemically dirty living environments" and chronic outbreaks of plague and other epidemic diseases, it is, of course, not at all surprising that Puritan bodies would fare better once settled in New England.[9]

During the fall and winter of 1630/1, more than two hundred persons—20 percent of the Winthrop group—had died of scurvy and several other diseases. In marked contrast to the experience of English planters in the Chesapeake, however, this incidence of high immigrant mortality proved to be a single episode rather than a chronic risk. Ironically, New Englanders' brief episodes of seasoning mortality served to amplify rather than undermine colonial Puritans' belief in their favored providential medical status. Those settlers who had the most reason to doubt medical providentialism either died from their sicknesses or became disillusioned and returned to England. Those

---

Angelical Conjunction, 12–23; Ann Leighton, *Early American Gardens: "For Meate or Medicine"* (Boston, 1970), 113–138.

9. The absence of medical doctors and formal medical institutions does not imply that New England's leaders were unconcerned about providing adequate medical care. Even during initial settlement, the Massachusetts Bay Company took pains to assure that the Winthrop fleet was well provided with medical assistance. Among the roughly 1,000 Puritans who emigrated with John Winthrop in 1630 were nine medical practitioners: two physicians (university-educated medical practitioners who had not received formal medical degrees), two educated medical practitioners whose medical training had occurred outside the university, four barber-surgeons (who specialized in treating wounds, fractures, and dentistry), one apothecary, and two or more midwives. Excluding the midwives (for comparative purposes with existing data), there was one trained medical practitioner for each 140 passengers in the Winthrop fleet. This was comparable to or better than the per capita numbers of trained medical practitioners available in the best-served parts of England. Perhaps this reflects a conscious awareness on the part of the Massachusetts Bay Company leaders that there was likely to be a seasoning period of high disease pressure following initial settlement. Pelling and Webster argue that the ratio of medical practitioners to population in England during the sixteenth century was probably around 1:200 ("Medical Practitioners," in Webster, ed., *Health, Medicine, and Mortality in the Sixteenth Century,* 165–235).

who remained interpreted their avoidance of or recovery from illness as an intentional providential blessing. Many New Englanders came to think that the seasoning mortality had been a divine winnowing of godly wheat from fallen chaff, which set the stage for intensification of medical providentialism among the godly.[10]

A decade after the seasoning period had ended, John Eliot underscored for English readers how sharply New England's providential healthfulness contrasted with the medical status of other English plantations, notorious for their high mortality.

> Thus farre hath the good hand of God favoured our beginnings. . . . In blessing us generally with health and strength[,] as much as ever (we might truly say) more then ever in our Native Land; many that were tender and sickly here [in England], are stronger and heartier there [in New England]. That wheras diverse other Plantations have been the graves of their inhabitants and their numbers much decreased: God hath so prospered the climate to us, that our bodies are hailer, and Children there born stronger, wherby our number is exceedingly increased.[11]

The hand of providence in human health was made manifest in far more telling ways than through a restorative climate. The most direct, powerful, and incontrovertible evidence of God's intervention in medical affairs was an event that provided indisputable proof of God's special medical blessings on New England's colonists. God sent the Puritans, through the selective administration of disease, a spectacular sign that they were his chosen people. During 1633 and 1634, just as native Americans were beginning aggressively to

---

10. Aug. 27, Sept. 20, 1630, in Richard S. Dunn, James Savage, and Patricia Yeandle, eds., *The Journal of John Winthrop* (Cambridge, Mass., 1996), 38–39. Within the first year, two hundred settlers abandoned the New England project and went home. Between death and repatriation, Winthrop's group was diminished by 40 percent (39n). In March 1631, Winthrop wrote his wife, who was preparing to sail over from England, confidently assuring her, "The Lord our God, who hath kept me and so many of my Company in health and safety among so many dead Corps . . . will . . . preserve us and ours still that we shall meet in ioye and peace" (John Winthrop to Margaret Winthrop, Mar. 28, 1631, in Samuel Eliot Morison et al., eds., *Winthrop Papers* [Boston, 1929–], III, 20 [hereafter cited as *Winthrop Papers*]). In the aftermath of the epidemic, Winthrop providentially linked the deaths of poorer settlers to their lack of commitment to New England's godly enterprise. He observed in his journal that it had "been alwayes observed heere, that suche [of the poorer sort] as fell into discontente and lingered after their former Conditions in Englande, fell into the skirvye, and dyed." Feb. 10, 1631, in Dunn, Savage, and Yeandle, eds., *Journal of John Winthrop*, 46.

11. [John Eliot], *New Englands First Fruits; in Respect, First of the Conversion of Some, Conviction of Divers, Preparation of Sundry of the Indians, 2. of the Progresse of Learning, in the Colledge at Cambridge in Massacusets Bay: With Divers Other Speciall Matters concerning the Countrey* (London, 1643), 20, 21.

resist English territorial expansion, smallpox broke out in Indian villages in Massachusetts and Plymouth colonies, spreading along the Narragansett Bay and Connecticut River and up the Atlantic coast as far as Pascataqua, Maine. Indian populations with little or no prior immunity to English diseases were devastated while the English remained largely untouched. Mortality rates among the Indians were four to five times those the Puritans had experienced in 1630; some native groups were completely extinguished. This was not the first medical catastrophe suffered by New England's Indians. From 1616 to 1619 another pandemic had devastated tribal groups along the New England coast. In 1633, though, English settlers and Indians were living in proximity while the epidemic devastated one group, leaving the other virtually unscathed. Puritans witnessed firsthand the wide geographic horizon of the epidemic and its terrifying impact on natives. "The poore Creatures being very timorous of death," Edward Johnson reported, "would faine have fled from it, but could not tell how, unlesse they could have gone from themselves." Because the Indians were too ill and in some cases too fearful to tend those stricken, Puritan men and women cared for the victims. They buried the native dead, as many as thirty in one day in a single village. They took orphaned or at-risk Indian children into their homes and hastened to convert those of the dying who had concluded that the God of the Puritans was more powerful than their own gods.[12]

---

12. Edward Johnson underscored the increasing tension over English expansion. "The Indians, who had all this time held good correspondency with the English, began to quarrel with them about their bounds of Land . . . but the Lord put an end to this quarrell also, by smiting the Indians with a sore Disease, even the small Pox; of the which great numbers of them died." J. Franklin Jameson, ed., *Johnson's Wonder-Working Providence, 1628-1651* (New York, 1910), 79; Dunn, Savage, and Yeandle, eds., *Journal of John Winthrop*, 106–107 n. 40; Neal Salisbury, *Manitou and Providence: Indians, Europeans, and the Making of New England, 1500-1643* (New York, 1982), 101–109; William Cronon, *Changes in the Land: Indians, Colonists, and the Ecology of New England* (New York, 1983), 85–91; Sherburne F. Cook, "The Significance of Disease in the Extinction of the New England Indians," *Human Biology*, XLV (1973), 485–508. Average native population declines as a result of English diseases are estimated to be as high as 90 percent (Kathleen J. Bragdon, *Native People of Southern New England, 1500-1650* [Norman, Okla., 1996], 25–28). The 1616 epidemic left the site that the Mayflower passengers settled in 1620 virtually uninhabited. The Separatists of Plymouth also drew important providential implications from the native diseases. E[dward] W[inslow], *Good Newes from New-England . . .* (London, 1624), 10–11, 52; Thomas Morton, *The Essential "New English Canaan"* (1637), ed. Jack Dempsey (Scituate, Mass., 1999), 19–20; John Winthrop to Sir Simonds D'Ewes, July 21, 1634, *Winthrop Papers*, III, 171–172; Jameson, ed., *Johnson's Wonder-Working Providence*, 80.

Given the degree to which person-to-person contact was considered a risk factor for contagion during epidemics, the readiness of the English to assist the natives highlights their faith in medical providentialism (Georges Vigarello, *Concepts of Cleanliness: Changing Attitudes in France since the Middle Ages*, trans. Jean Birrell [Cambridge, 1988], 7–8). "It wrought muche with them, that when

This decimation of native populations while the English remained all but unscathed confirmed for the Puritans their favored medical and providential status and the power that God could exert in manifesting his will via medical means. Colonial leaders hastened to inform English supporters of the extraordinarily selective contagion and to underscore the important providential meanings of the pandemic. John Winthrop confidently asserted that God had sent the diseases specifically to terminate Indian resistance to English territorial expansion. "For the natives, they are neere all dead of the small Poxe," noted Winthrop, "so as the Lord hathe cleared our title to what we possess." Ever afterward, the legitimation of Puritan expansion in New England would rest, at least in part, on the divine assurances received early in colonization through medical providentialism.[13]

The awesome demonstration of selective contagion of the Indian epidemics transmitted an unforgettable lesson in the presence and power of God's hand in medical matters from New England's founding generation to its successors. It helped heighten the sense of God's agency in medicine that permeated seventeenth-century New England culture. Alchemical medicine, because of its godly theoretical underpinnings, was inherently compatible with this heightened medical providentialism, and elite Puritan alchemical practitioners such as John Winthrop, Jr., operated well within, and strongly reinforced, the mainstream of Puritan New England cultural expectations. New England alchemical practitioners provided their Christian medical services and administered their spiritually derived medicines within a culture that attached spiritual significance to medical matters of all kinds.[14]

Throughout the century, a searching inquiry into the meanings and messages of medical occurrences occurred at every level of colonial society. Ministers exhorted their congregations to see illness as chastisement and never to forget the connections between spiritual and physical corruption. During periods of widespread disease among the colonists, they located its cause in New Englanders' declension from their mission to be a spiritual example to a sinful world. Puritan political leaders also employed providential medical interpretation during times of unrest to underscore claims of divine sanction

---

their owne people forsooke them, yet the Englishe came dayly and ministered to them, and yet fewe <onely 2 famyles> tooke any Infection by it." Dunn, Savage, and Yeandle, eds., *Journal of John Winthrop*, 105. Winthrop reported in November 1633 that "many of their children escaped and were kept by the English." By the following February, all but three of the children had died (101, 110, 105).

13. John Winthrop to Sir Nathaniel Rich, May 22, 1634, *Winthrop Papers*, III, 167; the claim is reiterated in John Winthrop to Sir Simon D'Ewes, July 21, 1634, 171–172.

14. Watson, *Angelical Conjunction*, 13, 14, 20.

for their judgments or policies: the elder Winthrop's exhumation of Mary Dyer's "monstrous birth" and his exploitation of it as a providence condemning Anne Hutchinson's heresy is merely the most memorable example of a relatively common occurrence. Early New Englanders' surviving letters, journals, and diaries are permeated with detailed accounts of the writers' own illnesses and those of persons close to them; invariably such accounts note the providential implications of these bouts of sickness or make explicit reference to God's authority over the diseases' outcomes. Many saints gained the certain knowledge of grace requisite for joining New England's churches only after grappling with "salvation panic" during the course of an extended or particularly acute illness. Prayer itself became, for many Puritans, the first therapy resorted to in cases of illness, the one many considered most effective.[15]

New Englanders' emphasis on divine agency in medical affairs differed from that of English Puritans in degree rather than in kind. As a result of the intensity of New England Puritans' belief in the link between divine providence and personal health, they welcomed alchemical medicine as a highly effective form of medical treatment. Medical providentialism and alchemical healing never achieved the uncontested acceptance in England that they received in the Puritan colonies. Given the highly contentious medical and religious pluralism of English society, specific providential interpretations of medical events and even the very concept of medical providentialism were almost always contested matters. Furthermore, English Puritans never received the dramatic affirmation of medical providentialism through selective contagion that New Englanders did during the Indian epidemics. In England, large disease outbreaks were ecumenical; saints and sinners (or saints and sectaries) were equally affected, and interpretations of whose sins had precipitated a crisis were subject to endless disputation. Moreover, in England's elite medi-

---

15. Wigglesworth's *Meat out of the Eater* was a widely popular example of spiritual-medical exhortation; other examples included James Allin, *Serious Advice to Delivered Ones from Sickness* . . . (Boston, 1679), 26–27; Peter Tha[t]cher, *The Alsufficient Physician Tendering to Heal the Political and Spiritual Wounds and Sickness of a Distressed Province* (Boston, 1711), 4, 12. "Gods rods are teaching, o[u]r epidemical sicknesse of colds, doth rightly by a divine hand tell the churches what o[u]r epidemical spir[itua]l disease is. Lord help us to see it." John Eliot, in William B. Trask, "Rev. John Eliot's Records of the First Church in Roxbury, Mass.," *New England Historical and Genealogical Register*, XXXIII (1879), 236–238.

See also Sacvan Bercovitch, *The American Jeremiad* (Madison, Wis., 1978), 84; Dunn, Savage, and Yeandle, eds., *Journal of John Winthrop*, 253–255; Kenneth P. Minkema, "The East Windsor Conversion Relations, 1700–1725," Connecticut Historical Society, *Bulletin*, LI (1986), 27–54; John Dane, "John Dane's Narrative," *New England Historical and Genealogical Register*, VIII (1854), 147; Douglas Winiarski, "Lydia Prout's 'Dreadfullest Thought': Female Piety and Maternal Bereavement in Provincial Boston," MS, presented at University of Connecticut Humanities Institute, Apr. 9, 2008; Hall, *Worlds of Wonder*, 197–200.

cal circles divine agency was often considered of little practical significance in the treatment of disease.[16]

Many members of the professional English medical community, including the eminences of the Royal College of Physicians, rejected the concept of direct divine intervention in medical matters as fundamentally false and theoretically unfounded. The Galenic medical tradition that dominated English medicine during New England's Great Migration held that God's will and wisdom with regard to medical matters had been permanently encoded directly into Creation. Elite medical practitioners in England strongly opposed the fusion of medical practice and religion. Such hostility was not new; the standard medical works of the Elizabethan era gave only perfunctory reference to divine agency in disease while excluding any mention at all of the supernatural from actual medical analysis. "The readiness of medical men to ignore the spiritual side of illness," Keith Thomas has noted, "had long gained them a reputation for atheism."[17]

Given that dismissal by the formal medical establishment and the criticism by Anglican physicians such as Robert Burton and Richard Napier, many English patients also rejected medical providentialism or, more commonly, minimized its significance. Richard Napier's medical account books give evidence that his patients rarely placed disease in a divine etiological framework. "In practice both doctors and laymen often regarded disease as a purely natural phenomenon." Most historians agree that the majority of English people retained some belief in an association between divine will and physical cor-

---

16. On English providentialism, see Keith Thomas, *Religion and the Decline of Magic* (New York, 1971), 79–102; William Birken, "The Dissenting Tradition in English Medicine of the Seventeenth and Eighteenth Centuries," *Medical History,* XXXIX (1995), 197–218. David N. Harley notes that English Calvinist ministers adopted a "distinctive attitude towards medicine, quite unlike that of Catholics or formalist Protestants" ("Medical Metaphors in English Moral Theology, 1560–1660," *Journal of the History of Medicine and Allied Sciences,* XLVIII [1993], 396–435, esp. 404). As Keith Thomas succinctly summarized, "What was an obvious providence to one man might be only a case of bad luck to another" (*Religion and the Decline of Magic,* 104). While conflicting interpretations did little to stem the flood tide of polemical providentialism in England prior to the Restoration, conflicts over meanings ultimately helped undermine confidence in providentialism itself. Michael Winship discusses the three-pronged assault on Puritan providentialism made by Anglicans in post-Restoration England and notes that the roots of this antiprovidentialism can be found much earlier in seventeenth-century England. Importantly, the assault on providentialism in England took hold during a period when, as Perry Miller found, providentialism in New England was experiencing a period of intensified interest (Winship, *Seers of God,* 31–52; Miller, *The New England Mind,* 227–228). Thomas describes the variety of theories propounded about epidemics in England in *Religion and the Decline of Magic,* 86–88, 104–105.

17. Kocher, *Science and Religion in Elizabethan England,* 263–266; Michael Hunter, "Boyle versus the Galenists: A Suppressed Critique of Seventeenth-Century Medical Practice and Its Significance," *Medical History,* XL (1997), 322–361; Thomas, *Religion and the Decline of Magic,* 85.

ruption, but at least through the first half of the seventeenth century English society as a whole did not attribute to divine agency the centrality in medical affairs uniformly accorded by New England's Puritans.[18]

New Englanders' wholesale acceptance of medical providentialism was instrumental in shaping the Puritans' medical culture. It helps explain, for instance, why Puritans unequivocally embraced the practice of having ministers provide medical care, despite the fact that the practice was highly contested in England. Historians have often assumed that ministers commonly practiced medicine in early modern England, particularly in rural areas, but the frequency of that practice is questionable, and the medical knowledge possessed by ministers seems to have been quite limited.[19]

---

18. On Napier, see Sawyer, "Patients, Healers, and Disease in the Southeast Midlands," 242–243; Thomas, *Religion and the Decline of Magic*, 85 (quote); Watson, *Angelical Conjunction*, 20–21. Failure to claim a divine etiology for disease did not, however, mean that Napier's patients were irreligious. Michael McDonald found that almost three hundred of Napier's psychologically troubled patients suffered from religious anxiety or spiritual problems. Although McDonald's study does not consider medical providentialism directly—whether Napier's patients thought of God as the *source* of their diseases—he concludes: "Few ordinary people . . . thought it was illegitimate to marry physic and astrology, medicine and divinity, in spite of the efforts of professional physicians to distinguish these arts" (*Mystical Bedlam: Madness, Anxiety, and Healing in Seventeenth-Century England* [Cambridge, 1981], 32, 220). Efforts to pinpoint the degree of commitment to medical providentialism among non-Puritans are murky at best. Thomas, for example, argues, "The doctrine of providence was always less likely to appeal to those at the bottom end of the social scale than the rival doctrine of luck" (*Religion and the Decline of Magic*, 111). Doreen Evenden Nagy, on the other hand, argues that popular belief in the connection between spiritual and physical health was so firmly established that it "encouraged the retention of traditional medicine at the expense of professional medicine" (*Popular Medicine in Seventeenth-Century England* [Bowling Green, Ohio, 1988], 35–42). David Harley fuses these two positions, claiming that the elite had worked for more than a century to instill medical providentialism among the laity: "Devout Calvinist ministers and physicians throughout the country attempted to convince the lay public that the two professions [medicine and divinity] were almost exactly homologous" ("Pious Physic for the Poor: The Lost Durham County Medical Scheme of 1655," *Medical History*, XXXVII [1993], 148–166, esp. 154). The most balanced view perhaps is found in Andrew Wear, who concludes, "Religion . . . not only allowed the practice of medicine but was also an integral part of the debate between rival medical systems" (*Knowledge and Practice in English Medicine, 1550–1680* [Cambridge, 2000], 34).

19. Nagy, *Popular Medicine in Seventeenth-Century England*, 38–40; Charles Webster, *The Great Instauration: Science, Medicine, and Reform* (New York, 1976), 255; MacDonald, *Mystical Bedlam*, 9; James Hart, *Klinike; or, The Diet of the Diseased* (London, 1633), 12–13; John Cotta, *A Short Discoverie of Severall Sorts of Ignorant and Unconsiderate Practisers of Physicke in England* . . . (London, 1619), 89; Peter Elmer, "Medicine, Religion, and the Puritan Revolution," in French and Wear, eds., *Medical Revolution of the Seventeenth Century*, 15n. Cotton Mather, who never visited England, noted that ministerial medicine had been commonly practiced since the days of Luke but supposed "the Greatest Frequency of the Angelical Conjunction, has been seen in these parts of America" (*Magnalia Christi Americana*, III, 151). Watson argues that minister-physicians were common in seventeenth-century England (*Angelical Conjunction*, 39–40). She also argues that economic necessity forced ministers into dual professions in New England.

English medical professionals sharply criticized the pattern of ministers' practicing medicine. The medical doctor James Hart claimed that clerics who practiced physick "doe wrongfully and injuriously, both contrary to the Law of God and man, intrude upon another weighty profession." Such intrusions were not only unsound medically; they violated the natural order of society as expressed in vocational hierarchies. John Cotta, fellow of the Royal College of Physicians, agreed. In *A Short Discoverie of Severall Sorts of Ignorant and Unconsiderate Practisers of Physicke in England,* Cotta called the practice of medicine by ministers "offensive to God, scandalous unto religion and good men, and injurious unto commonweales." He ridiculed the incompetence of minister-physicians as well, citing one whose ill-advised faith in his own medical abilities led him to overdose fatally on opium and another whose "pride and conceit of his [medical] knowledge" led him to claim there was no distinction between stibium (antimony) and ratsbane (arsenic). "His confidence herein so farre bewitched him," Cotta claimed, "that he made triall thereof in himselfe, and as a just execution upon himselfe, was the same day poisoned."[20]

George Herbert, whose *Priest to the Temple; or, The Country Parson, His Character, and Rule of the Holy Life,* is often cited as evidence for the widespread practice of medicine by ministers in rural England, actually specified a limited medical role for clerics. His work called for rural parsons, or their wives, to provide elementary medical services. Such care was to be given only to members of the parson's own congregation. A parson could acquire sufficient medical knowledge by "seeing one Anatomy, reading one book of Phisick, having one Herbal by him." His pharmacopoeia was to exclude alchemical remedies; it was limited to simple herbal ingredients from his own kitchen garden. Herbert recommended that the parson provide himself with ready access to more professional assistance. Should the parson or his wife have no medical skill at all, he was to keep "some young practitioner in his house for the benefit of his Parish, whom yet he ever exhorts not to exceed his bounds, but in ticklish cases to call in help." And in recognition of those moments when "all else fail[s]," a country parson "keeps good correspondence with some neighbour Physician, and entertains him for the Cure of his Parish." Herbert clearly distinguished between the rudimentary medical care he thought country ministers should be prepared to provide and the more formal medical services provided by trained physicians. Even with that

---

20. James Hart, *Klinike,* 12–13. Hart objected that "God is the God of order, not of confusion; and never did allow of this confused *Chaos* of callings" (*The Arraignment of Urines: Wherein Are Set downe the Manifold Errors and Abuses of Ignorant Urine-monging Empirickes* [London, 1623], [xiii/xv]). John Cotta, *A Short Discoverie of Severall Sorts of Ignorant and Unconsiderate Practisers of Physicke in England . . .* (London, 1619), 78, 89.

distinction, the number of minister-medical practitioners in England seems to have been circumscribed. Analyzing seventeenth-century English medical culture through the Civil War and Commonwealth period, Peter Elmer suggests English ministers rarely engaged in medical practice. "Although it would be wrong to suggest that Puritan clergymen never combined the functions of priest and physician, it was nonetheless very rare for such ministers to practise both professions simultaneously." Of 1,766 nonconforming ministers deprived of their church positions between 1660 and 1662, fewer than 3 percent turned to medicine as a calling.[21]

No such division between ministry and medicine existed in New England, where intensified medical providentialism and a lack of professionally licensed medical practitioners favored practice of medicine by divines. In colonial New England, 126 ministers who also served their communities' medical needs have been identified. Judging from the wide variety of medical texts found in the extant library inventories of these minister-healers, it is clear that many of them sought to acquire a medical expertise that far exceeded Herbert's one anatomy, one book of physick, and one herbal. Cotton Mather recognized New Englanders' high dependence on ministerially provided medical care as one of the distinctions between New England and Old. While noting that ministers had provided medical services since the days of Luke, Mather affirmed that "the Greatest Frequency of *Angelical Conjunction,* has been seen in these Parts of *America.*" To New England's Puritans, uniting the role of spiritual leader and healer made perfect sense. In a culture that emphasized the providential meanings of every aspect of quotidian experience, few questioned the inseparability of what Patricia Watson called the link between theology and pathology.[22]

Culturally sanctioned and reinforced medical providentialism also helps explain why colonial Puritans embraced learned healers who incorporated

---

21. Geo[rge] Herbert, *A Priest to the Temple; or, The Country Parson: His Character, and Rule of the Holy Life* (London, 1671), 79–80; Elmer, "Medicine, Religion, and the Puritan Revolution," in French and Wear, eds., *Medical Revolution of the Seventeenth Century,* 10–45, esp. 15n. In arguing that New England's medical culture mirrored that of Old, Patricia Watson argues that minister-physicians were common in seventeenth-century England (*Angelical Conjunction,* 39–40). She also argues that economic necessity was the primary motive drawing ministers into dual professions in New England. See also Birken, "The Dissenting Tradition in English Medicine," *Medical History,* XXXIX (1995), 198.

22. Watson, *Angelical Conjunction,* 3, and table 3.1, "Medical Authorities Most Popular among Preacher-Physicians," 76–77. Watson's analysis of the surviving library inventories of nineteen of these minister-physicians shows them to have relied on many medical authorities, including Osvald Croll, Nicholas Culpepper, Van Helmont, William Salmon, Lazarus Riverius, and Hermann Boerhave. Mather claimed that England had commonly produced eminent physician-ministers, but not to the same degree as in New England (*Magnalia Christi Americana,* III, 151).

the mysteries of alchemy into their treatment regimes. Alchemical medicine might have been readily accepted in New England for the same reasons it was criticized overseas: many alchemists claimed to be the recipients of divinely granted revelations into healing. They offered their patients a spiritually empowered form of medical care whose potency was manifest in the strong physical effects produced by the new chemical medicines they administered.

While it is hard to evaluate fully, and impossible to quantify, the overall importance of alchemical medicine within the broader spectrum of colonial New England's medical culture, Winthrop and presumably the many other alchemical healers who can be identified there derived personal power and status from their knowledge of this scientific and spiritual art.

Alchemical medicine had gained widespread attention in early modern Europe through the medical theories of Paracelsus. In place of traditional Galenic medicine, which he considered pagan and theoretically bankrupt, Paracelsus offered a highly religious medical philosophy that advocated treating illnesses primarily with medicines derived from metals and minerals rather than plants and animals. Paracelsians asserted that medicine was fundamentally a spiritual occupation—that physicians were called by and received their power from God.

> The physician is he who in the bodily diseases takes the place of God and administers for Him, and therefore he must have from God that of which he is capable. For in the same way as physic is not of the physician, but of God, so is the physician's art not of the physician, but of God.

The Reverend John Davenport of New Haven recognized exactly that same relationship between God and the physician in a letter to Winthrop expressing appreciation for healing his son:

> Many hearty thancks being praemised, to God, and you; to God as to the principal efficient, who stirred up your heart, and guided your minde to pitch upon such meanes as his blessing made effectual; and to yourselfe, as to a blessed Instrument in God's hand, for our recovery, my sons especially, from that weaknes, and those great paines, wherewith he was lately and long afflicted, unto this measure of strength, whereby he was enabled to come into the publick assembly, the last Lord's day, to bless God the Author of all blessings upon your endeavours: which I pray, may be stil continued, for the good of many![23]

23. Allen G. Debus, *The Chemical Philosophy: Paracelsian Science and Medicine in the Sixteenth and Seventeenth Centuries* (New York, 1977), 124–126; Pamela H. Smith, *The Body of the Artisan: Art and Experience in the Scientific Revolution* (Chicago, 2004), 82–181; Ole Peter Grell, *Paracelsus: The Man*

In addition to asserting a direct connection between divine will and healing, Paracelsus propounded elementary theories and models of disease causation that challenged almost every aspect of traditional Galenism. He replaced the four Galenic-Aristotelian elements (earth, air, fire, and water) with three (mercury, sulfur, and salt). He argued that diseases were caused, not by imbalances of the four Galenic humors (blood, yellow bile [choler], phlegm, and black bile [melancholia]), but rather by problems with an entity he named the archeus, or life force, unique to every object in the universe. In opposition to the Galenic method of curing diseases by applying a medicine with qualities opposite of those thought to cause a disease (treating a warm, moist disease with a medicine whose qualities were cold and dry, for example), Paracelsus argued that medicines with qualities similar to the disease agent brought about cures. At the level of physical causation, almost every aspect of Paracelsianism seemed incompatible with or even antithetical to Galenism. Nevertheless, many leading alchemical authorities whose works were part of the medical libraries of New England's alchemical practitioners—men such as Osvald Croll, Johannes Wimpenaeus, Michael Sendivogious, and Basil Valentine—managed to reconcile or gloss over the theoretical differences between Galenism and Paracelsianism, positing a variety of etiological bridging mechanisms that allowed them to practice spiritually focused alchemical medicine within a Galenic, humoral context.[24]

As a result of the profusion of often idiosyncratic syntheses between Galenism and Paracelsianism, early modern alchemical medical theory as developed by post-Paracelsian alchemists was characterized by an astonishing syncretism. (Some historians argue that Paracelsianism was too loosely defined in the sixteenth and seventeenth centuries, and remains so today, to constitute anything like a defined movement.) Those alchemical physicians who continued to reject completely all aspects of Galenic medical theory

---

and His Reputation, His Ideas, and Their Transformation, Studies in the History of Christian Thought, LXXXV (Leiden, 1998); Henry E. Sigerist, ed., C. Lillian Temkin, trans., *Four Treatises of Theophrastus von Hohenheim, Called Paracelsus* (Baltimore, 1996), 15; Rev. John Davenport to John Winthrop, Jr., June, 14, 1666, MHS, *Collections*, 3d Ser., X (1849), 59–62.

24. On humoralism, see "The Humoral Body," in Trudy Eden, *The Early American Table: Food and Society in the New World* (DeKalb, Ill., 2008), 9–22. On the synthesis of Galenic humoralism and Paracelsian theory, see "The Humoral Paracelsian Body," in Margaret Healy, *Fictions of Disease in Early Modern England: Bodies, Plagues, and Politics* (New York, 2001), 18–49; Debus, *The Chemical Philosophy*, 136, 143. Other alchemists continued to follow the Paracelsian model of completely rejecting Galenism. Their rigid opposition to Galenic influences represented an extreme position, which has sometimes been represented as typical of all alchemical medical practice. Nevertheless, there was enough overlap between alchemical and Galenic medical therapies to allow many committed alchemical practitioners to blend the two schools' approaches, even while maintaining the fundamental superiority of alchemical medicine.

took an extreme position, which has sometimes been represented as typical of all alchemical medical practice. For many alchemical practitioners, however, there was enough overlap between their understanding of alchemical and Galenic medical theory to allow them to blend aspects of the two schools' approaches to healing even while maintaining that alchemical medical practice was fundamentally superior.[25]

While the nosological systems adopted by alchemists displayed an astonishing degree of syncretism and idiosyncrasy, two of the primary appeals of alchemical medicine to many practitioners, and certainly to many of New England's Puritans, were the intensely spiritual focus placed on both their medical research and iatrochemical medical practices, and the demonstrable effects of alchemical medicines. "Medicine is a serious, and hidden thing, I had almost said sacred," noted the New England-trained alchemist George Starkey, "nor doth it come to the knowledge of any, but by the special gift of the most high." Christian alchemists' unwavering insistence that medicine was fundamentally a spiritual calling and that their medications were divinely granted gifts (regardless of the theoretical basis behind their efficacy) resonated powerfully in a colonial culture that sought evidence of God's providence in all quotidian events.[26]

Christian medical alchemy shared in the enthusiasm for general reformation that characterized the seventeenth-century pansophic moment. Some alchemical healers hoped to be in the vanguard of a new divine dispensation: their newfound remedies, like the discovery of the New World, would herald the end of a period of great corruption and the approach of the millennium. Other alchemical healers did not believe that they were providing medical innovations as much as that they were reclaiming a spiritual birthright.

---

25. Steven Pumfrey, "The Spagyric Art; or, The Impossible Work of Separating Pure from Impure Paracelsianism," in Grell, ed., *Paracelsus: The Man and His Reputation*, 21–52. There were times and places when the tension between alchemical and Galenic positions flared into open conflict. In Commonwealth and Restoration England, for example, the opposition of alchemical extremists such as Noah Biggs and George Starkey to the medicine of the Royal College of Physicians sparked a heated pamphlet war and led to a brief move to set up a separate Alchemical College distinct from the Royal College of Physicians. For treatments focusing on the contention between the alchemical and Galenic schools, see Wear, *Knowledge and Practice in English Medicine*, 353–398; Allen G. Debus, "Paracelsian Medicine: Noah Biggs and the Problem of Paracelsian Reform," in Debus, ed., *Medicine in the Seventeenth Century* (Berkeley, Calif., 1974), 37–47; P. M. Rattansi, "Paracelsus and the Puritan Revolution," *Ambix*, XI (1963), 24–32, and "The Helmontian-Galenist Controversy in Restoration England," XII (1964), 1–23; Antonio Clericuzio, "From van Helmont to Boyle: A Study of the Transmission of Helmontian Chemical and Medical Theories in Seventeenth-Century England," *British Journal for the History of Science*, XXVI (1993), 303–334.

26. George Starkey, *Natures Explication and Helmont's Vindication . . .* (London, 1657), 200.

They believed God was allowing them to recover, through arduous and inspired labor at the chemical furnace, medical knowledge that had been lost to humanity at the fall of Adam. Adam, the first alchemical magus, had known all the secrets of alchemy. He had possessed the ability to produce both the alkahest that could cure all diseases and the philosopher's stone. Critics ridiculed the pursuit of the alkahest as a form of medical quackery and the search for the philosopher's stone as evidence of foolish greed. For many alchemists, however, such pursuits represented simultaneously spiritual and eminently practical quests—efforts to overcome through faith and science the physical corruption in both humans and metals that mirrored the spiritual corruption of fallen man.[27]

Although individual alchemists had no assurances of research success, the alchemical community as a whole produced an array of perceived medical advances that persuaded proponents their methods represented a vast improvement over traditional medical practice. Few alchemists ever claimed to have attained the panacea alkahest, but the lesser achievements of alchemical research—cures for specific diseases and medicines that improved upon traditional remedies—seemed remarkable in their own right. The concept of divinely granted medical advances found natural acceptance among the Puritans of New England, and Christian alchemists claimed to have achieved a plethora of them.

New medicines were not the only practical aspect of alchemical research that was inherently beneficial to patients in New England. In their quest to find the alkahest, alchemists gained a wealth of practical knowledge about distilling and refining that enabled them to produce more carefully refined traditional medications. Alchemists claimed that even their herbal medicines were more effective than the medicines produced by Galenic apothecaries. Simeon Partlicius underscored the contradiction between the outward appearance of the brightly packaged commercial Galenic medicines and their dubious healing qualities.

> When you look upon one of *Galens* Apothecaries Shops, you see fine painted Boxes and curious pots, that it would dazle your eyes to look upon them, they are so finely painted, That if there be a paradice upon Earth you would think it were there: yet in the inside is nothing but filth and the very Carkeises and Dung of all Medicines. . . . How can they ease the sick without calling the help of an Alchymist to resolve, seperate, and exhale what is obnoxious, thereby producing the hidden

27. Debus, *The Chemical Philosophy*, 96.

Natures of things for use (for God hath vayled the greatest and most wonderful things, that so he may stir up man to search after them.

Thomas Palmer, a New England alchemical physician, echoed Partlicius. "Then surely a true Physitian must be an Imitatour of Nature in the separation of all the excremoniall (worthless) and unprofitable part of his medicine," he wrote in his *Vade Mecum* ("Go with me"), a commonplace book recording his observations on diseases and remedies. "But how few have the Knowledge of the secret mysterys of physick."[28]

Many of England's elite doctors, the members of the Royal College of Physicians, eschewed manual preparation of medicines as a point of honor. They considered sweating at the hearth demeaning physical labor unworthy of their education or status. Those who let others prepare their medicines while they themselves remained essentially ignorant of refining processes were, from many alchemists' perspective, engaging in a kind of malpractice. Basil Valentine, whose *Triumphant Chariot of Antimony* might have been the source for Winthrop's heavy reliance on antimonial medications, claimed:

> A Doctor, who knows not how himself to prepare his own Medicines, but commits that Business to another . . . that good Man knows not what Medicines he prescribes to the Sick; whether the Colour of them be white, black, grey, or blew, he cannot tell; nor doth this wretched man know, whether the Medicament he gives be dry or hot, cold or humid; but he only knows, that he found it so written in his Books. . . . Good G O D, to what a state is the matter brought! what goodness of mind is in these men! what care do they take of the Sick! Wo, wo to them! in the day of Judgment they will find the fruit of their ignorance and rashness.

In New England, hands-on knowledge of how to fabricate medicines was advantageous for both the healer and those provided for. Even though apothecaries were present in New England from the earliest settlement and even though alchemical healers such as Winthrop purchased ingredients from them, the medicines he himself produced were the foundation of his superior medical reputation.[29]

28. Simeon Partlicius, *A New Method of Physick* . . . , trans. Nicholas Culpeper (London, 1654), 9; Thomas Palmer *The Admirable Secrets of Physick and Chyrurgery*, ed. Thomas Rogers Forbes (New Haven, Conn., 1984), 165, 167.

29. Basil Valentine [Basilius Valentinus], *Triumphant Chariot of Antimony* (London, 1678), 33. Starkey: "As for the preparation of Medicaments, that the Doctor little acquaints himself with it,

Many English defenders of traditional medicine—the medical doctors New England lacked—rejected fundamental tenets of alchemical medicine, though what emerges in any study of the period is what Andrew Wear has called "a picture of confused and overlapping identities, and paradoxically of sharp and polemical conflict over apparently well-defined views." Even though the Royal College of Physicians included chemical remedies in their 1618 *Pharmacopoeia* as *auxiliaries* to Galenic medicine, and had gone so far as to build a chemical laboratory and appoint a staff chemist in 1648, most of its members were said to have remained fundamentally Galenical and antialchemical in their outlook. To gain institutional respect for their practices and emphasize their differences from the members of the Royal College of Physicians, proponents of the new alchemical approaches cataloged an array of unjust criticisms the Galenic medical doctors employed against them. The Galenists, they said, mistakenly derided alchemists' enigmatic texts and the alchemical insistence on protecting their secret knowledge. They also argued that the effects produced by chemical medicines were evidence that they were too dangerous for human consumption. They called alchemical physicians mere medical amateurs and dismissed outright the idea of a universal panacea. Some even claimed such a medicine, if possible, must be diabolical in origin. The Paracelsian George Thomson defended the alchemical physicians:

> They have villified and slandered true Artists, perswading the World falsly, that Chimical Medicines . . . were dangerous, received an Empyreuma, or hurtful impession from the Fire; that they did either Kill or Cure in a short time; yea, some of them have been so impudently audacious to assert, That whosoever took Chymical Medicines, although he were cured for the present, yet in the revolution of a year it would cost him his life.

> So blinde and stupid are they in most needful Phylosophy, that what their Brains cannot conceive, they presently reject as Diabolical or impossible. Tell them of the *Alkahest,* or universal *Menstruum,* of *Lapis Chrysopeius, Lapis Butleri,* of a *Panacaea,* they will but deride and shout at it.

George Starkey expressed similar outrage at the criticism of the traditionalists:

---

his Theory consisting only in turning over of leaves, and his Practise in tossing of Pisse-pots and writing of Bils" (*Natures Explication and Helmont's Vindication,* 88).

And as for the dangerousnesse of my Medicaments, which I know they will insinuate; that is but a meer Bug-bear, by which ignorant people are frighted without cause, or ground, as the Jesuites are reported to affright their deluded Catholisks, by telling them that the *English* since the casting of the Popes Supremacy, are turned into Monsters, which those who know our Nation see to be but an invention.[30]

Such objections, reported at the height of the contest between the Galenists and Paracelsians, might or might not have reflected the views of the majority of England's licensed medical doctors, but they were never raised, at least publicly, in early New England. There, in the absence of a professionalized medical community and in the presence of a culture closely attuned to providentialism, godly alchemical physicians were praised for their powerful medicines, admired for their Christian service, and encouraged to search for Christian-chemical cures. That is not to imply that all physicians practiced alchemical medicine; undoubtedly, many male practitioners in New England followed exclusively Galenic and herbal approaches to healing. Women's household medicine, the first and most common form of medical treatment in New England, was almost exclusively herbal unless performed in consultation with an alchemical physician such as Winthrop. Yet, unlike England, where polemical debates between alchemists and Galenists ebbed and flowed, one is hard pressed to find even a single instance in which a New England physician, woman practitioner, or patient spoke out against the godly remedies provided by the alchemists. When alchemists' views came under public scrutiny, as occurred surrounding Robert Child's remonstrance in the 1640s, it was because of their politics and suspect confessional leanings, not their medical practices or their medicines. Indeed, a major factor helping young Winthrop transition from being perceived as an intruder into the Pequot country to becoming a much-honored and sought-after candidate for governor of Connecticut was his provision of alchemical medical services to so many of that colony's settlers.

HERMES CHRISTIANUS

It is in the context of New England's intensified belief in medical providentialism and its cultural predisposition to favor spiritually imbued medical

---

30. George Thompson, *Galeno-pale; or, A Chymical Trial of the Galenists, That Their Dross in Physick May Be Discovered* . . . (London, 1665), 17, 31–32; Starkey, *Natures Explication and Helmont's Vindication*, epistle to the reader, B–C; Wear, *Knowledge and Practice in English Medicine*, 428. Wear analyzes the power struggle between the proponents of Helmontian alchemy and the physicians of the Royal College in the 1660s (353–433).

treatments that we can understand the importance of alchemical medicine in New England and the leadership in medical treatment of John Winthrop, Jr. Of all Winthrop's many services to colonial New England, those for which he perhaps came to be most revered by ordinary New England women and men were his services as an alchemical healer.

Although Winthrop had been providing some medical care since his initial emigration to New England, his 1641–1643 journey to Europe marked a turning point in his approach to rendering Christian service through medicine. His exposure to Paracelsian physicians such as Augustus Tanckmarus in Hamburg and Johann Moraien and Johann Glauber in Amsterdam was formative. When he left Amsterdam and journeyed to the Hague to recover manuscripts and books stolen by Dunkirk privateers in the fall of 1642, his letter of introduction from the English diplomat Sir William Boswell described him as a "student of physique." Inspired, perhaps, by the examples of Moraien and Glauber, who each conducted voluntary-payment clinics for those in need of medical care as part of their alchemical service to God, Winthrop returned to New England with new medical knowledge, and, as soon as his new harbor town was established, he began to practice medicine on a much-expanded scale.[31]

When, exactly, he began to distribute his alchemical medicines widely is unknown, but the reputation of his alchemical cures preceded him to New London. One of the first acts he and the minister Thomas Peters did while laying out the New London townsite was to go to the fort of the Mohegan chief Uncas and provide medical assistance to many of the more than thirty warriors who had been wounded in a daylong battle with the Narragansetts. Soon thereafter, he received a letter from Roger Williams in Rhode Island requesting him to send an alchemical powder for his sick daughter. Williams's request makes clear that the value of Winthrop's medicines was already well established.

> If youre powder (with directions) might be sent without trouble I should first wait upon God in that way: however 'tis best to wait on him. If the Ingredients be costly I shall thanckfully account. I have books that prescribe powders etc. but yours is probatum in this Countrey.[32]

31. Winthrop had arrived in New England in 1631 with a surgical chest and a large supply of chemical glassware. Dunn, *Puritans and Yankees*, 81; Sir William Boswell to the Chevalier De Vic, Nov. 1, 1642, *Winthrop Papers*, IV, 362–363.

32. Thomas Peters to John Winthrop, May 1645, *Winthrop Papers*, V, 19–20; Roger Williams to John Winthrop, Jr., June 22, 1645, in Glen W. LaFantasie, ed., *The Correspondence of Roger Williams*

While supervising the array of tasks involved in settling the new town, Winthrop also undoubtedly assumed the duties of providing medical services to the residents of his new community. It is equally likely, based on the number of Indian patients that are listed in his surviving medical account books (from 1657 to 1669), that he served as a physician to at least some of the three hundred to five hundred Pequots under Robin Cassacinamon, next to whom the English town was located. Patients in other places received treatment, too, such as the wife of Edward Hopkins of Hartford, who received medicinal waters from Winthrop, and Samuel Symonds of Ipswich, who sought medical advice regarding his wife's stomach pain. The level of demand for Winthrop's services in the late 1640s, however, was only a prelude to what was to come. In the early 1650s, as the hard labor of settlement subsided and the crisis over Pequot custody was resolved, the demand for Winthrop's medical assistance increased dramatically.[33]

Between 1650 and 1654 (based on surviving medical records) Winthrop received requests for medical advice and medicine from patients in Hartford, Windsor, Boston, New Haven, Farmington, Ipswich, Springfield, Southampton, Stamford, Stratford, Southold, Saybrook, Rehoboth, Branford, Wethersfield, Watertown, Middletown, and Pawcatuck. The requests came from both the elite, such as William Goodwin of Hartford, whose wife was suffering from paralysis, and the desperately poor, such as Jonathan Sergeant, Jr., of Branford, who suffered a malady that left him heavy and benumbed and who sent thanks through the literate Abraham Pierson to the physician who "had compassion on him in his misery." Most of Winthrop's medical correspondence came from men—a reflection of gendered differences in the ability to write and male dominance within the social hierarchy. Of more than sixty surviving written requests to Winthrop for medical assistance during the period surveyed, male correspondents outnumbered female correspondents by a ratio of twelve to one. Winthrop did receive correspondence directly from women, such as Sarah Rood, who wrote him about her prolapsed uterus, and Katherine Doxey, who reported on her "wasted and Consumed" infant

---

(Providence, R.I., 1988), I, 219. *Probatum:* "A thing proved; a demonstrated conclusion or fact; *esp.* a means or remedy that has been tried and found efficacious; an approved remedy" *(Oxford English Dictionary).*

33. John Winthrop, Jr., "Medical Account Books," in Winthrop Family Papers Bound Volumes (Boston, 1657–1669), vol. 20a (1657–1660), vol. 20b (1660–1669), Massachusetts Historical Society, Boston; John Winthrop, Jr., "Transcript of Medical Account Books, 1657–1669," Connecticut Historical Society, Hartford; Edward Hopkins to John Winthrop, Jr., May 5, 1647, *Winthrop Papers,* V, 156, Samuel Symonds to John Winthrop, Jr., Feb. 24, 1648, 199.

who was "nothing but skin and bones," but correspondence from women was relatively infrequent.³⁴

On the other hand, though men did the writing, they often were requesting medicines for their wives or children, and occasionally one can hear the voice of a wife in the background telling her husband what to say. Men regularly served as the intermediary between female healers and physicians in face-to-face relationships, so, in this regard, the predominance of men in Winthrop's medical correspondence is also a translation into text of an existing social practice. Women's illnesses were issues of concern twice as often as men's illnesses (thirty versus fifteen). This may reflect a difference in the general state of health between women and men, based on the twenty- to thirty-month childbearing cycle of colonial women. When children's health was at issue, girls were only slightly more likely to be the prospective patients than boys.³⁵

The problems for which people sought Winthrop's medical help ran the gamut from relatively minor problems (though serious enough to warrant long distance consultation) to almost certainly incurable diseases. John Haynes of Hartford wrote about his wife's "soare paine one her backe, betweene her shoulders, streming downe her left Arme," for which he requested plasters. Edward Wigglesworth of New Haven, father of Michael Wigglesworth, the future alchemical physician, minister, and popular poet, was encouraged by his son's description of a visit to Winthrop to write and ask him for help with a condition now known as amyotrophic lateral sclerosis, or Lou Gehrig's disease. After twelve years of living with the disease the senior Wigglesworth reported:

> It is come up to the head, in so much that I have not ability to move one joint in my body, save onely my neck a little but though all motion is quite gone, yet sense remaineth quick in every part. And

---

34. William Goodwin to John Winthrop, Jr., Dec. 16, 1651, *Winthrop Papers,* VI, 153, Abraham Pierson to John Winthrop, Jr., Jan. 29, 1653, 243–244, Katherine Doxy to John Winthrop, Jr., Aug. 10, 1652, 214–215, Sarah Rood to John Winthrop, Jr., Spring 1653, 271–272. Kenneth A. Lockridge found that fewer than two women in six in early-seventeenth-century New England could sign their names. More than 60 percent of men were literate. *Literacy in Colonial New England: An Enquiry into the Social Context of Literacy in the Early Modern West* (New York, 1974), 39.

35. John Davenport to John Winthrop, Sept. 19, 1654, *Winthrop Papers,* VI, 426–428; Rebecca Jo Tannenbaum, "A Woman's Calling: Women's Medical Practice in New England, 1650–1750" (Ph.D. diss., Yale University, 1996), 90–91; Tannenbaum, *The Healer's Calling: Women and Medicine in Early New England* (Ithaca, N.Y., 2002), 60–62; Laurel Thatcher Ulrich, *Good Wives: Image and Reality in the Lives of Women in Northern New England, 1650–1750* (New York, 1982), 135.

thorough the goodnes of God, my understanding, memory, with my eyesight and hearing remaine untouched: neither is my stomach apt to be offended with food, but a small quantity suitable to my weaknes it can close with. I do not find any sicknes within save onely the paine of weariness thorough sitting and lying. I am not sensible of any obstruction in my inward parts. My flesh is much fallen which began first in my lower parts, and now is in my upper parts: but my complexion remaineth pretty ruddy in my face. My age is about 49 yeares.

Wigglesworth hoped Winthrop "would seriously consider this my condition, and if it shall please God to discover to you any cranny of hope of any degree of cure; that you would be pleased to send mee your thoughts in a few lines."[36]

In most cases, Winthrop served the function of referral physician. Other healers directed patients to him after they had been unable to treat an afflicted person's condition. A family member or the patients themselves, concerned at another healer's ineffectiveness, might decide to seek Winthrop out on their own. William Peck, a barber-surgeon in New Haven, wrote Winthrop requesting medicine to help his wife after a midwife had confirmed that a growth in her abdomen was the source of her extreme pain. Samuel Stone, the minister at the Hartford church, contacted Winthrop about his infant son's yellow jaundice. The boy had been treated by Mrs. Hooker (widow of the founding minister of Hartford), who gave the boy saffron and turmeric, two yellow herbs that, because their yellow color corresponded to the yellow skin of the jaundiced boy, were supposed to draw out the jaundice through a sympathetic action. The boy had improved for a short time but then had become sick again. Stone had turned to Mrs. Haynes, wife of the governor, who had given the baby three-quarters of a grain of Winthrop's purging powder, and he slept well but seemed to be worse in the morning. Stone wanted Winthrop's advice on whether the boy should take more of Winthrop's powder and asked him to send his counsel by an Indian runner.[37]

---

36. John Haynes to John Winthrop, Jr., May 31, 1653, *Winthrop Papers*, VI, 296-297, Edward Wigglesworth to John Winthrop, Jr., July 18, 1653, 310-312. (The diagnosis of ALS was made by Norman J. Selverstone, M.D., 312 n. 3.)

37. Winthrop was consulted on cases that had formerly been treated by midwives, cunning women, and other physicians: William Peck to John Winthrop, Jr., Sept. 15, 1651, ibid., 141-142, Samuel Stone to John Winthrop, Jr., Feb. 28, 1653, 256-257, William Andrews to John Winthrop, Jr., Mar. 24, 1653, 272-273, John Pynchon to John Winthrop, Jr., May 22, 1654, 383, Elizabeth Chute to John Winthrop, Jr., Oct. 10, 1653, 341. Other persons, unable to come or too ill to travel, sent pleas for diagnoses, medicine, and advice. Multiple requests came to Winthrop from patients in Hartford, New Haven, Farmington, Windsor, Ipswich, Pawcatuck, Milford, Boston, Saybrook,

Such referrals provide insight into some of the categories of medical practitioners found in colonial New England. Barber-surgeons like William Peck had practices that were generally limited to surgical procedures and dentistry, employments considered to require less medical knowledge and skill. Surgery was not limited to the barber-surgeons: Winthrop himself performed surgery, as did female healers. Joanna Swift, a healer of Sandwich who wrote Winthrop seeking return of a runaway servant she believed had made his way to New London, told Winthrop, "If he should deny his name he maybe knowne by his wrist wher I took out aboane when I cured it." Most seventeenth-century women were medical practitioners to some degree; they served as caretakers of family health within the home and participated in women's community medical networks. Those who developed a reputation for exceptional medical skill or who possessed special talents, as Joanna Swift seems to have, became recognized as doctresses. They treated chronic and difficult illnesses and filled a second gradation in the medical hierarchy. The elite occupied a special place in healing by virtue of their position as community leaders. Women like Susanna Hooker, widow of the Reverend Thomas Hooker, a founder of Hartford, and Miss Haynes, daughter of the Connecticut governor John Haynes, shared with their husbands and fathers a social obligation to render medical services to their communities. One of the responsibilities of colonial leadership was demonstrating medical charity. Both husbands and wives rendered these services, though elite women were more commonly found at the bedsides of the ill than were their husbands.[38]

Correspondents made efforts to offset the disadvantages of seeking aid at a distance by providing thorough and accurate descriptions of the maladies afflicting their family members. Richard O'Dell's description of his daughter Lydia's paralysis provides a history of the progression of the disease as well as a complete description of her current state.

> The cause is it hath pleased the Allmightie to laye his afflickting hand upon a child of mine by a disease which most do thinke to be the palsey. The child is about 5 years of age. Shee hath ben aflicted with an ichey humer all over her body especially in her head ever since she

---

and Long Island. See, for example, Richard O'Dell to John Winthrop, Jr., Nov. 16, 1652, 229–230, Richard Smith to John Winthrop, Jr., Feb. 21, 1653, 251–252, William Peck to John Winthrop, Jr., Sept. 15, 1651, 141–142, Samuel Stone to John Winthrop, Jr., Feb. 28, 1653, 256–257.

38. Joanna Swift to John Winthrop, Jr., 1650, ibid., 7–8; Tannenbaum, *The Healer's Calling*, 22–84, 114–133; Mather, *Angel of Bethesda*, ed. Jones, 289; Watson, *Angelical Conjunction*, 40–41; Laurel Thatcher Ulrich, *A Midwife's Tale: The Life of Martha Ballard, Based on Her Diary, 1785–1812* (New York, 1990), 63.

was aquarter old [1 year] with great Cirnells [enlarged glands] in her neck which hath allways continued till this tyme ebing and flowing some tymes biger and then againe leser. As the humers run som have thought it to be the kings evell. Shee hath been neer death severall tymes with the striking in of the humers. About the begining of October last her head began to run exceedingly more than ever it did before. So it continued for the space of 3 weekes tyme and stoped by degrees but before it was quit stoped shee began to be very plesent and joccondt and spoke very plesently. The same morning that shee was thus struken about the 20 day of october last about 11 or 12 as shee was sitting by the fier shee rise up hastily to fech some vittells shee fell downe upon on side. Been set up againe she was nether able to stand nor speak and so she hath continued till this Instant haveing lost the use of the right side from the head to he foot. She hath gained lately som strength in her leg. She can stand with alitell helpe and is able to sit up right with her body with out support but cannot sepeak a word. She is sencable of feeling through out all her laime part. It is somthing swelled and very sore in the flesh and hath an exceeding payne in ther other part as in her leg and arme I meane that which shee hath use of. She hath allso hath a stopage in her throat or below the swallow when shee drinks that she can scarsely induer to drink or take any Liquid matter.[39]

Such detailed descriptions were necessary for attempting to make a meaningful diagnosis, but only of limited effectiveness in serious cases such as Lydia O'Dell's paralysis. As Winthrop wrote back to her father with a wide variety of suggestions that might offer some relief, "The cure depends upon the knowledge of the right cause, and not only that but the constand and due aplication of such things as may conduce thereto, which is difficult to doe at a distance."[40]

The desire to receive "the constand and due aplication" of the treatments that would help them recover from their maladies led patients from all over New England to make the long, sometimes painful journey to New London to be where Winthrop could provide them continuous medical care. Such journeys imposed a real burden on the patients and their families, yet the desire to seek relief from chronic illness or unremitting pain was sufficient to

---

39. Richard O'Dell to John Winthrop, Nov. 16, 1652, *Winthrop Papers*, VI, 229–230. The "King's Evil" is the name given to scrofula, which is characterized by enlarged and degenerating lymph glands.

40. John Winthrop, Jr., to Richard O'Dell, Nov. 27, 1652, ibid., 230–231.

make the visit to New London a desirable option. Roger Newton of Farmington, writing about his wife's chronic swelling in her body and legs and shortness of breath, listed the difficulties that the family would face in her absence. His commitment to his wife's recovery underscores the emotional cost of separation that would also be a burden of her residency at Winthrop's plantation. He asked Winthrop to

> acquaint mee with your apprehension about my wives disease and whether it may not be necessarie to convey her to your plantation where shee might bee in view and phisick applied more sutable to all possibilie variations in her bodie then by absence can be administred or directed which though it be a hard thing to weane her child before a convenient time leave a famili of small children and the ordering of the affaires of a familie besides our absence one from another in all which respects I know my wife will be backward to yeeld . . . unless you expresse your judgment for it yet to mee I could look over all other difficulties as not considerable in comparison of my wives health.[41]

Newton's wife was only one among many suffering patients for whom New London offered the hope of physical relief. Patients from Massachusetts, Connecticut, Rhode Island, and Long Island undertook grueling pilgrimages to Winthrop's remote plantation for physick. The Winthrop correspondence for 1653 and 1654 shows ten patients in residence at the plantation, with requests from at least five more to come for treatment, and this probably significantly underrepresented New London's patient population, who came for an array of acute and chronic illnesses. The descriptions of their conditions indicate the degree to which illnesses, low-grade infections, dental caries, worm infestations, and a variety of serious physical discomforts were pervasive factors of early modern life, whether in Europe or America. New Englanders—even though their longevity was greater than their European counterparts'—endured a host of chronic conditions, which reminded them that the corruptions of the flesh were all too real.[42]

---

41. Roger Newton to John Winthrop, Jr., June 15, 1653, ibid., 299–300.

42. Elaine Forman Crane directs our attention back to the time when "pain controlled people." See "'I Have Suffer'd Much Today': The Defining Force of Pain in Early America," in Ronald Hoffman, Mechal Sobel, and Fredrika J. Teute, eds., *Through a Glass Darkly: Reflections on Personal Identity in Early America* (Chapel Hill, N.C., 1997), 370–403.

Towns of residence represented among patients include Hartford, New Haven, Windsor, Farmington, Ipswich, Springfield, and Southhampton, Long Island. John Haynes to John Winthrop, Jr., Sept. 12, 1652, *Winthrop Papers*, VI, 220–221, John Pynchon to John Winthrop, Jr., May

Daniel Clark of Windsor, Connecticut, sought in midwinter of 1653 to bring his four-year-old son and one-year-old daughter to New London to receive treatment for extremely abscessed teeth.

> On the outside of the upper jaw betwixt his lipp and his goome just uppon the foure decaying teeth there would foure smal pimples arise very redd in coulour and after a day or two they would breake and coruption would issue forth and thus they continue for about two yeares space.[43]

Clark had exhausted all the local Windsor medical resources for assistance and had even tried to remove his son's rotted teeth with a small silver hook and a pair of pincers, but to no avail. Clark wanted to bring both his children to Winthrop if "there may be any meanes used for the preservation of his goome and getting the teeth out ... And for preventing the same disaster in my yongest child," and he was willing to pay for Winthrop's services in silver. William Goodwin of Hartford sought permission to bring his daughter to New London "in the fittest Season" for treatment of some chronic but unspecified illness that was already known to Winthrop. When Goodwin made his request, the daughter of William Andrews, the town clerk of Hartford, was already at the New London plantation receiving treatment. In March, he wrote asking Winthrop to send her home by means of a pinnace up the Connecticut River. Andrews was one of the many people who sought medical care from Winthrop but could not afford to pay him. Andrews thanked Winthrop for his "love in the cost you have been at and means you have used for the recovery of the health of my daughter" and, acknowledging his family's indebtedness, regretted that they "cannot recompense it as is meet wee should." A different William Andrews, "a poore man home the Lord hath bene pleassed to veset with a distemper: a fluckes and loosnes: one yeare and 7 months for the most part," requested Winthrop to send him some medicine, or "if you would have me com to you if the Lord enable me I pray ser let me understand your mind in it." Elizur Holyoke of Hartford didn't ask. He sent his wife and son to New London with a note saying he did not doubt Winthrop's concern for his son's health and "therefore both Mother and child are come to yow a while: we much desire if it maybe Gods good pleasure Some benefitt may accrue to the use of meanes or at least that God would help us with Contentedness and

---

22, 1654, 383. Mrs. Pynchon arrived accompanied by the wife and ill son of Elizur Holyoke, also seeking medical aid. During her lengthy course of treatment, she learned from her husband of the death of her infant son, William, back in Springfield. John Pynchon to John Winthrop, Jr., June 20, 1654, 393–394, July 26, 1654, 410–411, before and on Sept. 12, 1654, 422, 423–424.

43. Daniel Clark to Hugh Caulkins, Feb. 11, 1653, ibid., 245–246, Mar. 17, 1653, 263–264.

quietness under the cross." He hoped Winthrop would find his wife, who was "wholly a stranger at your plantacon . . . some Convenient house neere yours to abide awhile."[44]

New London became a hospital town, where patients, including Governor Haynes's wife, the wife of Springfield's founder John Pynchon, and family members of other notables, resided through extended periods of treatment, sometimes lasting up to a year. Along with the elite came the poor, whom Winthrop refused to turn away. Housing became so scarce that, when New Haven's minister John Davenport sought to visit New London in 1653 to consult Winthrop about a life-threatening medical condition he faced, there was no available lodging.[45]

Patients arrived in numbers that strained the resources of both the physician and the town. Both the pansophic agenda and the Rosicrucian tradition held that alchemical reformers were obligated to use their knowledge to provide Christian service. Richard Blinman, New London's minister, observed the drain on Winthrop and his personal resources and wrote to the Reverend Samuel Stone of Hartford seeking help addressing the problem. Stone replied: "I have spoken to some, of that which you write concerning Mr. Winthrop. It is certain it will be an insufferable burden to him, onlesse some way be taken to prevent it. It is evident to me, that if poore men want physick, and are not able some way or other to make allowanc The towne where they live must supply them." John Wilson of Boston sought to interest Winthrop and the Connecticut authorities in providing incentives to attract another physician, William Snelling of Boston, to New London to help alleviate Winthrop's caseload. This never came to fruition, and Winthrop remained heavily burdened with those seeking his medical services.[46]

Several factors account for why so many people were prepared to endure severe hardship in order to obtain Winthrop's medical services. His social and political status as a scion of New England assuredly reinforced his medical

---

44. Daniel Clark to Hugh Caulkins, Feb. 11, 1653, ibid., 245–246, William Goodwin to John Winthrop, Jr., Feb. 26, 1653, 253, William Andrews of Hartford to John Winthrop, Jr., Mar. 11, 1653, 260, William Andrews of New Haven to John Winthrop, Jr., Mar. 24, 1653, 272–273, Elizur Holyoke to John Winthrop, Jr., May 3, 1654, 381.

45. Ronald Sterne Wilkinson, "John Winthrop, Jr. and the Origins of American Chemistry" (Ph.D. diss., Michigan State University, 1969), 212. William Peck of New Haven apologized for having nothing fit to send in payment for his wife's treatment, enclosing women's clothing and a pair of shoes. William Andrews to John Winthrop, Jr., Mar. 11, 1653, *Winthrop Papers*, VI, 260, William Peck to John Winthrop, Jr., Dec. 11, 1653, 350–351, John Davenport to John Winthrop, Jr., Aug. 20, 1653, 322–323. Winthrop subsequently traveled to New Haven to attend Davenport.

46. Samuel Stone to Richard Blinman, June 12, 1653, *Winthrop Papers*, VI, 298–299, John Wilson to John Winthrop, Jr., Feb. 2, 1654, 357.

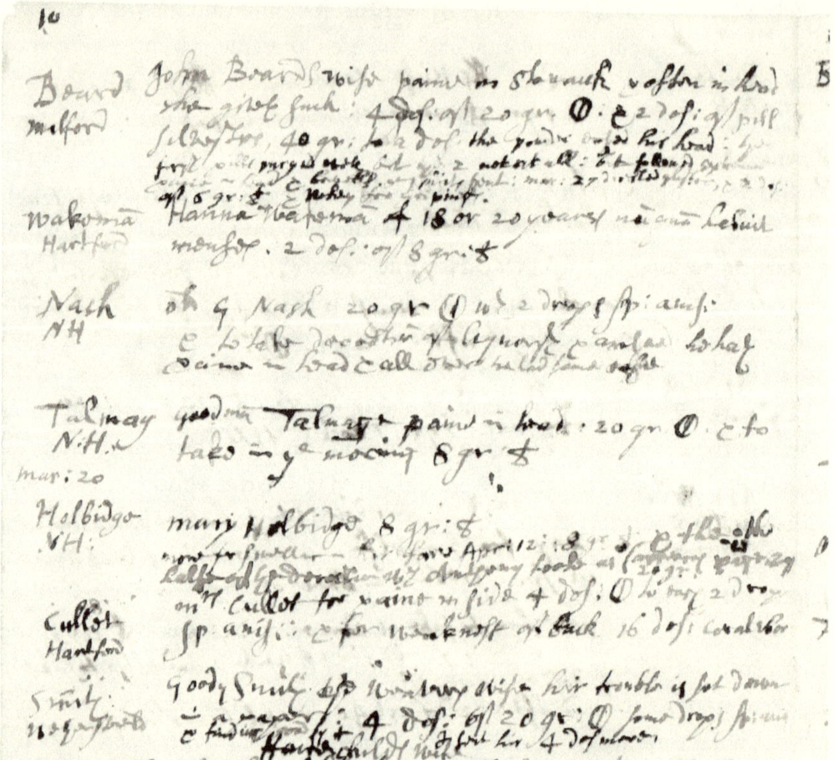

FIGURE 6. Medical Records, March 18, 1656, by John Winthrop, Jr. Manuscript, Winthrop Family Papers, Massachusetts Historical Society, Boston. In a two-day period, Winthrop treated patients from Milford, Hartford, New Haven, Wethersfield, and Stratford. *Courtesy of the Massachusetts Historical Society*

reputation and lent additional credibility to the alchemical medicines he provided. The Winthrop name carried a special authority and gravitas in New England, despite the suspicion it produced in Connecticut's leaders in the early days of New London's settlement. To be treated by Winthrop was to receive medical care from one of the most important and most honored men in all New England. Over time Winthrop's status increased, as he became Connecticut's governor in 1657 and a member of the Royal Society in 1661. "At the patient's bedside," Richard D. Brown has noted, "the chief distinction [between elite and popular medicine] might be social rather than medical."[47]

A second factor reinforcing Winthrop's medical reputation was his personal style. Winthrop, by almost every account, was a gentle, affable, and likable human being. People on three continents reveled in his company; patients, too, assuredly benefited from what Robert Black has called "the sickroom influence of a vibrant and kindly man." John Bishop of Stamford wrote of Winthrop's benevolent nature, "You are caried with a delight of doing good and lay out your selfe that way."[48]

A third factor was his commitment to providing medical charity. Although providing medical care at the level he did strained his resources at times, Winthrop never made an occupation of medicine, and he never obligated people to pay him. As a result, his example of serving God through medical care gained extra credibility. People paid Winthrop in all sorts of ways, and undoubtedly he priced his services to those who could afford them at a level equal to their perceived value. Those who could not pay were not asked to pay, however; nor were they turned away or, from what the existing records reveal, given a different level of care. Cotton Mather's assertion that the poor flocked to Winthrop as if he were the Angel of Bethesda focuses on one of the important distinctions between Winthrop and many other healers. It was a distinction that reinforced his own reputation and the reputation of his alchemical medicines.

Winthrop's alchemical medicines might have been the primary force behind Winthrop's preeminent reputation, for they often provided in his absence ready and direct confirmation of God's particular blessing on his means. Winthrop's medical account books list the type of medicines Winthrop prescribed and the frequency with which he prescribed them, and he falls firmly in the camp of the medical alchemists. Most of the medicines Winthrop issued to his

---

47. Brown, "The Healing Arts in Colonial and Revolutionary Massachusetts," in *Medicine in Colonial Massachusetts*, 43.

48. Robert C. Black III, *The Younger John Winthrop* (New York, 1966), 169; John Bishop to John Winthrop, Jr., Mar. 2, 1657, Winthrop Family Papers, Mss. W.11.11; Wilkinson, "John Winthrop, Jr. and the Origins of American Chemistry," 211–212.

patients were chemical, as opposed to herbal, medicines. Almost half of his prescriptions were for two remedies, antimony, which was actually antimony sulphide, and nitre, or common saltpeter. Antimony and saltpeter, powerful purgatives, were the foundations of Winthrop's medical practice. Both were alchemical products that tested the abilities and purity of the alchemist, even as they promised special rewards to the patient. Unlike Johann Moraien, who claimed to have knowledge of forty-two antimonial medicines, Winthrop used six antimonial preparations: crocus antimonii, cerusa antimonii (a diaphoretic, that is, sweat-inducing antimonial preparation); white antimony (the oxide); sulphur auratum antimonii (an acetate); and antimony of copper. Each of these was a powerful cathartic and potentially dangerous. Basil Valentine, whose *Triumphant Chariot of Antimony* provided both spiritual and intellectual foundations for medical reliance on the chemical, recognized antimony's dangers but saw it also as a particularly powerful and divinely blessed source of restoration. He argued that alchemical research into the qualities of antimony ought to be alchemists' first priority; it

> ought above all things to be sought, that (being brought to Light) the wonderful Works of our *GOD* may be made manifest, and the Glory given to him, with great thankfulness. It is not to be denied, but that more of Riches and Health may be found in it than either you all, or I my self, can believe: for I profess my self no other than a Disciple in the Knowledge of Antimony.

Valentine argued that a vivifying spirit in antimony affected whatever it came in contact with in the same invisible way that magnetism affected iron. Although antimony in its unpurified form was "Venome and a most swift poyson," it could become, through the work of the faithful alchemist, "voyd of Venome and a most excellent Medicine." Theodore Kerckringe also praised antimony as a beneficial medicine. "Of Antimony I can affirm, that being duly prepared it is as harmless a Medicine as *Cassia* or *Manna*."[49]

Antimony became pure medicine when the alchemist was able to separate its impurities from its active healing agent. Not everyone could do that; but, for those whom God enabled to understand the secrets of antimony preparation, great things were in store. Valentine claimed: "No Man unto this day could ever experience all its Virtues. We have seen many of its Effects, and many new Effects are daily found by curious Searchers, yet many more remain unknown." In relying on antimonial medicines, Winthrop was assert-

---

49. Wilkinson, "John Winthrop, Jr. and the Origins of American Chemistry," 194–212; Valentine, *Triumphant Chariot of Antimony*, 8, 19, 20.

ing his own abilities as an alchemist and providing his patients with divinely blessed medicine of remarkable powers. Similarly, Paracelsus, Robert Fludd, and Johann Glauber had all associated nitre with the vivifying spirit, and it is natural that Winthrop would, in coordination with his use of antimony, focus on the medicinal properties of saltpeter.[50]

Winthrop prescribed antimony in doses of up to ten or more grains for adults; nursing infants received a single grain. Nitre was given in doses of up to twenty grains. What made these medicines so essential to Winthrop's medical practice was that they were adaptable to a number of medical needs. The forces within them, because they were fundamental, could be applied against a wide array of medical conditions. Thus, Winthrop prescribed antimony as a vermifuge, an antidiarrheal agent, and an eyewash, among many other uses. Nitre was applied to problems from toothache, to stomach problems, to urinary blockages.[51]

Winthrop combined antimony and nitre into a medicine that became identified with the Winthrop family. Rubila, according to Oliver Wendell Holmes, who found the recipe for this medicine in the medical account books, was composed of four grains of antimony with twenty grains of nitre and a little salt of tin added, with some rubifying ingredient to give it the distinctively red color by which rubila became known. (Rubila was not John Winthrop's most frequently prescribed medication, or even nearly so, but, because it was widely distributed by his sons after his death, many historians incorrectly assumed that it was Winthrop's primary prescription.)[52]

Winthrop prescribed many other alchemical medicines, though not nearly as frequently as he did the antimonial or nitre medicaments. Depending on the illness of his patient, Winthrop would prescribe sal prunellae, alum, burnt alum, sal ammoniac, blue vitriol, iron, and iron derivatives such as aqua martis, distillation of vitriol, flowers of sulphur, balsam sulphuris, and mercury dulcis (or calomel). He prescribed the most primal of alchemical elements, salt; and the purest, gold. Winthrop also made use of a variety of chemical plasters.[53]

While serving a patient base estimated to eventually included nearly half the population of Connecticut, Winthrop also pursued alchemical research

---

50. Valentine, *Triumphant Chariot of Antimony*, 20.
51. Wilkinson, "John Winthrop, Jr. and the Origins of American Chemistry," 196.
52. Oliver Wendell Holmes, "The Medical Profession in Massachusetts," in Holmes, *Medical Essays, 1842-1882* (Boston, 1911), 331-335; Wilkinson, "John Winthrop, Jr. and the Origins of American Chemistry," 198-199; Black, *Younger John Winthrop*, 170; Walter R. Steiner, "Governor John Winthrop, Jr. of Connecticut as a Physician," *Johns Hopkins Hospital Bulletin*, XIV (1903), 294-302, esp. 301; Dunn, *Puritans and Yankees*, 82.
53. Wilkinson, "John Winthrop, Jr. and the Origins of American Chemistry," 204.

into the alkahest and other cures. He was especially interested in the weapon salve, a sympathetic magic medicine that was a running subject of debate among seventeenth-century intellectuals. This occult balm was thought to enable its possessor to heal wounds at a distance, not by treating the wound itself, but by rubbing the ointment on the sword that had inflicted it. The essential benefit of such a medication was, of course, its ability to heal at a distance, something a hard-pressed frontier physician such as Winthrop would have found of immense value. Proponents of the salve, including Paracelsus, Fludd, and Winthrop's friend Sir Kenelm Digby, debated whether it worked through magnetism or by capturing the sympathetic harmony that existed between the macrocosm and the microcosm. Critics claimed that, if it worked at all, it could be only as a result of diabolic intervention.[54]

As far-fetched as twenty-first-century readers find the concept of the weapon salve, it is important to remember that Winthrop's interest in the cure was deeply pragmatic. The weapon salve as described by some authors could cure not just wounds but diseases as well. Given the degree to which he was continually solicited by sick people from distant plantations and the limitations that trying to diagnose and treat illnesses from afar imposed, finding a medicine that could heal from a distance would have been a remarkable asset. Winthrop possessed many books on sympathetic cures and borrowed another from Sir Henry Moody of Gravesend, Long Island, Nicholas Papin's *De pulvere sympathetico dissertatio*. Given the caseload and the crowding at New London, it is understandable that Winthrop might make such a medicine a focus of his research.[55]

In special cases, Winthrop sought to develop new medicines for specific clients' maladies. New Haven's minister John Davenport, ready to embark for England to seek relief from a life-threatening renal-urinary condition, canceled the trip after receiving assurances that Winthrop would try to prepare "a special Arcanum" for him.[56]

Like most alchemical physicians, Winthrop also made use of many herbal and animal remedies. His prescriptions for such medicines derived from more than forty different herbal ingredients, including guaiacum, china root, cyclamen, oregano, sarsaparilla, birch, parsley, verbena, wormwood, elder leaves, unguent of elder, raisins, turpentine, anise, saffron, aloes, horseradish, senna, nutmeg, rhubarb, ointment of tobacco, betony, agrimony, mugwort, sage, roses,

---

54. Dunn, *Puritans and Yankees*, 83; Thomas, *Religion and the Decline of Magic*, 190.
55. Thomas Adams to John Winthrop, Jr., Mar. 5, 1654, *Winthrop Papers*, VI, 368.
56. *Arcanum:* "One of the supposed great secrets of nature which the alchemists aimed at discovering; *hence*, a marvellous remedy, an elixir" *(Oxford English Dictionary)*. John Davenport to John Winthrop, Jr., Sept. 11, 1654, *Winthrop Papers*, VI, 419–420.

and turnips. From the animal kingdom, Winthrop made relatively frequent use of coral and ivory. Other faunal ingredients include deer horn, kermes, spermaceti, ambergris, castor, pearls, and millipedes. Winthrop collected mice, probably in order to use their feces, which were a cathartic and diuretic. He also made use of unicorn (probably narwhal) horn and recommended seahorse pizzle (the penis of a seahorse) as an ingredient to treat kidney stones.[57]

Diseases, and patients' beliefs about the effectiveness of their treatments, are located within social systems of meaning, deeply informed by culture as well as physiology. Given our modern understanding of the limited physical therapeutic value of most early modern medicines, it is particularly important to locate Winthrop's alchemical medicines within the cultural expectations of New England's Puritans, in order to understand their widespread popularity.

Compared to most other medicines, Winthrop's antimony and nitre compounds produced strong purgative effects. Patients who took them frequently vomited profusely and emitted numerous sets of stools. They were careful to record the number of times they purged both above and below for future reference. Because of the humoral views of physiology that underlay most healing, purging, the desired effect of "taking physick," was the fundamental characteristic by which most seventeenth-century English people measured the effectiveness of a medicine. Medicines that produced multiple vomits or bowel movements were particularly well regarded. The alchemical physician Thomas Palmer's *Vade Mecum* helps us understand why purging might have been seen as the first form of medical treatment. "Observe by the way that ordinarily diseases begin in the Stomach," Palmer wrote, "and in the beginning may be easily expelled either by Vomit or Siege." Ridding the body of corruption through purges was also symbolically and metaphorically consistent with Puritan conceptions of spiritual salvation. The physical corruptions of the body mirrored the corrupt nature of the human condition. Just as spiritual corruption enervated faith, physical corruption enervated the body. The alchemist, by providing chemical medicines that had been revealed through divine inspiration and that were purified of their own physical corruptions, brought a divinely sanctioned purity into the body that helped eject the body's physical corruption just as divine grace cleansed the soul of the saint.[58]

---

57. Wilkinson, "John Winthrop, Jr. and the Origins of American Chemistry," 207–211; Henry Wolcott to John Winthrop, Jr., July 11, 1653, *Winthrop Papers*, VI, 306–307; John Winthrop, Jr., to Samuel Hartlib, Aug. 25, 1660, Hartlib Papers [32/1/6A–7B], University of Sheffield.

58. Lucinda McCray Beier, *Sufferers and Healers: The Experience of Illness in Seventeenth-Century England* (London, 1987), 169–170; Palmer, *The Admirable Secrets of Physick and Chyrurgery*, ed. Forbes, 72; Charles Lloyd Cohen, *God's Caress: The Psychology of Puritan Religious Experience* (New York, 1986), 205–206, 218.

Many New Englanders saw alchemical medicines as simultaneously physically and spiritually powerful. The Rhode Islander Samuel Gorton captured both those qualities in a letter he wrote to Winthrop thanking him for medicine Winthrop sent to treat a recent affliction. Gorton was filled with admiration "that a thing so little in quantity, so little in sent, so little in taste, and so little to sence in operation, should beget and bring forth such effects." "Whether it may be a cure, God knoweth, or an intermission," Gorton wrote, "it is no lesse marvellous." Gorton recognized God as the source of healing but saw Winthrop as God's agent, administering his wisdom. "The Lord can make it a cure . . . let him do as seemes good in his eyes. He hath given me a bodily experiment of his power, together with the administration of wisdom given unto his servant."[59]

The belief that the purgative power of Winthrop's medicines reflected a special godly "blessing on the means" kept them in demand all over New England. To help assure that they would be properly administered, Winthrop packed his chemical medicines in color-coded packets so that different medicines could be identified by whether they came in white, blue, brown, or printed papers. To distribute these medicines, Winthrop created alliances with local female healers, who served as agents distributing his powders and medical advice in their communities. Ephraim Child of Watertown, Massachusetts, wrote Winthrop in May 1653: "My wife would Intreate you, to send her a parcell of your physick devided into portions, for young and ould. She hath had many ocasions to make use thereof, to the help of many." Likewise, John Davenport of New Haven asked for more of Winthrop's powders, as "the supplye you left in [Elizabeth Davenport's] hand is spent." In distributing Winthrop's medicines, Elizabeth Davenport observed the same rules of Christian charity that Winthrop did himself. John Davenport wrote Winthrop that, when she acted as the agent for distributing Winthrop's medicines, "my wife is but your hand, who neither receiveth, nor expecteth any recompense." The wife of William Leete of Guilford, another colony leader, served a similar role. Through her husband's hand (but in her own wry voice) she sent a request for a new supply of Winthrop's medicines. "My wife entreats some more of your phisick," Leete wrote, "although she feareth it to have very contrary operations in Mr. Rossiters stomach." Rossiter was the local Guilford physician.[60]

---

59. Samuel Gorton to John Winthrop, Jr., Aug. 11, 1674, MHS, *Collections,* 4th Ser., VII (1865), 605–606.

60. Wilkinson, "John Winthrop, Jr. and the Origins of American Chemistry," 204; Rebecca J. Tannenbaum, "'What Is Best to Be Done for These Fevers?': Elizabeth Davenport's Medical Practice in New Haven Colony," *New England Quarterly,* LXX (1997), 265–284; Ephraim Child to John

Winthrop's decision to distribute medicines through a channel of women practitioners—in New Haven, Guilford, Norwich, Milford, and Watertown—reflects a general confidence in women's capacities as healers and a gender-collaborative approach to the exercise of medical authority. Equally significant is the fact that most of the female healers known to have distributed Winthrop's medicines were the wives of leading men, who as part of their elite status incurred a social and moral obligation to provide medical services to the poor. Distributing gratis medicines through these leaders' wives was not only consistent with Winthrop's commitment to the Christian alchemical principle of using alchemical knowledge to render Christian service (and his own personal practice); it helped Winthrop weave webs of familiarity and social obligation with leading Connecticut families, including the family of Captain John Mason, who had perhaps been Winthrop's fiercest critic during the early years in New London. This undoubtedly was a factor helping Winthrop in his transition from an unwelcome newcomer in the mid-1640s to his election as Connecticut's governor (without even standing for office) in 1657.[61]

Winthrop's appreciation for women as healers, while politically and socially useful, was also genuine and long preceded his arrival in Connecticut. It was an attitude he had acquired as a young alchemical student in England. Winthrop's cousin, Elizabeth Fones, noted for her skill as a surgeon, was invited to assist Winthrop and his friend Edward Howes with alchemical experiments in the 1620s. Howes later wrote that he not only appreciated her assistance; he praised her "more quick apprehension to discerne." Elite women were also included in the communications networks associated with the circle around Samuel Hartlib, of which Winthrop was a member. In America, Winthrop worked naturally and well with and through female healers, though his elevated medical status and reputation led to Elizabeth Davenport's noted deference to him.[62]

Winthrop, Jr., May 23, 1653, *Winthrop Papers*, VI, 292–293; John Davenport to John Winthrop, Jr., Aug. 4, 1658, in Isabel McBeath Calder, ed., *Letters of John Davenport: Puritan Divine* (New Haven, Conn., 1937), 125–126; Tannenbaum, *The Healer's Calling*, 79; William Leete to John Winthrop, Jr., Apr. 11, 1661, MHS, *Collections*, 4th Ser., VII, 546–548.

61. Identified high-volume distributors of Winthrop medicaments include Elizabeth Davenport of New Haven; Anna Mason, wife of Connecticut military leader John Mason; Mrs. Roger Newton, wife of the minister of Milford; the wife of future governor William Leete of Guilford; and the wife of Ephraim Child of Massachusetts. Tannenbaum, "'What Is Best to Be Done for these Fevers?': Elizabeth Davenport's Medical Practice," *New England Quarterly*, LXX (1977), 274–275, 278.

62. Edward Howes to John Winthrop, Jr., Mar. 7, 1632, *Winthrop Papers*, III, 66–67; Lynette Hunter, "Women and Science in the Sixteenth and Seventeenth Centuries," in Judith P. Zinsser,

ALCHEMICAL NETWORK

In addition to overseeing the widespread distribution of his pharmaceuticals and serving as a consulting physician to patients throughout New England, Winthrop supported the development of the alchemical talents of younger alchemical physicians, such as Jonathan Brewster, Gershom Bulkeley, and George Starkey, and he became the hub of a New England alchemical information exchange network. At the same time, he maintained contact with many European alchemists, including Samuel Hartlib, Robert Boyle, Benjamin Worsley, Johann Moraien, Johann Glauber, and the Kuffler brothers. At times, Winthrop's network seemed on the verge of major discoveries, as when Jonathan Brewster believed he had acquired the "knowledg of such a great mistery"—how to make the alkahest. Given the support for alchemy demonstrated at the upper echelons of colonial society and a colonial culture as conducive to the providential implications of medicine as was found in New England, it is clear why many other New Englanders would investigate "God's secrets" and become practitioners of alchemy.[63]

New England's identified alchemists—it is likely many practitioners left no records of their practices—were, like Winthrop, almost all members of the Puritan elite. Although healing was an important service they provided their communities, most also held other leading positions, as ministers, political leaders, or successful merchants. Being recognized as someone to whom God had granted special healing abilities was a significant status marker. In a culture that focused great attention on the spiritual connections between this world and the next and that saw the human body as a central metaphor defining the hierarchical organization of society (the body social), political authority (the body political), and the community of saints (the body ecclesiastical), the power to heal the body corruptible, especially by means of divinely blessed healing agents, provided a strong affirmation of one's right to serve at the head of the social, political, or religious hierarchy.[64]

At least two colonial governors, two presidents of Harvard and one president of Yale, twenty-two Puritan ministers, and nineteen elite practitioners are known to have practiced or studied alchemical medicine in colonial New England. As with Winthrop, alchemical knowledge increased these practi-

---

Men, Women and the Birthing of Modern Science (DeKalb, Ill., 2005), 123–140, esp. 123–124; Tannenbaum, *The Healer's Calling*, 82.

63. Jonathan Brewster to John Winthrop, Jr., Moheken, last January 1656, MHS, *Collections,* 4th Ser., VII, 77–80.

64. Robert Blair St. George, *Conversing by Signs: Poetics of Implication in Colonial New England Culture* (Chapel Hill, N.C., 1998), 116–203.

tioners' real value to their colonial communities. Because of their knowledge of medicines, metals, and minerals, alchemists were often prime actors in colonial settlement ventures and leading undertakers of economic development schemes. New settlements always needed good physicians, and ones who could read the landscape for its sanative qualities *and* its potential for economic development through metal or mineral extraction were highly sought after.

Alchemical healing attracted many New England divines. Making providentially focused medical treatment available was a logical extension of their ministry and a further confirmation of God's active control over quotidian affairs. Edward Taylor, minister and physician to the Massachusetts frontier town of Westfield, treated patients with a regimen that combined alchemy and astrologically propitious herbal remedies with reflection and repentance for sin. In his medical "Dispensatory," a commonplace book of recipes for useful medicines, Taylor painstakingly transcribed notes from John Webster's *Metallographia,* praising Paracelsus as the physician who "did really understand . . . the art of Medicines." Like Winthrop, Taylor sought knowledge of the weapon salve, recording Oswald Croll's recipe for an alchemical balm that could cure wounds at a distance.[65]

Minister-physicians in Puritan New England incorporated divine influence into every aspect of pathology and its remediation. All physicians performed in this world, through God's blessings on the means they used, small demonstrations of the great physical transmutation to come. Alchemical physicians, particularly when they participated in the quest to transmute base materials into the panacea alkahest, linked Christian healing to the Christian promise of the Resurrection. In the pervasively religious culture of early colonial New England, it was natural for a Puritan minister inclined to medicine and natural philosophy to practice Christian alchemy and for the best alchemical healers to be revered by their patients as specially blessed men. This reputation for piety seems to have remained intact, even when a number of New England's alchemists were known to be also studying magical arts related to, and sometimes conflated by the uneducated with, those practices considered beyond the realm of acceptable Christian practice.[66]

Leonard Hoar, who assumed the presidency of Harvard College in 1671, studied both alchemy and natural magic. While presiding over the college,

---

65. Karen Gordon-Grube, "Evidence of Medicinal Cannibalism in Puritan New England: 'Mummy' and Related Remedies in Edward Taylor's 'Dispensatory,'" *Early American Literature,* XXVIII (1993), 185–221, esp. 190, 197.

66. Mather, *Angel of Bethesda,* ed. Jones, 189; Watson, *Angelical Conjunction,* 100; Debus, *The Chemical Philosophy,* 117–122; Webster, *The Great Instauration,* 280–284.

PRACTITIONERS AND STUDENTS OF MEDICAL ALCHEMY

Minister Alchemists

John Allin	Samuel Danforth	Charles Morton
Moses Bartlett	Samuel Danforth, Jr.	Thomas Palmer
Gershom Bulkeley	Phineas Fiske	Ezra Stiles
John Bulkeley	Leonard Hoar	Edward Taylor
Charles Chauncey	Samuel Lee	Richard Treat
Elnathan Chauncey	Cotton Mather	Nehemiah Walter
Isaac Chauncey	Increase Mather	Michael Wigglesworth
Israel Chauncey		

Governors

John Winthrop, Jr. (Conn.)   William Stoughton (Mass.)

College Presidents

Leonard Hoar, Harvard   Charles Chauncey, Harvard   Ezra Stiles, Yale

Elite Practitioners

John Alcocke	William Barkeley	Richard Palgrave
George Alcocke	Jonathan Brewster	George Starkey
George Alcocke (grandson)	Robert Child	William White
	Richard Leader	Nathaniel Williams
Humphrey Atherton	Christian Lodowick	John Winthrop, F.R.S.
Jonathan Avery	Aeneas Munson	Wait Still Winthrop
William Avery	James Oliver	

Alchemists compiled from data found in Newman, *Gehennical Fire*, 39–50; Watson, *Angelical Conjunction*, 97–121; Ronald S. Wilkinson, "New England's Last Alchemists," *Ambix*, X (1962), 128–138. The qualification about a practitioner's practicing, or studying, alchemical medicine is necessary for two reasons. It is not possible to prove that a healer who owned a variety of alchemical medical texts actually practiced alchemical medicine. I have included several practitioners whose libraries contained alchemical medical works, even when that is the only evidence available linking them to alchemical practice.

Hoar wrote Robert Boyle expressing his desire to install an alchemical laboratory at the school so students could gain firsthand experience of chemical processing. Hoar owned standard works on medical alchemy, such as Van Helmont's *Ortus medicinae;* his interests also encompassed occult natural magic. His library included magic books such as Giovanni Battista della Porta's *Natural Magick* and Cornelius de Agrippa's *Occult Philosophy,* whose 1651 translation into English by John French was dedicated to Robert Child. Della Porta's treatise on natural magic covered subjects ranging from demonology to the occult power of magnetism, and Agrippa's work was a comprehensive exploration of magical practices ranging from geomancy and astrology to the angel magic of the Hebrew Cabala.[67]

In seeking such knowledge, Hoar was not unusual among the Puritan elite. Despite a common assertion that Puritan leaders were totally opposed to the practice of magic, Puritan intellectuals made clear distinctions between "natural magic" (manipulation through natural means of the occult or unseen forces at work in the world) and diabolical magic (manipulation of those same forces with the aid of the devil). Study of natural magic was cautiously acceptable; engaging in any form of diabolical magic was anathema. The potential problems associated with pursuing natural magic were that it could become the pathway of temptation leading the seeker into making a Faustian bargain with the devil in return for greater magical knowledge and, tangentially, that some of the effects produced by natural magic were indistinguishable from those thought diabolical in origin. On the other hand, one of the benefits that accrued to the elite who had studied natural magic was an enhanced ability to make wise judgments about whether suspected instances of magical deployment were natural or preternatural in origin. In practice, while the study or use of folk magic and other magical arts by New England's common folk was sharply censured by New England's divines, members of the Puritan elite, who could be trusted to use such knowledge wisely and in the service of God, were not reproached for their quest for acceptable occult knowledge. Thus, a sample of nearly a third of the medical libraries of New England's minister-physicians included books on the occult arts. Those Puritan leaders who studied natural magic did so with confidence that their piety and intentions were sufficient to shield them from both censure and diabolical temptation.[68]

---

67. Watson, *Angelical Conjunction,* 108, 114; John Baptista [della] Porta, *Natural Magick* (London, 1658); Henry Cornelius Agrippa, *Three Books of Occult Philosophy,* trans. John French (London, 1651).

68. Watson, *Angelical Conjunction,* table 3.1, "Medical Authorities Most Popular among Preacher Physicians," 76–77. "The established Calvinist doctrine of an all-powerful divinity left no sanc-

Leonard Hoar's interest in building an alchemical laboratory at Harvard might have represented an effort to institutionalize a course of study informally begun by Charles Chauncey, his predecessor. Chauncey, physician and minister, was also a patriarch of alchemical minister-physicians. Four of his sons, to whom he taught medicine, and a son-in-law were known for their practice of alchemical physick. While at Harvard around 1660, Elnathan Chauncey (1639–1684) transcribed alchemical notes from the medical works of Van Helmont and the English adept Thomas Vaughan. Elnathan practiced medicine in Boston and in England before finally settling in Barbados. His brother Israel Chauncey (1644–1702/3) collected alchemical medical books and studies of occult arts and magic while serving as minister and physician in Stratford, Connecticut. His 1703 inventory included an alembic and three "alchemy spoons." Isaac Chauncey (1632–1712), who sought his medical and ministerial fortunes among Puritans in England during the Civil War and Commonwealth, corresponded about alchemical matters with his New England kin from abroad. Their brother Ichabod Chauncey also practiced medicine in England, though the degree to which he practiced alchemical healing is not known.[69]

Charles Chauncey's son-in-law Gershom Bulkeley became one of New England's most successful alchemical minister-physicians and a close associate of Winthrop. Bulkeley, like his father-in-law, also became an alchemical patriarch. With Winthrop, Bulkeley would be instrumental in changing the nature of witchcraft prosecution in Connecticut. His knowledge of the occult practices, gained through his studies of alchemy and natural magic, helped him create a definition of diabolical magic in a way that would permanently end witchcraft executions in Connecticut and stop all witchcraft trials there for a generation. As a minister, Bulkeley attracted hundreds of parishioners to his congregation in Wethersfield before stepping down from the pulpit to devote full time to medicine. His library contained most of the essential texts of Christian alchemy, among them works by Paracelsus, Comenius, Glauber, Valentine, Croll, and Van Helmont as well as della Porta's *Natural Magick*. Bulkeley's nearly thirty surviving volumes of personal medical and

---

tioned room for human manipulation of supernatural power, and the Puritans inherited a century-long tradition of the suppression of magic." John L. Brooke, *The Refiner's Fire: The Making of Mormon Cosmology, 1644–1844* (Cambridge, 1994), 34–36.

69. Watson, *Angelical Conjunction*, 53, 147–148; John Langdon Sibley, *Biographical Sketches of Graduates of Harvard University*, I (Cambridge, Mass., 1873), 302–309, 389–402, II (1881), 80–87; Thomas W. Jodziewicz, *A Stranger in the Land: Gershom Bulkeley of Connecticut*, American Philosophical Society, Transactions, LXXVIII (Philadelphia, 1988), 1–23. Ichabod obtained his Doctor of Medicine degree at Leiden in 1884.

account books reveal his own belief in the power of astral influences and his experiments at perfecting the philosopher's stone.[70]

Two of Bulkeley's children, John and Dorothy, followed in their father's footsteps. John Bulkeley became a minister and alchemical physician to the people of Colchester, Connecticut, and was praised as a healer to whom "GOD had given . . . a peculiar Insight into Medicine." From Colchester, he actively sought news of "noble discoveries in the Arcana of the chymists" from alchemical correspondents and was likewise consulted by physicians from the metropolis.[71]

Dorothy Bulkeley Treat shared the family interest in alchemical healing, as attested to by the contents of the two medical receipt books that bear her signature and the inscription "Her book." One was written in the hand of her father, Gershom, but the other, "Medical Cabinet," may be in Dorothy's own writing. Replete with medical recipes and alchemical citations from Van Helmont, Thomas Willis, and Robert Boyle, the manuscript emphasizes the secrecy that was the sine qua non of alchemical research. "Keep this Cabinet in your Cabinet and let it not ly comon for every eye to look into and read it, whereby you would not do others so much good as you will do hurt yourself: therefore keep this book close and ——— to yourself." At his death, Gershom Bulkeley left his alchemical books and medical equipment to Dorothy's son Richard, who continued the family tradition of serving as a minister-physician.[72]

Other elite New England families whose members transferred knowledge of alchemical healing from generation to generation included the Alcock, Oliver, and Avery families and perhaps the most noted of all New England families, the Winthrops. George Alcock of Roxbury, a surgeon and physician of the original Winthrop fleet, passed on his medical interests to his son John, who conducted alchemical experiments with George Starkey while both

---

70. Jodeziewicz, *Stranger in the Land*, appendix A, 73–92.

71. Watson, *Angelical Conjunction*, 149; Clifford K. Shipton, *Biographical Sketches of Those Who attended Harvard College* . . . , IV (Cambridge, Mass., 1933), 450–454; John Bulkeley to John Winthrop, F.R.S., MHS, *Collections*, 5th Ser., V, 302–303; Eliphalet Adams, *A Discourse Delivered at Colchester* . . . *after the Funeral of . . . John Bulkeley* . . . (New London, Conn., 1734), 35.

72. Gershom Bulkeley, "Medical Receipt Book," in Bulkeley Collection MSS, Watkinson Library Manuscript Collection, Trinity College, Hartford; Patricia A. Watson, "The 'Hidden Ones': Women and Healing in Colonial New England," in *Medicine and Healing*, The Dublin Seminar for New England Folklife, Annual Proceedings, XV (Boston, 1990), 25–33. The transfer of the medical library to Richard Treat provoked a legal battle between John and Dorothy over the validity of their father's will. John claimed that Dorothy had forced her son to feign interest in medicine in order to gain possession of the library. Watson, "The 'Hidden Ones,'" 28–33; Jodeziewicz, *Stranger in the Land*, 1–2.

were students at Harvard in the 1640s. Starkey and Alcock worked together under the direction of the Charlestown alchemical physician Richard Palgreave, whose daughter Sarah John subsequently married. Sarah was also adept in physick and surgery, though her alchemical interests, if any, are unknown. Sarah and John's son George carried the family interest in alchemical medicine into a third generation. George died from smallpox in England four years after graduating from Harvard in 1673, but his library already included the works of key alchemical writers, including Van Helmont and the English alchemist Noah Biggs.[73]

Thomas Oliver, Boston surgeon and elder of John Cotton's Boston church, was grandfather to James Oliver, who was mentored in his alchemical studies by Christian Lodowick, a late-seventeenth-century immigrant physician some called "the most skillful Chymist that ever came to these parts of America." Oliver, like Lodowick, was noted for alchemical medical skill. He married the daughter of Samuel Bradstreet, physician son of Anne Bradstreet (whose poetry, like Edward Taylor's, incorporated a sizable amount of alchemical imagery). At his death, Oliver willed his books, papers, medicines, and apparatus to his nephew Nathaniel Williams, who had studied alchemical medicine under him.[74]

The medical skills of the alchemical physician William Avery were praised by Samuel Lee, an English minister and physician, who resided in Rhode Island from 1687 to 1691. Lee was a member of London's Royal College of Physicians, and his biases in favor of a regulated system of medical licensure and institutional control of medical practice show clearly in his letters. Lee was critical of the nonlicensed medical community he encountered in New England; on the other hand, he seems to have possessed an interest in and appreciation for medical alchemy. Lee called William Avery "a man of pretty ingenuity" who had developed "some notable skill in physick and midwifery and invented some usefull intrumts for that case. And besides was a great inquirer [after the alkahest?] and had skill in Helmont and chemicall physick." Avery, who served Dedham as healer and was deputy to the Massachusetts General Court and lieutenant in the militia, corresponded with Robert Boyle about alchemical matters in 1682 and 1684, apologizing in one of his letters for not *yet* achieving the full status of adept, that is, perfecting transmuta-

---

73. Newman, *Gehennical Fire*, 48–49; Sibley, *Graduates of Harvard*, I, 124–126, II, 420–422; James Savage, *A Genealogical Dictionary of the First Settlers of New England*, I (Boston, 1860), 21.

74. Darrett B. Rutman, *Winthrop's Boston: Portrait of a Puritan Town, 1630-1649* (Chapel Hill, N.C., 1965), 105, 219, 241; Newman, *Gehennical Fire*, 51; Sibley, *Graduates of Harvard*, III (1885), 198–199; biographical notes, *New England Historical and Genealogical Register*, I (1847), 64, VII (1853), 312–314, XIX (1865), 101–102.

tion. Avery's interests included both alchemical medicine and the alchemical study of metals and minerals, as evidenced by his investments in multiple mining projects. Lee recorded one of Avery's theories about shipbuilding, which demonstrates Avery's belief in the sympathetic and occult relationships among objects in the natural world. According to Lee, Avery hypothesized, "If a ships planks and boards be laid from sterne to head in the graine as it grew from root to top: it were a great facilitation to its quicker moti[on thro the water]." Avery's son Jonathan, to whom he bequeathed "all my Physick books and instruments," became, like his father, "an assiduous labourer at the chemical fire."[75]

Cotton Mather, colonial New England's leading medical theorist, is best known medically for his promotion of inoculation for smallpox in the 1720s. This has earned him a reputation among some historians of medicine as a proto-Enlightenment rationalist. His support for alchemy in New England, however, has received little attention, nor has the highly religious basis of Mather's support for inoculation been analyzed. Alchemy was closely aligned with Mather's own medical beliefs; he penciled on the frontispiece of his copy of the *Transactions* of the Royal Philosophical Society the location of Robert Boyle's recipe for a particularly effective Helmontian medicine, and he made medicines that were distributed by New England alchemists. In his most important medical work, *The Angel of Bethesda,* Mather revealed his voluminous knowledge of both traditional and alchemical medical authorities. Though he did not argue in favor of one school over the other, he frequently cited alchemical authors with approval. Most important, Mather's most original medical theory—positing the existence of the Nishmash Chajim, a part-physical, part-spiritual element physically present in every body, mediating between the physical and spiritual worlds and regulating the body both before and after the Resurrection—was derived largely from Van Helmont's theory of the archeus, a spiritual life force said to be responsible for the well-being of all bodies. Alchemical authors also influenced Mather's physico-theology, and the spiritual scientific views Mather espoused in his most important scien-

---

75. Lee, one of only three medical doctors licensed by the Royal College of Physicians to come to New England in the seventeenth century, privately and censoriously observed New England's lack of a regulating medical corporation like the Royal College, the failure to offer a formal medical training at Harvard, and the absence of official regulation of practitioners. Kittredge, "Letters of Samuel Lee and Samuel Sewall," Colonial Society of Massachusetts, Publications, XIV, *Tranasctions,* 143–187, esp. 145–147; Savage, *Genealogical Dictionary,* I, 21; Newman, *Gehennical Fire,* 292 n. 177; William Avery to Robert Boyle, Nov. 9, 1682, May 1, 1684, in Thomas Birch, ed., *The Works of the Honourable Robert Boyle* (London, 1772), VI, 610–614; Watson, *Angelical Conjunction,* 53, 148; Sibley, *Graduates of Harvard,* II, 82–88.

tific work, *The Christian Philosopher,* look back as much to European pansophic movement of the middle seventeenth century as they look forward to the rational Enlightenment of the eighteenth.⁷⁶

Mather, like Leonard Hoar, united interests in occult chemistry with the magical arts. Though he vociferously condemned the use of astrological fortune-telling as diabolical, his writing reveals he had a thorough knowledge of the complexities of chart casting. Mather also espoused natural astrology, measuring the influences of stars and planets on natural processes such as planting and harvesting crops, to members of the Royal Society as late as 1717. He had an expansive view of the range of occult studies permissible to the Puritan elite, and he alluded to his own knowledge of darker arts in his dedication of *Memorable Providences Relating to Witchcrafts and Possession* to the alchemical physician Wait Still Winthrop:

> Had I on the Occasion before me handled the Doctrin of Daemons, or launched forth into Speculations about magical Mysteries, I might have made some Ostentation, that I have read something and thought a little in my time; but it would neither have been Convenient for me, nor Profitable for those plain Folkes, whose edification I have all along aimed at.

Discussions of magic and demons were not suitable for the masses to whom Mather targeted his book. Winthrop, however, was different. As far as Mather was concerned, Winthrop was specially qualified to pass judgment on the discussion of magical portents about which Mather had written. "Knowledge has Qualified You to make those Reflections on the following Relations, which few can Think, and tis not fit that all should See."⁷⁷

---

76. Louise A. Breen, "Cotton Mather, the 'Angelical Ministry,' and Inoculation," *Journal of the History of Medicine and Allied Sciences,* XLVI (1991), 333–357; Royal Philosophical Society, *Transactions,* IX (1674), frontispiece, in Mather Family Library, American Antiquarian Society, Worcester, Mass.; Mather, *Angel of Bethesda,* ed. Jones, 28–37; Margaret Humphreys Warner, "Vindicating the Minister's Role: Cotton Mather's Concept of the *Nishmath-Chajim* and the Spiritualization of Medicine," *Journal of the History of Medicine and Allied Sciences,* XXXVI (1981), 278–295.

For a description of Helmont's archeus, which in turn was derived from Paracelsus, see Walter Pagel, *Joan Baptista van Helmont: Reformer of Science and Medicine* (Cambridge, 1982), 96–102; Jeffrey Jeske, "Cotton Mather: Physico-Theologian," *Journal of the History of Ideas,* XLVII (1986), 583–594.

Ironically, Mather's discussion of febrifuges, which might have detailed the nature of his medicine, is one of the five chapters missing from the extant text of his most important medical work, *Angel of Bethesda.*

77. Cotton Mather, *Memorable Providences, Relating to Witchcrafts and Possessions,* in George Lincoln Burr, ed., *Narratives of the Witchcraft Cases, 1648–1706* (New York, 1914), 94; George Lyman Kittredge, "Cotton Mather's Scientific Communications to the Royal Society," American Antiquarian Society, *Proceedings,* n.s., XXVI (1916), 18–57. Additional discussions of Mather's interest

That alchemical knowledge helped provide the foundation for Winthrop's qualifications for passing judgments on magical practices seems clear. In the same dedication, Mather extended best wishes to Winthrop as a physician "in the name of the many hundred Sick people" to whom he had "freely dispens'd your no less generous than secret Medicines." Wait Winthrop had followed in his father's footsteps, continuing the tradition of alchemical medical charity that caused Mather to style him a Christian Hermes.[78]

The practice of alchemical medicine in early New England was pervasive. Though largely ineffective by modern medical standards, it resonated powerfully with Puritan cultural sensibilities. Patients and physicians believed in its therapeutic value and wrote glowing reports of the benefits they had received from it. Alchemists' insistence on hands-on experimentation helped lay the groundwork for modern medicine, and their studiously acquired knowledge of the occult led the way toward the transformation of witch-hunting.

---

in occult sciences: Michael P. Winship, "Cotton Mather, Astrologer," *New England Quarterly*, LII (1990), 308–314, "Prodigies, Puritanism, and the Perils of Natural Philosophy: The Example of Cotton Mather," *William and Mary Quarterly*, 3d Ser., LI (1994), 92–105, *Seers of God*, 105–109; David Levin, "Giants in the Earth: Science and the Occult in Cotton Mather's Letters to the Royal Society," *WMQ*, 3d Ser., LV (1988), 751–770.

78. Mather, *Memorable Providences*, in Burr, ed., *Narratives of the Witchcraft Cases*, 94.

{ SEVEN }

The Magus as Mediator
Witchcraft, Alchemy, and Authority in the Connecticut Witch-Hunt of the 1660s

Between 1647, when New England hanged its first witch, and the end of the Hartford witch-hunt in 1663, the Puritan elite prosecuted witches with zeal. Thirty-four persons were tried for witchcraft, and fifteen of them were convicted and hanged. Connecticut assumed leadership in Puritan witch-hunting. That colony convicted and executed each of the first seven witch suspects it tried and was responsible for eleven of the witch hangings in New England before 1663. Although Connecticut was initially New England's most aggressive prosecutor of witches, it subsequently dramatically reversed its attitudes toward witchcraft prosecution. From the end of the Hartford witch-hunt forward, Connecticut set the standard for moderation in witchcraft proceedings, eliminating executions permanently and witch trials themselves for more than two decades.[1]

Puritan magistrates' growing reluctance to prosecute and convict witches has frequently been attributed to two factors: ministerial insistence on obtaining proof that a suspected witch had made a compact with the devil, and increasing skepticism about witchcraft itself engendered by the scientific revolution. In Connecticut, the transformation was too sudden and dramatic—and took place much too early—to be explained by such factors. Before 1663,

---

1. Data are derived from John Putnam Demos, "List of Known Witchcraft Cases in Seventeenth-Century New England," appendix to *Entertaining Satan: Witchcraft and the Culture of Early New England* (Oxford, 1982), 401–409; Richard Godbeer, appendix A, "Witchcraft Trials in Seventeenth-Century New England (Excluding Persons Accused during the Salem Witch Hunt)," in Godbeer, *The Devil's Dominion: Magic and Religion in Early New England* (Cambridge, 1992), 235–237. During this period (1670–1691), Massachusetts conducted fourteen witchcraft trials. R. G. Tomlinson, *Witchcraft Trials of Connecticut: The First Comprehensive Documented History of Witchcraft Trials in Colonial Connecticut* (Hartford, Conn., 1978), 11–12, 26. Caught up, perhaps, in the climate of fear that accompanied the Salem witch panic, Connecticut initiated its final two witch trials in 1692. Neither of the accused (Elizabeth Clawson and Mercy Disborough) was executed. Richard Godbeer, *Escaping Salem: The Other Witch Hunt of 1692* (New York, 2005).

Connecticut ministers were aggressive witch-hunters, often instrumental in securing the evidence of the diabolical complicity they required for conviction. Furthermore, Connecticut's change in attitudes occurred decades before Enlightenment skepticism began to exert meaningful influence on New England culture. Rather ironically, cessation of witchcraft prosecutions in Connecticut occurred through the agency of occult practitioners who assumed leadership at the highest levels of colonial society.[2]

Analyzing how Connecticut was transformed so rapidly from the colony most likely to execute witches to a colony without witchcraft trials highlights the prominence in colonial society of a group of occult philosophers, Puritan ministers, and magistrates who pursued knowledge of the unseen forces at work in the natural world through the study of alchemy, alchemical medicine, and natural magic. Led by Connecticut Governor John Winthrop, Jr., and the alchemist-minister Gershom Bulkeley, these magi-among-the-elect possessed both specialized knowledge of magic and the political authority to play a decisive role in reversing elite attitudes toward witchcraft prosecution. Their leniency toward witchcraft suspects was based, not on skepticism regarding magic's efficacy, but, rather, on their intellectual understanding of the magical arts witches were accused of practicing. It is to Winthrop and his understanding of the role of occult forces in the operations of nature, not to the arrival of modern scientific thinking, that Connecticut's sudden and sustained reluctance to prosecute witches must be attributed.

Magic in colonial New England occurred in many guises, both licit and illicit, sanctioned and condemned. The distinction between acceptable and unacceptable magic praxis was often a fine one. The boundary was not merely unclear to the elites and laity: it was highly dynamic. Magical practice was, therefore, subject to continual analysis and review in the court of community opinion, where suspicions of someone's magic might ebb and flow with changing circumstances. On those occasions when public reputation for practicing unacceptable magic was transformed into the more serious and formal accusation of witchcraft, it became the duty of the ministerial and magisterial elite to arrive at an official position regarding whether a suspect had in fact

---

2. Richard Weisman, *Witchcraft, Magic, and Religion in Seventeenth-Century Massachusetts* (Amherst, Mass., 1984), 57–72, 96–114; Godbeer, *Devil's Dominion*, 153–178; Keith Thomas, *Religion and the Decline of Magic* (New York, 1971), 641–688; Chadwick Hansen, *Witchcraft at Salem* (New York, 1969), 284; Demos, *Entertaining Satan*, 392–394; David D. Hall, *Worlds of Wonder, Days of Judgment: Popular Religious Belief in Early New England* (new York, 1989), 100–102, 106–109. Carol F. Karlsen argues that concern about the elevated social status of the accused generated elite resistance to prosecution, in *The Devil in the Shape of a Woman: Witchcraft in Colonial New England* (New York, 1989), 27–31, 254–255.

engaged in witchcraft. Leaders like Winthrop—students of the occult with knowledge and experience to make the often-obscure distinctions between natural and diabolical magic—could be highly influential, even decisive in this arbitration. It is in this context, as a recognized and respected authority on how supposedly magical forces operated in nature, that we can best understand Winthrop's pivotal position in changing the tenor of colonial witchcraft prosecution.

Winthrop believed magic was such a complex and powerful array of forces that the use of diabolical magic by ordinary people was both overstated and overfeared, and his opinions carried extraordinary weight. As a consultant on witchcraft cases in New Haven prior to becoming Connecticut's governor and then as chief magistrate in Connecticut witchcraft trials, Winthrop consistently refused to endorse witchcraft charges, resisting both popular and ministerial pressure to convict witch suspects. He not only orchestrated acquittals in witchcraft cases; on at least two occasions he flatly refused to enforce convictions, ultimately overturning the guilty verdicts. He recognized, however, that formal witchcraft accusations reflected a social pathology affecting the entire community, and he managed witchcraft acquittals with an eye to validating the accusers and reintegrating the community, even while protecting suspected witches from execution. This political and social balancing act sometimes met with resistance, from both laity and some members of the ministry, producing conflicts that called into question the very nature of witchcraft and raised concerns about the acceptability of Winthrop's own pursuit of knowledge. The conflict came to a head in the popular revolt surrounding the 1669 witchcraft case of Katherine Harrison, in which Winthrop, in careful coordination with the alchemist and minister Gershom Bulkeley, simultaneously undermined the legal foundations for convicting witches in Connecticut and removed intellectual study of the occult from ministerial suspicion and official scrutiny. Their actions set precedents that influenced witch prosecution in New England for more than two decades. When historians examine the changing attitudes of the elite toward witchcraft prosecution in New England during the second half of the seventeenth century, they must include the actions of these students of the occult who themselves were members of the Puritan elite.

RESHAPING WITCHCRAFT PROSECUTION

Between 1647 and 1654, Connecticut Colony conducted seven trials for witchcraft. In each case, the accused witch was found guilty and subsequently executed. John Winthrop, Jr., had no involvement in any of those trials. Between

1655 and the summer of 1661, four more witches were prosecuted in New Haven and Connecticut colonies. All of them were acquitted. Winthrop was involved, formally or informally, in at least two of these trials, and probably all four, and through his participation he began reshaping how the Connecticut elite approached the prosecution of witchcraft.[3]

Winthrop assumed leadership in the affairs of Connecticut slowly and reluctantly. As discussed earlier, his initial relations with Connecticut's magistrates were severely strained and left hard feelings that only slowly subsided. Connecticut had sought several times to gain formal recognition from Winthrop of its dominion over his new plantation by electing him a colony magistrate in 1647, 1648, and 1649. Winthrop, however, declined even to take the Connecticut freeman's oath until 1650.[4]

By the middle years of the next decade, however, Winthrop's relationship with Connecticut's leaders had substantially improved as a result of Connecticut's growing appreciation for Winthrop's special skills as an alchemical physician, projector, Indian negotiator, political leader, and entrepreneur and for his ability to achieve the kind of accommodations with Connecticut over issues such as Pequot subjection to Uncas that he had sought from the beginning of settlement. As the many strengths he brought to the region became more apparent, Winthrop was, Connecticut's leaders realized, exactly the kind of person most desired by colonies and towns trying to establish stable and productive footholds in New England. As a result, in the mid-1650s Winthrop transitioned from being an unwelcome pariah on the Connecticut frontier to becoming the object of a bidding war between Hartford and New Haven over which could win him as a resident.

The competition to attract Winthrop began formally in August 1654, when eleven inhabitants of Hartford, led by the Reverend Samuel Stone, wrote offering to provide him a furnished house if he would live in Hartford on a trial basis during the ensuing winter. A month later, the Reverend John

---

3. Godbeer, *Devil's Dominion*, 235–237. The victims were Alice Young, Windsor (1647); Mary Johnson, Wethersfield (1648); John Carrington, Wethersfield (1651); Joan Carrington, Wethersfield (1651); Goody Bassett, Fairfield (1651); Goody Knapp, Fairfield (1653); Lydia Gilbert, Windsor (1654). During the same period, Massachusetts tried and executed three witches. Winthrop's attentions during this period were focused on New London. The accused were Elizabeth Godman, New Haven (1655); Goodwife Bailey, New Haven (1655); Nicholas Bailey, New Haven (1655); Elizabeth Garlick, East Hampton, Long Island (1658).

4. Richard S. Dunn, *Puritans and Yankees: The Winthrop Dynasty of New England, 1630-1717* (Princeton, N.J., 1962), 74. He was subsequently elected a magistrate in every Particular Court session in his own town in 1654 and 1655; he remained inactive in colonial government. *Records of the Particular Court of Connecticut, 1639-1663*, Connecticut Historical Society, *Collections*, XXII (Hartford, 1928), 129, 145 (hereafter cited as *Records of the Particular Court of Connecticut*).

Davenport of New Haven, reacting to the Hartford offer, upped the ante, offering to provide both a house and free transportation for Winthrop and his family and goods if he would spend the winter in New Haven.[5]

Hartford and New Haven were eager to acquire Winthrop as a resident in part because of his ability to provide two commodities then in very short supply in those places: political leadership and entrepreneurial skill. Connecticut's governor John Haynes had died suddenly the preceding January; New Haven's governor Theophilus Eaton was a venerable sixty-four. Winthrop, through his experience as an assistant in Massachusetts, governor of Saybrook, and founder of three towns, was an excellent candidate for political leadership. He was an entrepreneur of proven ability as well, an indefatigable and successful promoter of unique development schemes.[6]

These were, however, only some of the benefits to be derived from Winthrop's presence. Foremost among the talents he could bring to any town was his remarkable ability as a healer. A letter from Hartford's Matthew Allyn leaves no doubt of the importance Hartford placed on that aspect of Winthrop's talents. Allyn asked Winthrop to "consider of our want of the presence of such a one as your selfe to live at Hartford, whare you will have more optunitye to emprove your abilityes to greater advantage for Gods glorye and his peoples good: I mean as for fissicke so for government."[7]

---

5. The Hartford writers anticipated they would not be alone in making offers, for they urged "that what invitations soever may be presented to you for others else wher may not divert your thoughts from us who stand in soe much need of you." Samuel Stone and others to John Winthrop, Jr., Aug. 17, 1654, in Samuel Eliot Morison et al., eds., *Winthrop Papers* (Boston, 1929–), VI, 416–417 (hereafter cited as *Winthrop Papers*). Alluding to Winthrop's previous tensions with the Hartford magistrates (most of whom were not numbered among the signers of Stone's letter), Davenport noted, "I assure you that you have more hearts desirous of you here, then there are hands unto that [the Hartford] invitation." To back up that assertion, New Haven sent two additional solicitations: one from New Haven governor Theophilus Eaton personally, on behalf of the church and town; and a second from the governor with a formal invitation to Winthrop from the General Court. John Davenport to John Winthrop, Jr., Sept. 19, 1654, *Winthrop Papers*, VI, 426–428, Gov. Theophilus Eaton to John Winthrop, Jr., Sept. 28, 1654, 436–437, Oct. 4, 1654, 445. In addition to weighing the offers from Hartford and New Haven, Winthrop had received encouragement to resettle from Roger Williams in Rhode Island, Peter Stuyvesant in New York, and would-be migrants to the Delaware. Dunn, *Puritans and Yankees*, 76.

6. Dunn, *Puritans and Yankees*, 74, 76–77, 83–92; Robert C. Black III, *The Younger John Winthrop* (New York, 1966), 173; the magistrates of Connecticut to John Winthrop, Jr., Jan. 9, 1654, *Winthrop Papers*, VI, 354–355; E. N. Hartley, *Ironworks on the Saugus: The Lynn and Braintree Ventures of the Company of Undertakers of the Ironworks of New England* (Norman, Okla., 1957), 42–59; Ronald Sterne Wilkinson, "John Winthrop, Jr. and the Origins of American Chemistry" (Ph.D. diss., Michigan State University, 1969), 35–60, 89–135, 238–260; Samuel Eliot Morison, "John Winthrop, Jr., Industrial Pioneer," in Morison, *Builders of the Bay Colony* (Boston, 1930), 268–288.

7. Matthew Allyn to John Winthrop, Jr., Mar. 1, 1654, *Winthrop Papers*, VI, 366–367.

AUTHORITY, BOTH MEDICAL AND MAGICAL

This study rejects a fundamental assumption of most American historiography on witchcraft: that the Puritan elite categorically rejected the practice of magic. It further rejects two claims frequently made about the occult scientific pursuits of John Winthrop, Jr.: (1) that such practices were an insignificant part of his intellectual endeavors, and (2) that he practiced them so secretly they were unknown to his contemporaries. Winthrop's pursuit of occult knowledge, through alchemy, alchemical medicine, and natural magic, was neither secret nor unimportant, and the benefits derived from that quest were solicited by ministers and laity alike.[8]

Colonial New England, like the rest of the early modern world, was believed by its Puritan inhabitants to be permeated through and through with occult forces, influences whose *effects* were readily apparent but whose *causes* were hidden to the human intellect. Examples of these occult influences could be found everywhere: in the reaction of iron to a lodestone, the spread of contagious disease and epidemics, the workings of poisons and antidotes, the spontaneous generation of insects in putrefied matter, the effect of solid bodies in free fall, the influence of stars and planets on crops and health, the ability of purgation to cure a wide variety of ailments. No one was really sure *how* such things occurred, but *that* they occurred was self-evident. The Puritan world of wonders was chock-full of puzzling but powerful occurrences whose effects were manifest even if their causes were occult.[9]

Some of these happenings, effects such as the healing from a distance wrought by the "weapon salve" or the spectral apparitions reported in witchcraft cases, were so unusual or extraordinary that they bordered on the miraculous. Such "preternatural" phenomena, by definition *magical* in origin, were understood to be produced by purely natural means, even when produced by or through the aid of the devil or demons. Though their causes might have been occult—that is, unknown or hidden—they could potentially be explored, analyzed, replicated, and over time, perhaps, even understood.

As a result, the study of natural magic, like alchemy and astrology, was

---

8. Godbeer, *Devil's Dominion*, 5; Brooke, *The Refiner's Fire: The Making of Mormon Cosmology, 1644-1844* (Cambridge, 1994), 36; Weisman, *Witchcraft, Magic, and Religion*, 55. See also Jon Butler, *Awash in a Sea of Faith: Christianizing the American People* (Cambridge, Mass., 1990), 67, 70-73; but note David D. Hall's study, which focuses on the compatibility between lay magic beliefs and Puritan religion: *World's of Wonder*, 7. "Winthrop's alchemy was so secretive that it has only recently been rediscovered" (Brooke, *Refiner's Fire*, 37).

9. Stuart Clark, *Thinking with Demons: The Idea of Witchcraft in Early Modern Europe* (Oxford, 1997), 214-232.

considered by many to be a legitimate avenue of inquiry into the secrets of the natural world. Natural magic became "one of the most enduring enthusiasms of early modern natural philosophers." The source of the term "magic" itself came from the Persian *magia,* and it signified both universal science and universal wisdom. Although twenty-first-century hearers automatically associate magic with either the supernatural or the deceptive, early modern thinkers from Pico della Mirandola to Francis Bacon saw it as one of the most elevated and rewarding forms of natural philosophy, and it attracted many devotees. Far from being a solitary and secretive pursuit, intellectuals across Europe supplemented or focused their investigations into nature through the study of natural magic.[10]

Winthrop became a student of natural magic as a logical extension of his investigations into the occult properties at work in alchemy and alchemical medicine. Rooted in the exploration of the sympathies and antipathies at work between the macrocosm and microcosm, all three pursuits were intellectually compatible, theoretically interrelated, and at times overlapping. In addition to his extensive collection of alchemical texts, some of which dealt with magical topics, the Winthrop library contained major magical works such as Cornelius Agrippa's *Three Books of Occult Philosophy,* Giambattista della Porta's *Natural Magick,* and Marco Antonio Zimara's *Cave of Magic and Medicine, in Which There Is a New, Secret, and Overflowing Treasure of Magico-Physical Secrets, of Seals, Signs, and Magicall Images.* How Winthrop specifically made use of these texts and what role, if any, they played in his own studies is simply unknown. But Winthrop's knowledge of the occult, along with his medical expertise, helped provide him respected authority when he was called on to adjudicate New England's most troublesome encounters with the preternatural—witchcraft.[11]

As a self-professed Hermetic philosopher who spent his life seeking mastery over the hidden forces at work in the cosmos, Winthrop represents a type of inquirer into the nature of magic well known to colonial New Englanders (though not always their historians) whose exploration of the occult was not only acceptable to elites: it was lauded. Cotton Mather praised Winthrop as "Hermes Christianus," the Christian Hermes, and, as his reputation grew in the mid-seventeenth century, it helped encourage a younger generation of Puritan divines to combine occult medicine with their ministries. Eight of the thirteen Harvard graduates between 1650 and 1675 who became minister-physicians are known to have practiced alchemical medicine or to have col-

10. Ibid., 155.
11. Ibid., 217–218.

lected occult texts. Winthrop was only the prime exemplar of a class of elite occultists—minister-physicians, merchant-alchemists, and students—whose presence had an important but little-noted effect on New England's cultural understandings of magic practice.

Clearly, the binary opposition that has been frequently and often axiomatically posited between New England Puritanism and magic did not really exist. As David D. Hall has astutely noted, "We do better if we perceive an accommodation between magic and religion than if we regard magic as somehow the substance of a different tradition." Just as learned Puritans like Winthrop sought to achieve mastery over the occult through study and experiment, ordinary New Englanders sought the same kinds of controls over the unseen forces working in their lives through the application of a broad array of folk magic practices. Though New England's ministers worried that such practices, particularly among the uneducated, might seduce people into turning to the devil for greater control over the occult, they did not see magic as inherently alien to Puritan philosophy or practice.[12]

A perspective on New England's magic that incorporates the presence of sanctioned intellectual magic as a vital force in Puritan society would better serve history. This perspective should consider all occult practices, whether licit or illicit, as interrelated, based on their shared belief in manipulating natural phenomena through the human exploitation of hidden forces, and as occupying a position along a continuum of acceptability. The outer, unacceptable limits of this continuum would be direct diabolical manipulation of natural processes, and the other extreme the divine manipulation of natural processes. The occult disciplines employed by New England's Puritans—astrology, the occult aspects of alchemy, sympathetic magic, image magic, and possibly diabolical magic—appear in bands along that continuum, because each discipline encompassed a variety of occult practices spanning a range of legitimacy. Judicial astrology (star-based divination of future events applicable to specific persons or predictions), for example, was unacceptable to most Puritans, because it was assumed that knowledge of future events could come only from divine prophecy or diabolical magic. The natural astrology that allowed predictions about such things as weather and agriculture, however, was completely acceptable. Combining prayer and astral magic in the creation of alchemical medicines fell into the acceptable range, but using scripture and incantations to charm away illnesses did not. Use of a talismanic sigil (such as

---

12. David D. Hall, *Worlds of Wonder*, 7; Cotton Mather, "Hermes Christianus: The Life of John Winthrop, Esq.; Governour of Connecticut and New-Haven United," in *Magnalia Christi Americana: or, The Ecclesiastical History of New-England* . . . (London, 1702), book 2, chap. 11, 30–33.

Winthrop's use of John Dee's *monas hieroglyphica*) to focus the powers of the macrocosm on spagyric endeavors was sanctioned; use of a paper with symbols and liturgical expressions to cure toothache was proscribed.[13]

Dynamism was the normal state of affairs on this continuum of magical practice. The acceptance of particular magic acts and tolerance for magic in general could vary from person to person and over time. The debate over the weapon salve demonstrates the ambiguity inherent in many magical practices. Healing from a distance could be considered either acceptable deployment of sympathetic magic or totally diabolical, depending on the knowledge and perspective of the person analyzing it.

Within this fluid context of acceptability, individual magic practitioners occupied at any given time a specific area along the continuum. In determining where a practitioner was situated in the eyes of the community, many factors in addition to the kind of magic exercised came into play. Education, status, gender, character, religious conformity, public reputation, social adjustment—all helped a community evaluate the acceptability of a person's magic practices. Witches who harmed neighbors and brought unrest to communities were highly unacceptable. Magi like Winthrop and the physician ministers who manipulated the occult in the service of God were favored. As long as someone's magic fell within acceptable tolerances, that practitioner drew no formal censure from the elite or the laity. But, should one's magical or social reputation shift into the unacceptable range and remain there for any length of time, one ran a distinct risk that informal community gossip would lead to a formal accusation of witchcraft.[14]

Once that occurred, the ministerial and magisterial elite were obliged to sift through the accounts of a suspect's reported magical acts, determine to what degree those accumulated acts corresponded to unacceptable magical practice, and take an official position whether the suspect had actually practiced witchcraft. Precisely because magical practices were located on a dynamic continuum of licit and illicit actions, making such determinations was often difficult. Elite leaders like Winthrop, widely experienced in the work-

---

13. Patrick Curry, *Prophecy and Power: Astrology in Early Modern England* (Princeton, N.J., 1989), 9; Thomas, *Religion and the Decline of Magic*, 286–287; Jon Butler, "Magic, Astrology, and the Early American Religious Heritage, 1600–1760," *American Historical Review*, LXXXIV (1979), 328–332; Godbeer, *Devil's Dominion*, 127–128; John Hale, *A Modest Enquiry into the Nature of Witchcraft* (1702; Bainbridge, N.Y., 1973), 131–132; Increase Mather, *An Essay for the Recording of Illustrious Providences* . . . (Boston, 1684), 261.

14. On the importance of gossip networks in colonial America, see Mary Beth Norton, *Founding Mothers and Fathers: Gendered Power and the Forming of American Society* (New York, 1996), 20, 210–217, 232–236, 250–255.

ings of occult forces in the natural world could and did play a critical role, both formally and informally, in the deliberative process.

The same belief in the power of occult forces that led the Reverend John Davenport to postpone a European voyage in the hope of obtaining an arcane elixir from Winthrop led him and his fellow New Haven colonists to fear the unseen power of witches. In a world of wonders—where devils' snares, like God's providences, were omnipresent and occult magic could be a force for both good and evil—the inexplicable was always analyzed for its deeper providential or diabolical meanings. In such a world, too, the deviant, the difficult, and the contentious persons were commonly the objects of a special kind of scrutiny.

So it was that on May 21, 1653, Elizabeth Godman stood before the magistrates of New Haven seeking formal redress from ten of her neighbors, all of whom had, she claimed, wrongfully accused her of witchcraft. The evidence presented that day and at a second hearing held two days later, however, suggested that Mistress Godman had given ample cause for such accusations. Witness after witness told how, through heart, tongue, and eye, Elizabeth Godman had demonstrated the concealed malevolence, bitter words, and evil looks that many associated with witchcraft. Three things about Godman most concerned her neighbors: her foreknowledge of events, her association with sudden or inexplicable illness, and her seemingly deliberate self-representation as someone intimately familiar with the powers of witches and witchcraft.[15]

Several witnesses reported that Godman had unnatural knowledge of both distant and future events. The Reverend William Hooke, whose charge was corroborated by both his daughter and an Indian servant, said that Godman always knew immediately what had happened in church despite her exclusion (for reasons unknown) from church services. Even more serious were Godman's predictions regarding people's health. Godman had stated that Mary Bishop—betrothed to James Bishop, a man to whom she herself was attracted—would have many fits after she married him, a prediction that had proved correct. She had also accurately foreseen that Hooke's son, who was ill

---

15. Charles J. Hoadly, ed., *Records of the Colony or Jurisdiction of New Haven* . . . (Hartford, Conn., 1858), 31–36 (hereafter *New Haven Colony Records*). Karlsen recounts these cases from a different perspective in *Devil in the Shape of a Woman*, 125–128, 297–298n. Norton focuses on the Godman case as an example of the importance of gossip networks in witchcraft cases (*Founding Mothers and Fathers*, 250–252). Much of the testimony relevant to this case is reprinted in David D. Hall, ed., *Witch-Hunting in Seventeenth-Century New England: A Documentary History, 1638–1692* (Boston, 1991), 61–73.

"in such a strange manner as the docter said hee had not mett with the like," would recover.[16]

Through such predictions, Elizabeth Godman had accrued public suspicion of being a witch, a reputation she seemed to cultivate rather than try to dispel. Hooke reported, "She would be often speaking of witches and did rather justifye them then condemne them; she said why doe they provoake them, why doe they not let them come into the church." Hooke found particularly disturbing Godman's comment that "she had some thoughts, what if the Devill should come to sucke her, and she resolved he should not sucke her." A conversation with Mrs. Goodyear, who had said, "I never knew a witch [to] dye in their bed," brought a correcting response from Godman: "You mistake, for a great many dye and goe to the grave in an orderly way." Godman also habitually muttered to herself or conversed with unseen respondents, acts that made observers suspect she was communicating with a diabolic familiar (the diabolical spirit thought to accompany witches in the shape of animals).[17]

Godman's statements defending witches and her peculiar conversations with beings unseen reinforced the fears of her neighbors over the issue that most concerned them, Godman's close association with harms and illnesses. Hooke was nearly certain Godman had caused his son's mysterious illness, especially because, like a witch, "when his boy was sicke, she would not be kept away from him, nor gott away when she was there." Mr. Goodyear had asked her directly whether she was the cause of the boy's illness, and "she denyed it, but in such a way as if she could scarce denye it." Godman was also thought to be responsible for Mary Bishop's fits and, further, for bewitching to death Bishop's infant children. Her "workes of darkness" at the home of Betty Brewster had also supposedly sent Betty into fits. Hannah Lamberton, daughter of Mrs. Goodyear (by her first marriage), had been a third seizure victim. All three women's fits, once begun, had continued, unresponsive to medical treatment.[18]

---

16. *New Haven Colony Records*, 31–33. Godman's foreknowledge embraced mundane activities as well. She had once told Joshua Atwater's wife that Atwater had figs in his pocket, even though she couldn't see them; another time she had known Atwater served pease porridge when there was no natural way for her to know such a thing.

17. Ibid., 31, 34. Witches were said to have a special teat, an excrescence somewhere on their body, through which the devil or his familiar would nurse and gain nourishment. Elizabeth Lamberton had heard Mrs. Godman alone in her room saying to no one visible, "What, will you fetch me some beare, will you goe, will you goe." Henry Boutele said that early one morning he had "heard her talke to herselfe as if some body had laine with her." Godman was one of those Puritan women descried by Elizabeth Reis for whom the devil had a presence that was both physical and immediate: *Damned Women: Sinners and Witches in Puritan New England* (Ithaca, N.Y., 1997), 55–57.

18. *New Haven Colony Records*, 32, 33. Godman exhibited the same behavior when Mrs. Atwater

Godman's court appearance marked an important transition in her case that was central in any witchcraft case: the institutional formalization of previously informal witchcraft suspicion. Whereas, before, Godman's place on the continuum of magical acceptability was amorphous, subject to the ebb and flow of community opinion, now it must be officially located and fixed. The magistrates must decide whether the ten people named by Godman had falsely accused her of witchcraft or their suspicions warranted further action. They ordered Godman to gather witnesses and return to the next court to make her case.

Although she had been the one who initiated the proceedings against her neighbors, the evidence presented against Godman was compelling. She had an unusual, perhaps preternatural foreknowledge of events, potential association with the devil or a diabolical familiar, and was believed by her neighbors to be responsible for causing injury though occult means to at least four persons. Furthermore, the persons Godman had accused of slander, who subsequently testified against her, were among the most reputable in the colony. Their testimony was not just credible; for Hooke and Stephen Goodyear, it carried high authority.[19]

Central to evaluating whether Godman should be indicted for witchcraft was determining the cause of the fits that affected Mary Bishop, Elizabeth Brewster, and Hannah Lamberton. If their seizures were determined to be of predictable natural origin, then the case against Godman was greatly weakened. Were they determined to be of preternatural origin, possibly with diabolical assistance, as many suspected, then indictment and perhaps conviction would almost certainly follow.

It was precisely in important interpretive situations such as this that occult practitioners like Winthrop strongly but informally influenced witchcraft prosecution. Winthrop and his peers served as authoritative interpreters of magical activity. As a known student of the occult with a superior medical reputation, Winthrop could be relied on to provide expert testimony on

---

warned her away from her home on suspicion of being a witch. She returned the next evening asking for beer. Animals, too, were said to suffer from Godman's bewitchment. Goodwife Thorp told the court that, when she refused to sell Godman a chicken, the angered Godman had issued an implied threat, saying, "Will you give them all[?]" and went away. Soon after, her chickens died, disappeared, or were consumed from within with worms. Ibid., 35–36.

19. Stephen Goodyear was deputy governor and a member of the court, and Joshua Atwater was the colony treasurer. The Reverend William Hooke was the teaching elder, second only in ministerial rank to John Davenport. The unquestioned authority of the elite in judicial proceedings was even stronger in New Haven than in other colonies, because of New Haven's elimination of juries in judicial proceedings. Cornelia Hughes Dayton, *Women before the Bar: Gender, Law, and Society in Connecticut, 1639–1789* (Chapel Hill, N.C., 1995), 24–28.

suspected magical practices, especially when the supposed use of magic was thought to cause injury or illness. As Connecticut's leading physician and alchemist and a student of natural magic, Winthrop was in a better position than anyone else to determine whether illness came from natural or preternatural and possibly diabolical causes.

New Haven naturally turned to Winthrop for counsel in the Godman case. Between her first court appearance in May and her second official hearing in August, Winthrop was contacted by the New Haven physician Nicholas Augur, who requested help in analyzing the fits of the afflicted women. Augur's letter, which thanks Winthrop for his "redines to bee helpefull in doubt full Cases," is at least the third letter in their correspondence but the only one that has survived. Based on internal evidence, however, we can extrapolate critical information regarding Winthrop's response to the Godman accusation.[20]

Augur's initial letter presumably had informed Winthrop that Augur had been the primary physician treating the women. It described the chronic, acute fits they were experiencing and, probably without naming her, the New Haven person suspected of causing them. Augur had asked Winthrop to help him diagnose the cause of the fits, possibly suggesting that Augur anticipated Winthrop would confirm his tentative diagnosis of suspected witchcraft.

Winthrop, with some dispatch, provided an analysis that rejected the witchcraft theory, suggesting several likely physical causes. This drew a cautiously defensive response from Augur. "I must needs say the disease wheare with these wimen weare taken to me was and still is very dubious not so much in respect of the fitt[s] and manner of Takeing them as in the causes that went bee fore ever any of them was taken with such fitts as also in theire cure, which to mee is as strange and more strange than the fitts them selves." Augur reported that since his first letter to Winthrop the "partye suspected" (Elizabeth Godman) had been called before the magistrates and that, following Godman's examination, the afflicted women's "fitts left them, and they Never weare troubled with them since." The formerly afflicted women had "found such an alteration in the stat of theire bodyes that . . . [it] did Confirme them more . . . that it [their seizures] might arise from that suspected party." While Augur had formerly believed, perhaps paying a nod to Winthrop's diagnosis, the cause might be hysterical illness or the beginnings of epilepsy, the continued health of the women after Godman's hearing very strongly suggested to him witchcraft was involved.[21]

---

20. Nicholas Augur to John Winthrop, Jr., June 17, 1653, *Winthrop Papers*, VI, 300–301.
21. Ibid. Winthrop dispatched his most reliable Indian ally, Robin Cassacinamon, to New Haven

In resisting the diagnosis of witchcraft, Winthrop was responding with the cautious but unwavering rejection of witchcraft accusations that came to characterize his involvement in witchcraft cases. His skepticism regarding witchcraft was rooted in the occult philosophies from which his own magical practices were derived. Historians of science have long recognized that early modern Hermetic philosophers and their followers, men such as Paracelsus, Van Helmont, Agrippa, and della Porta, were among the most ardent critics of the wholesale prosecution of witchcraft. Precisely because they studied the operations of occult forces in the natural world and recognized that there were a plethora of ways to create occult effects without diabolical assistance, they were inherently skeptical of most witchcraft charges, though few if any of them doubted in theory the existence of witches or witchcraft. Paracelsus, for example, whose views Winthrop seems to have followed closely, believed in witches but denied that they made compacts with the devil or evil spirits. Paracelsus's medical theories had dramatically expanded the range of natural explanations for disease and drastically circumscribed the influence of diabolical agents in human endeavors. According to him, much of the harm attributed to the diabolical powers of the witch actually resulted from the influence of the imagination of one person on another. English Paracelsians like Winthrop denied concepts like the witch's familiar and witch's teats.[22]

Such skepticism about the likelihood of witchcraft was not, of course, shared by all of the elite who consulted Winthrop, as the testimony in the Godman case of leaders like Hooke and Mr. Goodyear indicates. Because they were professing an alternative understanding of magic and the workings of the occult, "advocates of Neoplatonism and Hermeticism needed to tread warily in order to avoid suspicion of heresy." Fortunately, though, for Winthrop to effectively influence the outcomes of witchcraft cases, he did not have to win converts to the Hermetic perspective. All he had to do was create sufficient doubt about the witchcraft accusations that the accused witch could not be convicted.[23]

---

with the message, a clue that both speed and confidentiality might have been issues. Because of the lengthy sentence construction, I have broken sentences and inserted punctuation to make the meaning more clear.

22. For an interpretation of the scientific revolution and its relationship to changing views of witchcraft, see Charles Webster, *From Paracelsus to Newton: Magic and the Making of Modern Science* (Cambridge, 1982), 88–100. On the relationship between Hermetic philosophies and demonology, Clark's work is essential, especially *Thinking with Demons*, 233–250. See also Charles Webster, *From Paracelsus to Newton*, 81–84, 97; John Webster, *The Displaying of Supposed Witchcraft* . . . (London, 1677), n.p. In no witchcraft case Winthrop ever adjudicated was a search for a witch's teat conducted.

23. Webster, *From Paracelsus to Newton*, 77.

Winthrop followed a consistent strategy in every witchcraft case with which he was involved. Recognizing that formal public witchcraft accusations reflected a communitywide social pathology, his priority was always reintegrating the community and restoring social cohesion. Winthrop refused to let a witch die, but he was not at all averse to coercing a witch to conform to social conventions. Often, from Winthrop's perspective, this meant that it was useful for witch suspects to be found, not completely innocent, but, rather, not exactly guilty. Such findings validated the grounds for the public suspicion that had produced the witchcraft charges, without empowering the accusers, thus defusing some of the fear and anger that underlay the charges. Simultaneously, it put witch suspects on notice that there would be strict expectations regarding their future behavior.

In the case of Elizabeth Godman, this meant that Winthrop, using discretion and acting informally, would have worked, not to assert Godman's innocence, but rather to sow seeds of doubt regarding her guilt and to promote an alternative solution to the community crisis, a via media that would end the social pathology without sending an innocent person to the gallows. At least, from the outcome of the case, that is what appears to have happened. When, six weeks after Winthrop's consultation with Augur, Godman was again before the New Haven magistrates, she was not formally indicted, despite newly incriminating evidence presented against her. She was instead warned "that her carriage doth justly render her suspitious of witchcraft," and she was admonished "not to goe in an offensive way to folkes houses in a rayling manner" but to "keepe her place and medle with her own buisnes." If new suspicions of witchcraft arose, she was warned, "these passages will not be forgotten."[24]

When new suspicions did arise in 1655, two years later, the magistrates did recall their previous suspicions. Their response, however, was once more designed to increase the coercive pressure on Godman to conform to community standards while still avoiding a formal witchcraft conviction. Godman was called before the New Haven town court in August to answer new witchcraft charges, including harming cows and pigs, hindering the churning of butter, and causing objects to mysteriously fly about the house of Mistress Yale. Considering these and her former miscarriages, the court ordered Godman imprisoned until she could face the colony magistrates the following October. This proved to be a kind of shock probation, for a month later,

---

24. The evidence included testimony that Godman had caused Deputy Governor Stephen Goodyear to faint through the use of an evil stare after he read a scriptural passage she thought was directed against her. *New Haven Colony Records*, 29, 30.

while reiterating that the "suspicion of her lewd carriages" was still "exceeding strong," the town court released her from prison, with the proviso that she still must appear before the Court of Magistrates in October and that she "must not go up and down among neighbors to give offense." In October, the magistrates continued her release, though not without an ominous warning. "Though the evidenc is not sufficient as yet to take way her life," the court decreed, "yet the suspitions are clear and many . . . therfore she must forebeare from goeing from house to house to give offence, and cary it orderly in the family where she is, wch if she doe not, she will cause the court to commit her to prison againe." Godman had received a final warning. In the eyes of the court, she was almost, but not quite, guilty of witchcraft.[25]

It was a decision Winthrop undoubtedly had influence over. He had by this time agreed to become at least a short-term resident of New Haven to pursue an ironworks, practice medicine, and (presumably) support John Davenport's "Illustrious College." Given New Haven's intense desire to attract him as a resident and John Davenport's respect for his medical abilities and knowledge, his opinions regarding Godman's second trial would carry unusual authority. Coupled with the decision in the case of Nicholas Bailey and his wife decided during the same period, a pattern of witchcraft adjudication was set in 1655 in New Haven that Winthrop would follow for the rest of his public life.[26]

The Godman case demonstrates Winthrop's strategy of healing social pathology through progressively coercing a suspected witch into social conformity. The treatment of Nicholas Bailey and his wife reveals the other option (short of execution) open to the skeptical adjudicator, banishment. The Baileys were brought before the New Haven town court in July 1655, charged with acts that rendered them "both, but especially the woman, very suspicious in point of witchcraft." Mrs. Bailey had "grossly miscarried" in "impudent and notorious lying," causing "discord among her neighbors," and making "filthy and unclean speeches," including one sanctioning coitus between a dog and a pig and another justifying the sexual misconduct of a neighbor whose "wife was weak." Her actions, said the court, rendered her "as one possessed wth the very devil." Witchcraft prosecution seemed warranted, but the court elected

---

25. New Haven Colony Historical Society, *Ancient Town Records*, I, *New Haven Town Records, 1649-1662*, ed. Franklin Bowditch Dexter (New Haven, Conn., 1917), 249–252 (hereafter *New Haven Town Records*); *New Haven Colony Records*, 151–152. Godman was also ordered to post a bond of fifty pounds to assure her good behavior.

26. Black, *Younger John Winthrop*, 173–177; Dunn, *Puritans and Yankees*, 93–94; Wilkinson, "John Winthrop, Jr. and the Origins of American Chemistry," 225–237; *New Haven Town Records*, 235; John Davenport to John Winthrop, Jr., Jan. 16, May 14, 1655, in Isabel MacBeath Calder, *Letters of John Davenport: Puritan Divine* (New Haven, Conn., 1937), 100–102, 103–105.

"not to proceed at this time." Rather, the magistrates declared the Baileys unfit to live among such neighbors and ordered them either to post a heavy bond for their good behavior (an unlikely possibility) or "consider a way how to remove themselves to some other place." In each of the next three monthly courts the town ordered updates on the Baileys' plans, reminding Goody Bailey in September that the charges "remain as full and the suspicions of witchcraft as strong as before." Finally, in October, the Baileys departed from New Haven's records forever. Winthrop is not listed as participating in the Bailey case, but, given his relationship with New Haven, it is likely he was consulted. "Urged removal" was a strategy of conflict resolution Winthrop frequently resorted to in future witchcraft cases.[27]

THE HARTFORD WITCH-HUNT

Winthrop became officially involved in witchcraft prosecution, not in New Haven, but at Hartford. Connecticut, without Winthrop's knowledge or consent, elected him governor of the colony in May 1657. Given his early relations with Connecticut officials at New London and the current focus of his efforts at New Haven, it was a position he was not eager to accept. In December, however, after repeated entreaties from Connecticut officials, he took up his duties in Hartford. The next May he presided over his first witchcraft trial.[28]

Like Elizabeth Godman of New Haven, Goody Garlick of East Hampton, Long Island, had long been suspected of being a witch. But not until Betty Howell, the daughter of East Hampton's leading settler, cried out from her deathbed that Goody Garlick had bewitched her did informal suspicion become public accusation. Howell, a new mother, had been suddenly afflicted with a terrible headache and fever. Bedridden, fearful, and in agony, she swore that she saw Goody Garlick at one side of her bed accompanied by a black ugly thing at the other side. Garlick, Betty cried, was "readie to pull me in peeces and she prickt me with pins she prickt me with pins." Then, delirious, she gagged, and a middle-sized pin was taken from her mouth, of a kind that had not been in the house before.[29]

27. *New Haven Town Records*, 245–246, 249, 256–258. The basis of most of the complaints against Bailey was her public utterances, underscoring Jane Kamensky's assertion of the relationship between witchcraft charges and Puritans' need to maintain social and political order by controlling unruly women's speech; see Jane Kamensky, *Governing the Tongue: The Politics of Speech in Early New England* (New York, 1998), 5–10.
28. Black, *Younger John Winthrop*, 178; J. Hammond Trumbull, ed., *The Public Records of the Colony of Connecticut*, I (Hartford, Conn., 1850), 297–308 (hereafter cited as *Connecticut Colonial Records*).
29. At that time, East Hampton fell under the jurisdiction of Connecticut. *Records of the Town of East-Hampton, Long Island, Suffolk Co., N.Y.* (Sag-Harbor, N.Y., 1887), I, 140, Depositon of Good-

Her death the next day was followed by three weeks of deposition gathering, in which a comprehensive communal account of the reasons for indicting Garlick for witchcraft was compiled. Howell's death was not the first Garlick had been suspected of causing. There had been a "neger child . . . taken away . . . in a strange manner" and a man who had died suddenly. Goody Edwards accused Garlick of making her daughter's breast milk dry up, which led to the death of Edward's grandchild. Goody Davis also suspected Garlick of bewitching her child to death. Animals, too, had suffered under mysterious conditions associated with Garlick. And, when the owner of an ill-fated sow had burned its tail, an act of countermagic used to draw the offending party to the scene of the fire, Goody Garlick had appeared.[30]

In an apparent effort to ward off his wife's prosecution, Garlick's husband, Joshua, brought a suit of defamation against one of the most vociferous deponents. It fell, however, on deaf ears. Four days after he filed, "a major vote of the Inhabitants of this Towne" ordered Elizabeth Garlick to be delivered into the hands of the authorities in Connecticut "for the trial of the cause of witchcraft which she is suspected for."[31]

It is fortunate for Elizabeth Garlick that Winthrop presided over her trial. Prior to her appearance at the Particular Court in May 1658, Connecticut had the harshest record regarding witchcraft of all New England colonies. Of the seven witch suspects tried there since 1647, all had been convicted, and all had been executed. To Garlick's detriment, four of the seven magistrates who heard her case had each previously participated in four or more of those fatal witchcraft convictions. And, as everyone knew, in witchcraft cases, as in all capital cases, the influence of the magistrates was paramount, even though a separate jury rendered the actual verdict. The magistrates collected evidence, conducted the prosecution, interrogated witnesses, poked holes in testimony, and generally used their privileged position to shape the jury's understanding of the case. If the jury subsequently returned a verdict with which the magistrates disagreed, they could even overturn it.[32]

---

wife Simons, Feb. 24, 1657, 132, Testimonies of Goody Burdsill and Goody Edwards, Feb. 27, 1657, 134 (hereafter cited as *East-Hampton Records*). (The Garlick case is reconstructed and analyzed by Demos in *Entertaining Satan*, 213–245.)

30. Testimony of Goody Hand, Feb. 27, 1657, *East-Hampton Records*, I, 134–135, Testimony of Goody Edwards, Mar. 11, 1657, 139–140, Deposition of Richard Stratton, Mar. 8, 1658, 154, Testimony of Goody Hand, Feb. 27, 1657, 134–135.

31. Mar. 15, 1657, ibid., 140. Garlick's case against Goody Davis was no doubt buttressed by the report that Lion Gardiner, Betty Howell's father, had claimed that Goody Davis "had taken an Indian child to nurse and for lucre of A little wompom had merely starved her owne child." Deposition of Goodman Vaile and his wife, Feb. 27, 1657, 136.

32. The execution of Lydia Gilbert following her conviction in 1654 is undocumented but most

Winthrop's skepticism about witchcraft accusations was at odds with the pattern of convictions the majority of his fellow magistrates had previously demonstrated. Yet, as in New Haven, Winthrop's knowledge of occult medical practices made him an expert witness in matters involving suspected magic and illness. As governor, he was also the leading judicial authority in the colony, expected to set a direction for the magistrates to follow. As the new, much-solicited, and long-awaited governor, he was in the midst of a honeymoon period too, during which he likely enjoyed both extraordinary influence and immunity from criticism. As presiding magistrate, Winthrop could use the courtroom as a theater of persuasion, combining expert knowledge with carefully chosen questions to mold the jury's opinion.[33]

Whatever his welcome, it is sufficient to note that, at the conclusion of the Garlick trial, a Connecticut jury decided, for the first time, not to convict a suspected witch. Predictably, it did not find her completely innocent, either. As Winthrop explained to the townspeople of East Hampton, "Tho there did not appear sufficient evidence to prove her guilty yet we cannot but well approve and commend the Christian care and prudence of those in Authority with you, in searchinge into thatt case, accordinge to such just suspicion as appeared." As in New Haven, the Connecticut decision performed a dual function: it validated the accusers' suspicions without empowering them, and it protected the suspected witch from conviction while maximizing coercion on her to conform to community expectations.[34]

---

likely (Godbeer, *Devil's Dominion*, 235). On the magistrates, see *Records of the Particular Court of Connecticut*, 56, 92, 93; *Connecticut Colonial Records*, I, 220; William K. Holdsworth, "Law and Society in Colonial Connecticut, 1636–1672" (Ph.D. diss., Claremont Graduate School, 1974), 402–403; Peter Charles Hoffer, *The Devil's Disciples: Makers of the Salem Witchcraft Trials* (Baltimore, 1996), 159–161, 170.

33. Governor Haynes's views on the need for harsh treatment of witches had undoubtedly influenced the earlier witchcraft proceedings. Holdsworth, "Law and Society in Colonial Connecticut," 403.

Winthrop could be uniquely effective in the art of persuasion in the Garlick case, because, more than anyone else on the tribunal, he had long-standing personal relationships with the central characters. Lion Gardiner, father of the girl whose deathbed accusation had led to Elizabeth Garlick's indictment, had been associated with Winthrop for more than twenty years, first as military commander of Winthrop's plantation at Saybrook, and later as a trading partner. Joshua Garlick, husband of the accused witch, had carried messages and goods between Winthrop and Gardiner across Long Island Sound. Winthrop could also probably count on the support of committed personal allies among both magistrates and jurors. Two of the magistrates at Garlick's trial and three of the twelve jurors had signed the original petition from Hartford asking Winthrop to relocate there. Samuel Stone and others to John Winthrop, Jr., Aug. 17, 1654, *Winthrop Papers*, VI, 416–417; *Records of the Particular Court of Connecticut*, 188.

34. *Connecticut Colonial Records*, I, appendix 5, 572–573. To ensure that the underlying objective of community reintegration was explicitly understood, Winthrop reminded East Hampton: "It is

With Winthrop's assumption of the office of governor, witchcraft prosecution in Connecticut changed dramatically. Automatic convictions ceased, as did the expectation that complaints would lead to trials. That Winthrop's presence acted as the brake and restraining influence on witchcraft accusations between 1655 and 1661 is strongly suggested by the events that followed Winthrop's departure for England in the summer of 1661 to secure Connecticut's royal charter.[35]

Winthrop's rather sudden journey to England was precipitated by the awkward political situation in which the Connecticut colony found itself upon the restoration of Charles II. As a Puritan colony, Connecticut was known to have supported those who had beheaded the new king's father. (Hugh Peter, one of the regicides who signed the king's death warrant, was Winthrop's father-in-law.) Moreover, Connecticut, like the other New England colonies, was rumored (correctly, at least for New Haven Colony and Massachusetts) to be providing refuge to some of the regicides who had fled England upon the Restoration to avoid prosecution for the king's death. To make matters much worse, Connecticut was extremely vulnerable to crown retribution because it had no validating charter acknowledging its existence as a political entity. The Connecticut River towns had been originally settled by groups of colonists who arrived essentially as squatters and only later worried about political legitimacy. As a result, a new king bent on taking retribution against the colony could legally declare the colony's current government illegitimate and impose a new royal government controlled from London rather than from Hartford. This was a matter of tremendous concern to Connecticut's leaders once Charles II took office, and in an effort to stave off such a move

---

desired and expected by this Court, that you should cary neighbourly and peaceably, without just offence, to Jos: Garlick and his wife, and that they should doe the like to you." Garlick, like New Haven's Elizabeth Godman, was ordered to post a high bond to assure her good behavior and to appear again at a fall court session.

35. Ibid., 338. Samuel Wyllys was sent by the General Court to Saybrook in 1659 "to assist the Major [John Mason] in examining the suspicions about witchery, and to act therein as may be requisite," a visitation that produced no further action. It appears that the hearing might have been an inquest into the affairs of Margaret and Nicholas Jennings, who were indicted two years later on witchcraft charges and who seem to have fled the colony thereafter (Demos, *Entertaining Satan*, 508 n. 59; *Records of the Particular Court of Connecticut*, 50, 223, 225, 227, 238, 240). Charles H. Levermore, a nineteenth-century historian, claimed the Jenningses were indicted in 1659, but there seems to be no evidence of such indictment ("Witchcraft in Connecticut, 1647–1697," *New Englander*, XLIV [1885], 806; Hall, ed., *Witch-Hunting*, 98). Also in 1661, John Robbins's wife of Wethersfield accused Katherine Palmer of causing her illness. After she died, her husband drew up a paper to prove the accusation, but nothing came of it. Wyllys Papers, Ann Mary Brown Library, Providence R.I., W-7 (hereafter cited as Wyllys Papers); Demos, *Entertaining Satan*, 353; Hall, ed., *Witch Hunting*, 157.

the colony hurriedly authorized funding to support a preemptive mission by Winthrop to try to secure a legitimate charter from Whitehall.[36]

Within months of Winthrop's leave-taking, witchcraft trials began again, and the pattern of prosecution reveals the rapid devolution of Winthrop's moderate witchcraft policies. The first trial, that of Nicholas and Margaret Jennings of Saybrook, followed the model established by Winthrop—neither convicting nor acquitting the accused. But a subtle yet fundamental difference in the verdict handed down by the court produced an altogether different effect on the process of community reintegration.

The Jenningses, a Saybrook couple, were indicted in Hartford in September 1661 on suspicion of murdering through bewitchment several persons and committing other unnamed sorceries. Details of their October trial were not recorded, but at its conclusion the court secretary noted, "The major part find him [Nicholas Jennings] guilty of the inditemt the rest strongly suspect it that he is guilty." Regarding Margaret Jennings: "Some of them find her Guilty the rest strongly suspect her to be." Whereas in Winthrop-influenced cases suspects had been found not guilty but highly suspicious, in the Jenningses' cases the accused were found probably guilty but not convicted. The trial outcome was the same—the suspects were released—but the shift in emphasis from high suspicion to probable guilt was, in terms of community reintegration, fundamentally damning. Rather than validating the accusers without empowering them—Winthrop's strategy—the Jenningses' verdicts actually fueled further suspicion and encouraged future unrest. Unintentionally, perhaps, it encouraged Saybrook residents to find new, conclusive evidence of the Jenningses' guilt. Faced, therefore, with an untenable and highly volatile situation, the Jenningses fled the colony.[37]

The following March, witchcraft accusations resurfaced when eight-year-old Elizabeth Kelly awoke in excruciating stomach pain. She screamed for her father and told him that Goody Ayres, a woman known for spreading stories of encounters with the devil, was tormenting her. Echoing the reported deathbed cries of Betty Howell in East Hampton four years earlier, Kelly had

---

36. In 1643, Connecticut had purchased the rights held by the Saybrook founders in a patent supposedly issued to the earl of Warwick, but that patent was of questionable validity and could not in fact ever be produced.

37. *Records of the Particular Court of Connecticut*, 188, 238, 240, 243. Two of the jurors, Edward Stebbing and John Moor, had participated in the trial of Elizabeth Garlick. Three of the four magistrates, Daniel Clark, Matthew Allyn, and Richard Treat, had served with Winthrop when the General Court ordered the previous inquiry into witchcraft at Saybrook two years earlier (*Connecticut Colonial Records*, I, 338). Two Jennings children were apprenticed by the court at the session following their trial (Demos, *Entertaining Satan*, 302–303, 508 n. 59). Demos believes the Jenningses fled to Rhode Island; for a contrary view, see Tomlinson, *Witchcraft Trials*, 25.

screamed: "Goody Ayres torments me, she pricks me with pins she will kill me. . . . Father why doe you not go to the magistrates." When Kelly died in agony saying, with her last words, "Goody Ayres chokes me," the magistrates were sent for, and the Hartford witch-hunt began.[38]

Initially, the court continued to exercise the judicious restraint espoused by Winthrop. To evaluate the claim that Elizabeth Kelly's death had been caused by unnatural forces, the court sought the advice of a medical expert. In Winthrop's absence, it turned to the physician Bray Rossiter, who conducted a graveside autopsy. Rossiter was a far less temperate man than Winthrop and, by reputation, not as skillful a physician. Unlike Winthrop, there is no record of his practicing alchemical medicine or studying occult philosophies. Winthrop had at one time considered relocating to Rossiter's hometown of Guilford, and patients there revered him. Just before his departure for Europe, Winthrop had received a letter from Guilford's William Leete saying, "My wife entreats some more of your phisick, although shee feareth it to have very contrary operations in Mr. Rossiters stomach."[39]

Called to consult on the cause of death of Elizabeth Kelly, Rossiter rendered an opinion it is inconceivable that Winthrop would have rendered personally or have accepted unchallenged from someone else. Rossiter provided the magistrates six detailed medical reasons why Kelly's death had in fact been from preternatural causes. The impact of his testimony was profound. Suddenly there was manifest physical evidence, in the corpse of Elizabeth Kelly, that occult diabolical magic was a palpable presence, killing and corrupting the innocent. The evidence was damning. When Goody Ayres heard the depositions against her read, she exclaimed, "This will take away my life." Before that could happen, however, Ayres and her husband, like the Jenningses before them, fled the colony, leaving behind an eight-year-old son.[40]

---

38. "Did you not see her last night stand by the bed side readie to pull me in peeces and she prickt me with pins, she prickt me with pins." Elizabeth Howell, from "Deposition of Goody Simon," Feb. 24, 1657, *East-Hampton Records*, 132. Also: "I asked [Betty Howells] what would you have with her and she said, 'I could tear her in peeces.'" Kelly's statements are in Wyllys Papers, W-6.

39. While Winthrop was in England, Rossiter and his son were arrested by New Haven authorities for attempting to cause insurrection there and bring New Haven under Connecticut's jurisdiction. They were released only after Rossiter signed a statement admitting "he had been very rash and inconsiderate, and, he could freely say, offensive" (*New Haven Colony Records*, 456, 513). Gurdon W. Russell, *Early Medicine, and Early Medical Men in Connecticut* ([Hartford, Conn.], 1892), 15–23; William Leete to John Winthrop, Jr., Apr. 12, 1661, Massachusetts Historical Society, *Collections*, 4th Ser., VII (1865), 546–548.

40. Testimony of Dr. Rossiter, Mar. 31, 1662, Wyllys Papers, W-5. Rossiter's preternatural signs were the absence of rigor mortis; the deep blue tincture of the decedent's skin; the presence of blood only in the throat, and that uncoagulated; the presence of fresh blood in the backside of

Rossiter's medical confirmation of the colonists' worst nightmares unleashed a torrent of witchcraft accusations. Hartford—deeply divided by many years of bitter contention over issues such as ministerial authority, standards of church admission, and personality clashes within the church—began a year of panic that produced eight witchcraft trials in as many months. New witchcraft victims screamed their presence, as social tensions amplified by the church controversy erupted into new pathology. As in the Godman case years earlier in New Haven, women were afflicted with violent fits of suspicious origin. This time, though, there was no willing authority powerful enough and knowledgeable enough to defend a naturalistic explanation of their condition. One young woman, "or rather the Devill . . . making use of her lips," according to the Reverend John Whiting of Hartford, spoke in a "very awefull and amazing" Dutch accent, accompanied by "extreamely violent bodily motions . . . even to the hazard of her life." As the ministers scribbled notes of her words, Ann Cole's devil accused her neighbor Rebecca Greensmith of injuring others through witchcraft.[41]

Greensmith was called before the magistrates. Under ministerial prodding she broke down, confessing not only that she had had meetings with the devil—who "had frequent use of her body with much seeming (but indeed

---

the arm; the gall bladder ruptured "and curded" with no discoloration to adjacent parts; and the gullet contracted like a "fish bone." They are consistent with those commonly found in corpses several days after death and reflect the period's—and Rossiter's—extremely limited knowledge of the subject area (Tomlinson, *Witchcraft Trials,* 30). Robert Blair St. George sees Rossiter's autopsy as a sign of the "enlightened" discourses of empirical medicine beginning to impose themselves on the Puritan world of revelation. Rather than showing a highly detailed knowledge of anatomy however, as St. George suggests, Rossiter's failure to understand the changing effects of rigor mortis and the manner in which blood pools within corpses under certain conditions seems rather to reflect a quite limited empirical understanding of anatomy. See Robert Blair St. George, *Conversing with Signs: Poetics of Implication in Colonial New England Culture* (Chapel Hill, N.C., 1998), 118–119; Tomlinson, *Witchcraft Trials,* 31; Report of Trial of Goodwife Seager, Wyllys Papers, W-2. Ayres's assets were confiscated by the colony to pay his forfeited bond. *Records of the Particular Court of Connecticut,* 258.

41. I follow Demos in arguing that the church crises of the years preceding the Hartford witch-hunt were a significant contributing factor to the outbreak. The Hartford controversy of the late 1650s had pitted church members favoring Presbyterian polity and the halfway covenant against strict Congregationalists who favored limiting membership only to the demonstrably elect. The tensions generated thereby had been bitter and split churches at Hartford, Wethersfield, and elsewhere along the Connecticut River valley. Demos, *Entertaining Satan,* 349–351, 364–365, 369, 382–383; Tomlinson, *Witchcraft Trials,* 27; Sylvester Judd, *History of Hadley . . .* (Springfield, Mass., 1905), 3–17; Paul R. Lucas, *Valley of Discord: Church and Society along the Connecticut River, 1636–1725* (Hanover, N.H., 1976), 43–50; Robert G. Pope, *The Half-Way Covenant: Church Membership in Puritan New England* (Princeton, N.J., 1969), 83–85, 96–97; Edwin Pond Parker, *History of the Second Church of Christ in Hartford* (Hartford, Conn., 1892), 18–50; John Whiting to Increase Mather, Dec. 4, 1682, MHS, *Collections,* 4th Ser., VIII (1868), 466–469.

horrible, hellish) delight to her"—but also that she had attended meetings with other local witches. Greensmith named seven additional witches from Hartford and Wethersfield and confirmed for the court that her husband, too, was a witch.[42]

More witnesses were deposed, suspects interrogated, and new trials set. As in England, the legal foundations for conviction were informed by the writings of William Perkins and Richard Bernard. Both authors emphasized the need to prove legally that an accused witch had a direct relationship with the devil, either by extracting a confession or through the testimony of multiple witnesses to the witch's diabolical acts. In practice, however, Perkins and Bernard provided guidelines only. The English witch-hunter Matthew Hopkins also seems to have influenced the case. Perkins and Bernard rejected use of the ordeal by water, a test of suspected witches that Hopkins had favored. In Hartford, a couple was subjected to that test: bound hand to toe and dropped into water to see whether they floated. If they floated, they were sure to be witches; if not, they were probably innocent. The buoyant couple failed the test, but, before their prosecution progressed further, they fled the colony. The theory behind the water ordeal was magical, with a spiritual foundation: just as a witch had rejected baptism, so, it was reasoned, the water would reject the witch, forcing the witch to float to the surface.[43]

More important than the legal guidelines for prosecution were the aggressive attitudes of the prosecutors. Ministers have usually been represented as a moderating force in witchcraft trials, gatekeepers who restrained the public

---

42. John Whiting to Increase Mather, Dec. 4, 1682, MHS, *Collections,* 4th Ser., VIII, 466–469. Greensmith demonstrates Reis's argument that Puritans' emphasis on women's moral inferiority convinced women themselves of their susceptibility to and involvement in witchcraft (Reis, *Damned Women,* 121–163). Greensmith's angry initial resistance turned to self-incrimination. "Testimony of Rebecca Greensmith," Wyllys Papers, W-1.

43. William Perkins, *A Discourse on the Damned Art of Witchcraft* . . . (London, 1608); Ric[hard] Bernard, *A Guide to Grand Jury-Men* (London, 1627); Godbeer, *Devil's Dominion,* 159–163; Weisman, *Witchcraft, Magic, and Religion,* 99; Hoffer, *The Devil's Disciples,* 142–144. The specific guidelines followed in witchcraft cases are unclear for this period. The undated "Grounds for Examination of a Witch," copied in the hand of William Jones, has often been cited as the legal grounds for convictions in Connecticut, but internal evidence suggests it dates from the end of the century, rather than the middle (Wyllys Papers W-38; Tomlinson, *Witchcraft Trials,* 57). Floating, a sign that the suspect had been rejected by water, was a confirmation of the witch's diabolical compact, though not alone considered legally sufficient evidence for conviction. It carried more weight in the court of public opinion; see Tomlinson, *Witchcraft in Connecticut,* 34; Holdsworth, "Law and Society," 684 n. 2; Godbeer, *Devil's Dominion,* 161; Increase Mather, *Essay for the Recording of Illustrious Providences,* 139. Matthew Hopkins, the English witch-finder who gained fame in the 1640s through his prosecution of witches in Essex, encouraged swimming, but never used it as evidence. Rossell Hope Robbins, *Encyclopedia of Witchcraft and Demonology* (New York, 1959), 492–494; Thomas, *Religion and the Decline of Magic,* 551.

impulse to convict through their insistence on rigorous evidentiary standards. Not in Hartford: ministers were especially forceful in prosecution. Despite the fact that the ministers involved were at the center of both sides of the continuing church disputes, they all agreed with and joined together in the prosecution of the accused witches. Sixty-year-old Samuel Stone, aided by his young assistant John Whiting, the even more youthful Reverend Joseph Haynes of Wethersfield, and the Reverend Samuel Hooker of Farmington became a prosecutorial tribunal, gathering evidence, recording notes, and vigorously interrogating witnesses. Their ability to extract damaging testimony from even hostile deponents was underscored in the trial of Rebecca Greensmith, whose confession came in direct response to ministerial confrontation. When, at her trial, Joseph Haynes began to read his notes of the allegations against Greensmith made by the possessed Ann Cole, Greensmith was infuriated. She "could have torne him in peices," she said, "and was as much resolved as might be to deny her guilt." But, as Haynes persisted in the evidentiary assault, Greensmith crumbled. She felt "as if her fflesh had been pulled from her bones . . . and so could not deny any longer." Greensmith confessed and was subsequently hanged.[44]

By the time John Winthrop returned to Hartford in June 1663, four people—Rebecca Greensmith, Nathanial Greensmith, Mary Sanford, and Mary Barnes of Farmington—had gone to the gallows. As many as five others had fled the colony in fear. Only two persons had been acquitted. Andrew Sanford, the first person indicted in the witch-hunt, had barely escaped conviction; his wife Mary had been executed. Elizabeth Seager had been even more narrowly acquitted of witchcraft charges in January 1663, and she already stood accused a second time.[45]

---

44. Godbeer, *Devil's Dominion*, 161–163; Weisman, *Witchcraft, Magic, anhd Religion*, 98–105. In 1662, Whiting was twenty-eight, and Haynes twenty-two. Stone had gained notoriety in the 1648 trial of Mary Johnson for pressuring her into a gallows confession and repentance. Haynes was the son of the former Connecticut governor, himself a stern witchcraft prosecutor. Cotton Mather, *Memorable Providences, Relating to Witchcraft and Possessions* (1689), in George Lincoln Burr, ed., *Narratives of the Witchcraft Cases* (New York, 1914), 135–136; Arthur Adams, "John Haynes," in Charles E. Perry, ed., *Founders and Leaders of Connecticut* (Boston, 1934), 65–66; John Whiting to Increase Mather, Dec. 4, 1682, MHS, *Collections*, 4th Ser., VIII, 466–469.

45. Those known to have fled were Goody Ayres and her husband and James Wakeley of Wethersfield. John Demos suggests, based on their disappearance from local records, that Katherine Palmer, another suspect, might have fled the colony with her husband, Henry (Demos, *Entertaining Satan*, 352–355). One more suspect, Judith Varlet, had been released only after her brother-in-law Peter Stuyvesant, governor of New Netherlands, protested the Connecticut court's "pretend accusation of witchery." "Peter Stuyvesant to the Hartford Magistrates," in Robert C. Winthrop Collection of Connecticut Manuscripts, 1631–1794, Archives, History and Genealogy Unit, Connecticut State Library, Hartford, 1, doc. 1a; Tomlinson, *Witchcraft Trials*, 37; Karlsen, *Devil in the*

Winthrop must have been both astonished and disheartened at how far the moderate approach to witchcraft prosecution had been superseded in his absence. But, as the case of Elizabeth Seager demonstrates, it would be more difficult to restore that approach than it had been to institute it.

REDEFINING WITCHCRAFT

Elizabeth Seager lived under formal suspicion of witchcraft for nearly three years. She had been named as one of the Hartford witches in 1662 and had come under suspicion earlier at the trial of Goodwife Ayres. Seager was quickwitted and sharp-tongued, and she fiercely rejected the charges against her. Unlike Goody Greensmith, who broke down under Haynes's carefully documented reading of the evidence against her, Seager called the notes taken by Haynes "a great deal of hodgepodge." When told that Ann Cole had said she was a witch, Seager shocked her audience by praying to Satan to tell them she was no witch. A stunned Margaret Garrett demanded to know why she had invoked Satan's aid. In response, Seager taunted, "Because Satan knew she was no witch." At her first trial, Seager had remained aggressive. She rejected the ordeal by water, suggesting that the devil himself could falsify its results. "The devil that caused me to come here can keep me up," she claimed. And, when pressed by magistrate John Allyn on the details of her relationship with the suspected witch Goody Ayres, Seager fumed: "I do not know that I am bound to tell you. . . . Nay I will hold what I have if I must die."[46]

Resistance might have helped Seager gain the barest of acquittals in January 1663, but the nature of her verdict virtually guaranteed she would be in court again. "Half the jury ore more did in their vote cast Goody Seger [guilty], and the rest . . . were deeply suspicious. . . . They were some times likely to com

---

Shape of a Woman, 25–26. "The Jury returne they cannot agree Some find Inditement against Sanford the rest strongly Suspect." *Records of the Particular Court of Connecticut*, 250, 251, 258–260; Wyllys Papers, W-2.

46. When Ayres had said in court, "This will take away my life," a witness noted that Goody Seager had "shuffled her with her hands and sd hold your tongue wt grinding teethe" (Wyllys Papers, W-2). "There was a meeting under a tree in the Green by the house and there was there James Walkely Peter Grants wife Goodwife Ayres and Henry Palmers wife of Wethersfield, and Goody Seager, and there we danced and had a bottle of Sack; it was In the Night, and something like a cat called me out of the meeting" (Testimony of Rebecca Greensmith, Wyllys Papers, W-1). Case of Goodwife Seager, Wyllys Papers, W-2; Testimony of Garrett and Watson, Wyllys Papers, W-4. Seager said she did so relying on Acts 19:14, where people spoke to evil spirits in the name of Jesus (Wyllys Papers, W-2). Seager's refusal to engage in self-incrimination might have reflected a transferal of this newly recognized right accorded the accused in English courts. Her protest, however, was of short duration. Under continued prodding, she answered Allyn's questions. Holdsworth, "Law and Society," 527.

up in their Judgments to the rest; whereby she was almost gone and cast, as the foreman expressed to her at giving in of the verdict." Gone was any trace of the community reintegration strategy formerly employed by Winthrop. Seager's accusers were both validated for making their accusations and empowered to continue their quest for a conviction.[47]

When court met in July 1663, the month after Winthrop returned from England, Seager was charged again with witchcraft and with additional charges of blasphemy and adultery. Winthrop faced the challenge of attempting to undo a work in progress. He undoubtedly rejected both the heavy-handed verdict of Seager's first trial and the intellectually discredited means, like the water ordeal, that had been invoked for obtaining evidence against her. Because of the number of witch prosecutions that had occurred in his absence, however, Winthrop took what seems a cautious though moderating stance toward Seager's second prosecution. Considering the adulation that had accompanied his return to Hartford with a spectacularly generous royal charter and the enhanced intellectual authority he possessed as one of the charter members of England's newly founded Royal Society, Winthrop might easily have been able to defuse the accusations of witchcraft. More likely, a compromise was arranged with his fellow assistants and pressed on an intractable jury. Whatever the tactical considerations might have been, at the end of Seager's second trial she was once again acquitted of witchcraft, though found guilty of adultery.

This verdict, too, proved only a temporary solution. In June 1665 Seager was indicted yet a third time, for "continuing to practice witchcraft." Witnesses seemed determined to get a conviction—by using Seager's own words against her. One said that Seager told her "it was very good to be a witch, and desired her to be one." Another claimed Seager told her that Seager's accusers "Missed there Mark," since the chief actor among the witches was still at large. Seager went on to say she would reveal this chief witch's name at a future time.[48]

---

47. Wyllys Papers: Supplement, doc. 53, printed in John M. Taylor, *The Witchcraft Delusion in Connecticut, 1647-1697* (Williamstown, Mass., 1984), 83–85.

48. Holdsworth, "Law and Society," 520; Hartford City Court Records, Connecticut State Library, Hartford, 35, 36, 38, 52; Testimony of Mrs. Migat, Wyllys Papers, W-3. The inscriptions on the verso of Migat's deposition suggest that this testimony might have been used in both Seager's 1663 and 1665 trials. In this third trial, the effects of countermagic were offered as testimony. Goodwife Garrett claimed that an especially good cheese she had made was found to be full of maggots. Cutting the infested part out and throwing it into the fire, Garret heard Seager suddenly "cry out exceedingly . . . after a short time Goodwife Seager came into the house, cried out she was full of Paine, and sat wringing of her body and crying out." Seager's pained reaction and sudden appearance, as all knew, indicated that her witchcraft had been the source of the maggot infestation." Wyllys Papers, W-4.

Perhaps anticipating the likely verdict, Winthrop refrained from participating in Seager's third trial, in which she was, predictably, found guilty. He did, however, act forcefully to challenge the decision. Calling together an extraordinary meeting of the governor and magistrates, he refused to enforce the jury's verdict. Instead, he used the authority of his specialized knowledge and political power to buy time: "The Governor declared that it was his desire that the matter might be respited to a further consideration for advice in those matters that were to him so obscure and ambiguous and the issue is deferred to the Quarter Court in September next." September came and went without the promised decision. Winthrop delayed acting until the following May to take advantage of major judicial reforms then in progress.[49]

The colonial charter Winthrop secured from Charles II significantly expanded the authority of the governor. Not only was he now formally eligible for indefinite reelection, but the governor, deputy governor, and assistants were also given explicit authority to wield unlimited administrative, legislative, and judicial powers. In court cases, this included the right to "impose, alter, change, or annul any penalty, and to punish, release or pardon any offender." Additionally, fundamental reform of the judiciary institutions took place during the time Elizabeth Seager's verdict was in hiatus. In October, the old Particular Court was superseded by a new Court of Assistants, consisting of seven or more assistants, which was to meet to hear capital, important civil, and divorce cases. In May, the governor or deputy governor and assistants became empowered to summon special Courts of Assistants ad hoc, to handle unique situations or sudden needs.[50]

The colony's first special Court of Assistants, called the week after the creation of such courts, rendered the long-awaited decision in the case of Elizabeth Seager: "This court considering the verdict of the jury and finding that it doth not legally answer the indictment do therefore discharge and set [Elizabeth Seager] free from further suffering or imprisonment." For the first time in Connecticut's history, a convicted witch did not die. Seager, like several other witch suspects before her, left Hartford and made her way to Rhode Island.[51]

---

49. Adjourned Court, June 15, 1665, Records of the Connecticut Court of Assistants, Connecticut State Library, Hartford, microfilm (hereafter cited as Court of Assistants); Court of Assistants, July 8, 1665.

50. *Connecticut Colonial Records,* II (1852), 3–11, 28–29: Holdsworth, "Law and Society," 438, 440, 446, 451, 452. Except in capital cases, such special courts met without juries. Private meetings of the assistants, which constituted a kind of executive session of the court, also met without juries. *Connecticut Colonial Records,* II, 38–39; Holdsworth, "Law and Society," 456.

51. *Court of Assistants,* 52, 56; Tomlinson, *Witchcraft Trials,* 39.

In the aftermath of the Seager trial, magistrates once again embraced Winthrop's moderate stance toward witchcraft prosecution. A February 1667 hearing regarding accusations that William Graves of Stamford had bewitched his daughter to death resulted in no indictment. A few months later, the magistrates not only rejected what they considered an unwarranted witchcraft accusation; they told the accuser that he "did greatly sin in harboring such jealousies in his heart." Many members of the polity, however, proved less eager than the magistrates to accept Winthrop's lenient witchcraft position. In the wake of the Hartford witch panic, some seemed reluctant to cede full power to determine questions of witchcraft only to the elite, even to someone with the occult knowledge and social stature of Winthrop. Many who felt Seager's witchcraft had harmed them might have laid the blame for her release directly at the feet of the governor. Winthrop, however, whose medical practice now encompassed virtually all of Connecticut, continued regularly to provide a public demonstration of his special credentials for assuming authority over the interpretation of suspected magic practice.[52]

Nowhere was lay resistance to Winthrop's occult authority more apparent than in the case of Katherine Harrison, whose trial marked a turning point in New England witchcraft prosecution. Harrison, a Wethersfield widow who had risen from low status to become a person of substantial means, had been suspected of witchcraft since her early years as a servant in the household of Hartford's John Cullick. Like many suspected witches, Harrison was a medical practitioner—a "cunning woman" or doctoress—others had turned to for relief from injuries and ailments. Working at the boundary between life and death, sickness and health, Harrison was particularly vulnerable to charges of witchcraft. As Richard Godbeer has noted, "Medical skill implied magical skill," and the boundaries between magical and nonmagical treatment were ambiguous. One who could heal could also potentially harm. When an illness lingered or worsened, there was always the possibility that it had been magically prolonged or inflicted by the supposed healer.[53]

---

52. Wyllys Papers: Supplement, 37–44; Holdsworth, "Law and Society," 711 n. 45.
53. Harrison's case is among the best documented of all witchcraft cases and has been analyzed and retold from several perspectives. Demos, *Entertaining Satan*, 355–365; Karlsen, *Devil in the Shape of a Woman*, 84–89; Tomlinson, *Witchcraft Trials*, 43–50; Taylor, *Witchcraft Delusion*, 47–61; Carolyn S. Langdon, "A Complaint against Katherine Harrison, 1669," Connecticut Historical Society, *Bulletin*, XXXIV (1969), 18–25. Hall, ed., *Witch-Hunting*, 170–184, reprints substantial primary source material; see also Godbeer, *Devil's Dominion*, 65–69, esp. 67. On female doctors' special vulnerability to witchcraft accusations, see Rebecca J. Tannenbaum, *The Healer's Calling: Women and Medicine in Early New England* (Ithaca, N.Y., 2002), 125–133; Karlsen, *Devil in the Shape of a Woman*, 141–144; Guido Ruggiero, *Binding Passions: Tales of Magic, Marriage, and Power at the End of the Renaissance* (New York, 1993), 147–166.

Many of the witchcraft charges made against Harrison involved failed cures or accusations of intentionally caused illness or death. Joan Frances accused Goody Harrison of magically causing the death of her young child. At least five other people connected her to the death of the child of Josiah Willard. Witnesses noted as proof of Harrison's agency the fact that, when she touched the Willard's child's corpse, it moved. Jacob Johnson was also thought to have met death through Harrison's medical malevolence. When Johnson had sought other medical advice for an illness Harrison had unsuccessfully treated, Harrison supposedly caused his nose to bleed extraordinarily, which continued until his death.[54]

In addition to being charged with causing illness and death, other witnesses testified to seeing Harrison's spectral apparitions. Mary Kercum saw Harrison "and her black dog" appear by moonlight in the house where she was staying. Thomas Bracy observed a hay cart on top of which was a calf's head, its ears standing straight up. As the cart drew near, the calf's head suddenly vanished, and Harrison was standing there in its stead. Twenty-year-old Mary Hale was in bed when an "ugly shaped thing like a dog, having a head such that I clearly and distinctly know to be the head of Caterin Harrison," walked to and fro in her chamber.[55]

Assuming special importance in Harrison's case was testimony of her abilities as a fortune-teller. Harrison not only related things "that did come to pass," but she "foretold many matters that in future times were to be accomplished." Samuel Martin claimed that Harrison had publicly predicted the deaths of two men, Josiah Willard and Samuel Hale. Mary Olcott testified that Harrison had correctly foretold the fact that Elizabeth Smith would not marry her intended spouse, William, but, rather, a man named Simon. Significantly, Harrison had asserted that her remarkable foreknowledge was grounded, not in folk magic, but in occult philosophy, particularly her knowledge of the works of the English astrologer William Lilly. William Warren called Harrison a "common and professed fortune teller," who "had her skill from Lilly." Thomas Waples, too, testified that she claimed "she had read Mr. Lilly's book in England." Elizabeth Simon went even further, saying Harrison "would often speak and boast of her great familiarity with Mr. Lilley."[56]

Harrison's claims to mastery of occult arts, especially the highly ambiguous realm of predictive astrology, were to make her case especially sensitive for

---

54. Wyllys Papers, W-10, W-16; "Harrison's Answer to Winthrop," 1669, Winthrop Family Papers, microfilm, reel 9, Massachusetts Historical Society, Boston.

55. Ibid.; Hall, ed., *Witch-Hunting*, 181.

56. For Lilly, see Thomas, *Religion and the Decline of Magic*, 307–319, 409–414; Wyllys Papers, W-11; Hall, ed., *Witch-Hunting*, 173–175, 177–178.

Winthrop and the magistrates, for it called into direct question the boundary between intellectual magic and popular folk magic in ways most witchcraft cases did not. Given the special volatility of the Harrison case, it raised a question of boundary definition the magistrates would ultimately feel compelled to address.

Long before people came forward to witness formally against Harrison, neighbors had engaged in acts of vandalism or charivari against her property. Harrison's ox had been maimed right at her door; a cow in the street had been wounded in the udder. One of Harrison's pigs had wandered into the woods and come home with its ears pulled out and its hind leg cut off. Many in Wethersfield and other towns had feared and hated Katherine Harrison for a long time. In the fall of 1668, Harrison lost a defamation suit to Michael Griswold, who had apparently been trying to organize a group to bring formal witchcraft charges against her. In response, Harrison had attacked Griswold in public, saying "he would hang her though he damned a thousand souls, and as for his own soul, it was damned long ago." For good measure, she had also called Griswold's wife "a savage whore."[57]

Whether or not Griswold was the instigator, evidence gathering continued, and in May 1669 Harrison was formally indicted for witchcraft at the Court of Assistants. As many as thirty and probably more witnesses offered evidence of Harrison's fortune-telling, spectral appearances, and malevolent healing. Winthrop and the assistants seem to have vigorously asserted their courtroom authority to resist this concerted effort to convict Harrison, for, despite the plethora of testimony presented against her, at the conclusion of the trial the jury could not "as yet" agree on a verdict. The magistrates ordered Harrison imprisoned until the next court in October, at which time the jury would be reconvened.[58]

Harrison's opponents were clearly displeased with the proceedings of the Winthrop court. They immediately began collecting new depositions against her for use in the next trial. When, before the October court date, the magistrates released Katherine Harrison from prison and allowed her to

---

57. "A Complaint of Several Greeuances of the Widow Harrison," Oct. 6, 1668, Wyllys Papers, W-11; Hall, ed., *Witch-Hunting*, 171–172. Griswold asked one hundred pounds for himself. Simultaneously, he entered a fifty-pound suit on behalf of his wife, Hartford County Court, Sept. 2, 1668, Probate Records, Connecticut State Library, Hartford; Demos, *Entertaining Satan*, 359.

58. The estimate of the number of persons providing evidence is taken from the dates of the depositions in the Wyllys Papers and the names of deponents indicated in "Harrison's Answer to Winthrop," Winthrop Family Papers, microfilm, reel 9. Since no formal record was made of oral testimony in court, depositions reflect testimony from witnesses who were exempted from coming to court because they lived more than twenty miles away or could claim illness, pregnancy, or a nursing child. Dayton, *Women before the Bar*, 5–6 n. 9.

return to Wethersfield, local resentment flared into angry resistance. Thirty-eight Wethersfield residents, including two ministers, the local physician, and Wethersfield's representative to the General Court, signed a petition protesting Harrison's release and urging her immediate reincarceration.[59]

The Wethersfield petitioners explicitly challenged the authority of whoever had granted Harrison's freedom: "Wee are not satisfied upon what rightfull ground she is at such libertie." They noted that "very evle, hurtfull and dangerous effects" had occurred since Harrison had been released, and they named new witchcraft victims. They further demanded that Harrison be ordered to endure the water ordeal and that she be rejailed before she could flee the colony. "She hath disposed of great part of her estate to others in trust, and it is believed that she entendeth to be gone."[60]

Three specific points in the petition suggest the petitioners felt the magistrates had slanted the former proceedings in Harrison's defense. "Theire are diverse witnesses more to be considered," they noted, "whose witnessing was not given in to the consideration of the Juerie." These witnesses' testimony had since been taken in writing, the petitioners reported, and some of them would be present at the next trial "to declare themselves more at large." The petitioners further requested that the prosecution be directed, not by the magistrates, but by John Blackleach, a Wethersfield merchant whose name headed the list of signers. Blackleach, the petitioners said, would "be the better fitted to answere diverse questions, and cleare up some matters nessessary." Furthermore, to make sure that obscure legal matters should not prove impediments to conviction (as they did in the case of Elizabeth Seager), they desired that William Pitkin, the colony's foremost lawyer, be ordered to assist Blackleach in the prosecution.[61]

Was Winthrop the particular target of criticism in the Wethersfield petition? Had he pressed the authority of his occult knowledge and political leadership further than his fellow Puritans would allow? As chief magistrate, Winthrop bore most of the official responsibility for both the conduct of Harrison's trial and her subsequent release. Circumstantial evidence suggests also that he might individually have been instrumental in setting Harrison free.

---

59. Wyllys Papers, W-17; Langdon, "Complaint against Katherine Harrison," CHS, *Bulletin*, XXXIV (1969), 20–25.

60. Langdon, "Complaint against Katherine Harrison," CHS, *Bulletin*, XXXIV (1969), 20. Caustically, the petitioners urged that, if Harrison should be allowed to flee "before shee be thorowly tryed and Cleard from the Guilt of such abhorred and cryingwichednesses . . . *(from wch wee beleave she is not yet cleared,)* wee feare the dredful displeasure of Allmighetie God" (italics mine).

61. Ibid. On Pitkin, see Holdsworth, "Law and Society," 478.

In the aftermath of her trial, Harrison, presumably with the assistance of a legal representative, had composed a lengthy "Answer to Mr. Winthrop" in which, point by point, she denied or refuted the testimony of more than thirty witnesses. Her defense, at least from the twenty-first-century perspective, seems compelling. In response to claims that she had appeared as a spectral dog, for example, Harrison wondered "how any person can affirm that by a small fire light they can clearly and definitely know my head on a dog." To the charge that the corpse of the supposedly bewitched child moved when she touched it, Harrison cited testimony of a woman at the wake who said she bumped into the table the body was on, which had caused the child's motion. One by one Harrison countered the evidence against her with highly plausible defenses. Only on the issues of fortune-telling and prophecy were Harrison's responses weak. To these charges she simply said, "I know nothing of it."[62]

Reading Harrison's plea to Winthrop, it is easy to understand how a sympathetic reader could empathize with her plight. Given Winthrop's advocacy of "urged removal" as a strategy for community reintegration and his previous actions in the case of Elizabeth Seager, it seems likely that he engineered Harrison's release with the intention of turning a blind eye to a planned Harrison flight from Connecticut. The Wethersfield petition, however, put an end to any such plans. The October trial took place with Harrison present to receive the verdict. Again, as in Elizabeth Seager's case, Winthrop chose to absent himself from the courtroom, though not from the trial's ultimate outcome.

On October 12, the reconvened jury found Katherine Harrison guilty of witchcraft. Her conviction, however, was challenged by the magistrates. Apparently, much of her trial had indeed centered on the fine points of law, witchcraft, and magic. Fundamental questions had been raised regarding the evidentiary standards necessary for conviction in witchcraft cases, the acceptability of spectral evidence, and the distinction between acceptable and diabolical practice of magic. Before passing sentence, therefore, the magistrates sought clarification on those points. Eight days after her trial, they summoned a group of ministers to offer them advice on three matters.

Given the public reaction to the court's previous release of Harrison, it was politic that any effort to moderate Harrison's guilty sentence be based on outside opinions rather than magisterial fiat. Still, it seems, at first glance, unusual that a Winthrop court would turn to ministers for advice in a witchcraft case. During the Hartford witch-hunt, Connecticut's ministers had been vigorous, aggressive prosecutors of witches. Since Winthrop's return from England, however, the political and ideological dynamics of ministerial authority in

---

62. "Harrison's Answer to Winthrop," Winthrop Family Papers, reel 9.

Connecticut had changed significantly. Samuel Stone, who had led the ministerial assault on witches during the Hartford witch-hunt, had died in 1663. His leadership in ministerial affairs had been assumed by someone of a much different persuasion, the minister—and alchemist—Gershom Bulkeley.[63]

Bulkeley, who had previously ministered at Winthrop's New London, was the rising young star among Connecticut divines. He had become minister of the Wethersfield church in 1667, and by 1670 the popular preacher brought in nearly 350 new members, all in a town of only 600 people. Bulkeley was one of four ministers appointed to help resolve the long and hotly contested dispute between Connecticut's Congregationalists and Presbyterians in 1668 and had helped fashion the compromise that led to acceptance of both forms of worship there in 1669.[64]

Of considerable importance for the Harrison case are the facts that, in addition to being a popular minister, Bulkeley was also a physician, an alchemist, and a friend of Winthrop. Bulkeley's alchemical and medical notebooks fill thousands of pages and cover key works in alchemy and alchemical medicine. Among his sermon notes from his years in New London, Bulkeley copied a lengthy Latin description of how to prepare the alkahest, one of the objects diligently sought at Winthrop's New London alchemical plantation. In other places, Bulkeley transcribed works on Paracelsus and Van Helmont and related his own attempts at developing the "philosophical wine."[65]

Bulkeley shared Winthrop's interest in occult sciences, and he fully embraced the governor's lifelong opposition to witchcraft convictions. In addition to his role in the Harrison trial, Bulkeley would be a vigorous opponent of the 1692 witchcraft trial of Mercy Disborough of Farmington, and, in the aftermath of the Salem trials, he would write: "I wish N.E. have not a great deale of innocent blood to answer for, both of former and later times. The good Lord pardon his people, and give them to see theire error." Bulkeley was, from the standpoint of someone opposed to witchcraft convictions, the per-

---

63. Holdsworth, "Law and Society," 525.

64. Robert G. Pope, *The Half-Way Covenant: Church Membership in Puritan New England* (Princeton, N.J., 1969), 105.

65. Bulkeley consulted Winthrop regarding both his mother's and wife's illnesses, and he was comfortable urging him to "speedily come over to my house" to attend his wife. Gershom Bulkeley to John Winthrop, Jr., Feb. 29, 1669, Sarah Bulkeley to John Winthrop, Jr., Apr. 22/3, 1669, and Gershom Bulkeley to John Winthrop, Jr., May 15, 1669, all in Winthrop Family Papers, reel 9; Thomas W. Jodziewicz, *A Stranger in the Land: Gershom Bulkeley of Connecticut,* American Philosophical Society, Transactions, LXXXVIII, part 2 (Philadelphia, 1988), 17–19; Bulkeley Collection, Hartford Medical Society; Gershom Bulkeley Mss.: III, Medical and Chemical Notes, 48; IV, Laboratory Notes; V, Laboratory Notes; X, 13a; Hartford Medical Society; Bulkeley Collection, Trinity College, VII, 11–.

fect choice to guide the ministerial deliberations, and, through his carefully worded response to the magistrates' questions about witchcraft and magic, witchcraft prosecution in Connecticut was dealt a nearly fatal blow.[66]

BULKELEY WROTE the response to the magistrates' questions in careful, though not unambiguous, prose. His document reflects concern with the Harrison case per se as well as with marking a clearly defined boundary between the acceptable and unacceptable practice of magic. The magistrates had expressed concern over three distinct issues, all of which had bearing on Harrison's particular case and important implications for all witchcraft prosecution.[67]

The first, from the standpoint of Harrison's conviction, was the most important: The magistrates asked, "Whether [in cases of witchcraft] a plurality of witnesses be necessary, legally to evidence one and the same individual fact?" As a capital crime in which one could lose one's life upon conviction, English law and tradition demanded that any conviction be based on evidence presented by at least two witnesses to the crime in question. In cases of witchcraft generally, and certainly in Katherine Harrison's case, however, the two-witness requirement presented a particularly thorny problem of interpretation. Many people had testified to witnessing Harrison's evil magic, but, as in other cases, almost all had encountered her maleficia while alone. At issue was whether in order to convict someone of witchcraft there had to be two witnesses to the exact same specific act, or whether the testimony of witnesses to different acts of witchcraft occurring on separate occasions constituted proof of guilt. Before 1662, both Connecticut and New Haven courts had employed a broad interpretation of the two-witness rule, allowing witchcraft convictions based on the testimonies of different individuals who had witnessed the suspect's acts of diabolical malevolence on separate occasions.[68]

Bulkeley's document called for a much stricter interpretation of the two-witness rule, holding that a witchcraft conviction was valid only if there were two witnesses to the exact same act at the exact same time. In finding that, "if the proof of the fact do depend wholly upon testimony there is then a necessity of a plurality of witnesses, to testify to one and the same individual fact," the ministers were redefining witchcraft evidentiary standards in Connecti-

---

66. Godbeer, *Escaping Salem*, 127–129; Jeffries Family Papers, Massachusetts Historical Society, V, 20, cited in Jodziewicz, *A Stranger in the Land*, 52, 53.

67. Taylor, *Witchcraft Delusion*, 57; Wyllys Papers, W-18.

68. Wyllys Papers, W-18; Hall, ed., *Witch-Hunting*, 184. An exception was the case where five witnesses had seen the corpse of Josiah Willard's child move when Harrison touched it, a manifestation Harrison had obtained countertestimony to refute. "Harrison's Answer to Winthrop," Winthrop Family Papers, reel 9.

cut in a way that dramatically contracted the grounds for witchcraft conviction. Much of the evidence against Harrison or any suspected witch was automatically negated, for the spectral apparitions, familiars, and other preternatural encounters people reported with suspected witches almost always happened when the witnesses were alone. In finding that the two-witness rule required multiple witnesses to a single act, the ministers undermined much of the evidence upon which Harrison (and all previous witches, for that matter) had been convicted. Not until 1692, when a whole group of afflicted girls publicly writhed together in agony and in unison in a Salem, Massachusetts, meetinghouse because of the pain being inflicted on them by a witch trial suspect's familiar (visible in the courtroom to all of the afflicted, who testified together to the apparition's presence) was this new, more-guarded interpretation of the two-witness rule fully met, a fact about the Salem trials that has gone largely unnoticed.[69]

Having addressed the standards of testimonial evidence necessary for conviction, the ministers shifted their focus to a second question posed by the magistrates, also of central importance in witchcraft trials: the validity of spectral evidence itself. The magistrates had requested clarification regarding, "Whether the preternatural apparitions of a person, legally proved, be a demonstration of familiarity with the devil." The validity of spectral evidence in witchcraft proceedings had been long contested. No one doubted that spectral apparitions occurred: many witnesses had testified to seeing a number of different spectral visions of Katherine Harrison (Harrison's face on the head of a dog and her specter transformed from a calf's head on a cart). Many considered such apparitions prima facie evidence confirming the suspect had entered into a covenant with the devil. Critics, however, challenged spectral evidence on the grounds that the devil could assume the shape of an innocent person as well as a witch and thus provide evidence that could convict an innocent person. Since the devil was the prince of liars, how could the specters he was instrumental in making appear be relied upon as evidence? Because spectral evidence had been an important part of Harrison's trial and, presumably, her conviction, the magistrates, before rendering a final decision regarding the legitimacy of the jury's verdict, sought clarification regarding how such evidence should be valued.[70]

The ministers' answer can be confusing on whether they accepted or re-

---

69. Wyllys Papers, W-18; Holdsworth, "Law and Society," 528–530; Weisman, *Witchcraft, Magic, and Religion,* 110. The other option for conviction apart from testimony was for the witch to confess.

70. Wyllys Papers, W-18; Hall, ed., *Witch-Hunting,* 184.

jected the validity of spectral evidence. Such confusion may be warranted, for Bulkeley's ambiguously stated answer in essence does both: it confirms the validity of spectral evidence and simultaneously makes it all but impossible to secure a conviction on spectral grounds. "Wee answer," Bulkeley wrote, "that it is not the pleasure of the most high, to suffer the wicked one to make, an undistinguishable representation, of any innocent person in a way of doing mischief, before a plurality of witnesses. The reason is because, this would utterly evacuate all humane testimony; no man could testify, that he saw this person do this or that thing, for it might be said, that it was the devil in his shape." What the ministers did is *confirm* the validity of spectral evidence, *but only provided* that there were at least two witnesses to the spectral apparition in question. While the devil might appear in the shape of an innocent person to one witness, the ministers argued God would not allow him to do so to several people simultaneously, because then no testimony regarding spectral evidence could ever be accepted. Spectral evidence was, therefore, only conditionally reliable. And, importantly, since apparitions almost universally appeared to only one person at a time, the conditions under which spectral evidence could be considered reliable for judicial purposes would henceforth almost never be attained. In a marvelous demonstration of saying one thing while doing another, Bulkeley and the ministers lent support to the suspicions of Katherine Harrison's accusers while protecting Harrison herself from conviction. It was Bulkeley's writing, but it was Winthrop's strategy of community reintegration—validating the accusers' suspicions without empowering them. As a result of this finding, the use of spectral evidence in Connecticut trials was for practical purposes thoroughly undermined without ever being discredited.[71]

The ministers' response to the magistrate's final questions hints at broader issues than Harrison's particular witchcraft practices that might have emerged through the Wethersfield petition for redress and Harrison's second trial. Although Bulkeley's answer is clearly intended to clarify whether Harrison's fortune-telling practices should be considered diabolical, it also seems painstakingly crafted to definitively specify the boundaries between what should be considered acceptable magic practice and what should fall into the range of diabolical magic. Given the fact that Harrison's fortune-telling was based on her knowledge of the works of the noted English astrologer William Lilly,

---

71. Tomlinson and Langdon both argue that the ministers' answer refutes spectral evidence; Godbeer and Weisman that it accepts it. Tomlinson, *Witchcraft Trials,* 49; Langdon, "Complaint against Katherine Harrison," CHS, *Bulletin,* XXXIV (1969), 25; Godbeer, *Devil's Dominion,* 172; Weisman, *Witchcraft, Magic, and Religion,* 110; Wyllys Papers, W-18; Hall, ed., *Witch-Hunting,* 183.

learned magic as well as folk magic had come under scrutiny at Harrison's trial. Bulkeley's carefully worded final answer to the magistrates' queries suggests that, even as the ministers were condemning the diabolical act of fortune-telling, they were making a concerted effort to remove from suspicion those occult practices engaged in by occult natural philosophers such as Winthrop and Bulkeley himself.

Both Bulkeley and Winthrop were students of astrology. Winthrop collected astrologically focused treatises, like the works of the alchemist-astrologer Robert Fludd, and the theory of the macrocosm and microcosm Winthrop embraced unquestionably accepted the power of the occult influences of the stars and planets on the terrestrial world and its denizens. Among Winthrop's papers are horary astrological charts (created by someone other than Winthrop) drawn up for various Londoners in the 1650s, and Winthrop's medical and alchemical correspondence reflects numerous instances in which astrological influences figure in the issues under discussion. One reason Winthrop kept his patients in New London for treatment for such long periods might have been to treat particular illnesses during the time when astrological conditions were thought to help effect their cures. Since different body parts and organs were believed to be ruled by specific zodiacal signs, particular months were considered most favorable for specific diseases. Bulkeley believed that gathering minerals and herbs at appropriate astrological moments increased their strength. He also advocated gathering and preparing medicines when the sun and moon were sympathetic to the particular planet astrologically associated with a medicine's ingredients.[72]

Bulkeley's and Winthrop's astrological practices were almost certainly limited to the realm of natural astrology, attempting to employ or focus the powers of stars and planets to enhance natural effects. Neither person is known to have engaged in the kind of divination purportedly practiced by Harrison, and, because of their belief in the power of occult knowledge and the need to keep it closely guarded, both presumably deplored the public revelation of occult mysteries engaged in by people like William Lilly. Lilly's

---

72. Horary astrological charts were cast to determine information about a subject based on the positions of the planets at the time the question was asked. These were more common than nativity charts, based on birth dates, since many people did not know the exact moment of their birth. C. A. Browne, "Scientific Notes from the Books and Letters of John Winthrop, Jr. (1606-1676), First Governor of Connecticut," *Isis,* IX (1928), 340-341. John Davenport, for example, asked whether he could expect any help from Winthrop in treating his illness during the autumn of 1654, since "the season [was] now beginning to suite the use of meanes" for his urinary ailment. John Davenport to John Winthrop, Jr., Sept. 11, 1654, *Winthrop Papers,* VI, 419-420; Bulkeley Collection, Hartford Medical Society Library, Gershom Bulkeley Mss. XIII, 117-124.

publication of England's first astrological textbook, *Christian Astrology*, in 1647, had not only enabled ordinary people such as Katherine Harrison to misuse occult knowledge; it had created an environment that threatened the entire spectrum of occult philosophical practices. When critics found the pseudointellectual occult practices of people like Harrison diabolical, the risk that men like Winthrop and Bulkeley could face similar accusations increased. During his lifetime, Winthrop's role model John Dee had faced charges of diabolism, and Robert Fludd had also faced charges of practicing diabolical magic. And during periods of heightened public fear over occult harms, such as during suspected witchcraft cases, the potential for the arrow of suspicion to turn toward those known to possess occult knowledge was undoubtedly heightened.[73]

The demonic magic thought to underlie all the preternatural effects at issue in witchcraft cases was understood, despite its diabolical origin, to be effected through natural magic. Far from being able to work via supernatural means (as only God could do), the devil was subject to the exact same limitations and natural boundaries as human magicians. He was, however, believed to be a much more skillful and much more knowledgeable magus than any human and, therefore, could work far more powerful magical effects. As James I noted, the devil was "farre cunninger then man in the knowledge of all the occult proprieties of nature." No matter how much more skillful he was, though, the devil performed the same *kinds* of magic known to students of the occult and natural magic. This created an elision of categories that at times made the distinction between natural and diabolical magic difficult to ascertain. As Stuart Clark has noted generally, "Natural magic had to be defended repeatedly from the accusation that it was the work of demons . . . while the devil himself could count as just another natural agent."[74]

Special knowledge of natural magic gave publicly accepted and sanctioned

---

73. Winthrop and his alchemical correspondents displayed an obsessive concern with secrecy and protecting occult information from disclosure to the unworthy who might misuse it. From Samuel Hartlib, who withheld news from Winthrop until "the method of communication . . . be better chalked out between us," to New London alchemist Jonathan Brewster, who entreated Winthrop "to keepe both my name and my worke secrett to your selfe," to the Dorchester inquirer Humphrey Atherton, who begged Winthrop to "not lat it be known to any but my self that I did ask such a question," desire for secrecy is a pervasive subtext of Winthrop's occult communications. Samuel Hartlib to John Winthrop, Jr., Mar. 16, 1660, MHS, *Proceedings*, LXXII (1963), 45; Jonathan Brewster to John Winthrop, Jr., Jan. 31, 1656, MHS, *Collections*, 4th Ser., VII, 78; Humphrey Atherton to John Winthrop, Jr., Oct. 30, 1648, *Winthrop Papers*, V, 273–274; William Lilly, *Christian Astrology, Modestly Treated of* . . . (London, 1647) (1985); Patrick Curry, *Prophecy and Power: Astrology in Early Modern England* (Princeton, N.J., 1989).

74. Clark, *Thinking with Demons*, 245.

occult practitioners such as Winthrop particular authority over the interpretation of magic, on the one hand, but such authority always carried with it vulnerability to suspicion. Especially in a case as contested as Harrison's, where the actions of Winthrop and the magistrates between trials in releasing Harrison from jail were so vociferously protested, one senses that a background issue in the case—never extending beyond community gossip but influencing the court proceedings nonetheless—was Winthrop's own relationship to magic and his apparently special sympathy for the accused.

That is at least one interpretation that helps explain the care with which Bulkeley's response to the magistrates' third and fourth questions was crafted. For it seems clear that the magistrates framed their questions and Bulkeley penned the ministerial response with the specific intention of avoiding, or, perhaps, counteracting, the dangerous comparison of elite with diabolical magic. Even in wording, the magistrates had directly sought to separate questionable occult practices from those of the elite. They asked whether "vicious"—that is, rude, depraved, and immoral—persons who divined future events or revealed occult knowledge were demonstrating familiarity with the devil.[75]

It is suggestive that the magistrates assumed that the social status of the practitioner was a qualifying factor in determining what constituted acceptable magical practice. Yet the history of accusations against Fludd and Dee, among others, had shown that status could be a thin reed against charges of diabolical complicity. So, as Bulkeley framed the ministers' response to the question of what constituted diabolical foreknowledge of events, he ignored the status of the accused and concentrated on the nature of occult knowledge itself. He took care to create a clear distinction between knowing things "which are indeed secret" and therefore truly diabolical, and knowing that which was only hidden, or occult. By creating a firewall between secret diabolical knowledge and occult natural knowledge, Bulkeley insulated almost the entire realm of occult natural philosophies from links to any connection to the diabolical.

> Wee say thus much, That those things, whether past, present, or to come, which are indeed secret, that is, cannot be known by humane skill in Arts, or strength of reason arguing from the course of nature, nor are made known by divine revelation either mediate or immedi-

---

75. "To the 3d and 4th Quests. Together: whither a vicious persons foretelling some future event, or revealing of a secret, is a demonstration of familiarity with the devil?" Wyllys Papers, W-18; Hall, ed., *Witch-Hunting*, 185.

ate, nor by information from man, must needs be known (if at all) by information from the Devil."[76]

On the surface, the ministers again appear to have affirmed the devil's aggressive engagement in human affairs by agreeing that the revelation of truly secret knowledge was definitely diabolical. But, having affirmed the devil's agency in the realm of human affairs, they defined secret knowledge in a way that confined Satan to a very narrow theater of operations. All the occult philosophies studied by members of the elite such as Winthrop and Bulkeley—alchemy, astrology, numerology, even natural magic—produced acceptable knowledge. The knowledge of the occult such studies produced, through which the philosophers' powerful magical effects were generated, were by definition obtained through "humane skill in Arts, or strength of reason arguing from the course of nature . . . or made known by divine revelation either mediate or immediate, or through information from man." Sounding like witch-hunters, the ministers were actually sealing the devil off from involvement in most magic and occult practices. Only in the case of divination and only under certain circumstances did the ministers grant the possibility of true diabolical influence:

> Hence the communication of such things, in way of Divination (the person pretending the certain knowledge of them) seems to us, to argue familiarity with the Devil, in as much as such a person, doth thereby declare his receiving of the Devils testimony, and yeeld himself as the devils instrument to communicate the same to others.[77]

Only when one displayed knowledge of future events that could not have been obtained in any other way were there grounds for suspicion that one had compacted with the devil, and even then those grounds were qualified. The person must pretend "certain knowledge" and communicate that knowledge to others. The ministers' definition of secret, diabolical knowledge, for all its aggressive posturing, relegated the archfiend, at least in terms of witchcraft convictions, to the status of fortune-teller. It was a demotion he would endure for a generation.

The importance of Bulkeley's document, both for the Katherine Harrison trial and the future of witchcraft prosecution in New England, was profound. It provided the magistrates grounds to overturn the jury verdict and set Harrison free. On May 12, 1670, Winthrop presided over a meeting of the General

---

76. Wyllys Papers, W-18.
77. Ibid.

Assembly in which Harrison's case was referred into the Court of Assistants. Eight days later, a special court freed Harrison and ordered her to do what the Wethersfield petitioners had feared she was preparing to do months before: leave the town. Winthrop's strategy of community reintegration had again dictated the witchcraft trial's ultimate outcome. Harrison was banished; Wethersfield, under the ministry of Gershom Bulkeley, was left to recover from its long-term social unrest.[78]

This was, however, to be the last time the strategy of social reintegration would be used in a Connecticut witch trial during Winthrop's lifetime, for there would be no more witch trials in the colony for twenty-two years. The effect of Winthrop's sustained resistance to Connecticut witchcraft convictions, combined with the new standards for witchcraft conviction created by Bulkeley's ministerial response, made obtaining witchcraft convictions in Connecticut virtually impossible. As a result, witchcraft cases ended there until 1692. Moreover, Winthrop and Bulkeley's managed resolution of the Harrison case provided a precedent followed in other witchcraft cases and helps explain why every witchcraft trial in New England from 1670 until 1688 produced only acquittals. When witchcraft charges did surface again in Connecticut, the magistrates, in sharp contrast to the proceedings at Salem during the same year, displayed a remarkable reluctance to exact punishment on the suspects.[79]

Did the decline of witchcraft in seventeenth-century New England result from the skepticism associated with the scientific revolution and the acceptance of mechanistic natural philosophies? Did the Puritan elite completely reject magical practice as diabolical? Maybe, and maybe not. In Connecticut, the decisive shift in witchcraft prosecution came, not from Puritan positivists rejecting magic, but from alchemical philosophers who believed that the practice of witchcraft by ordinary people was overstated, overprosecuted, and overfeared.

To bring the Hartford witch-hunt of the 1660s to a final conclusion, Winthrop used his governmental authority, elite status, specialized knowledge,

---

78. "Haueing considered the verdict of the jury respecting Katherine Harrison," the court noted it "cannot concur with them so as to sentence her to death or to a longer continuance in restraint; but do dismiss her from her imprisonment, she paying her just fees. Willing her to mind the fulfillment of removing from Wethersfield which is that will tend most to her own safety and the contentment of the people who are her neighbors." Court of Assistant Records, Connecticut State Archives, LIII, 7; Hall, ed., *Witch-Hunting*, 184.

79. See Weisman's discussion of the Harrison case's influence on the trial of Elizabeth Morse, in *Witchcraft, Magic, and Religion*, 111–112. Peter Charles Hoffer, viewing witchcraft from the perspective of Salem, concluded that reintegration was the primary objective of Massachusetts's witchcraft trials. Hoffer, *Devil's Disciples*, 127, 249 n. 89. See also Godbeer, *Escaping Salem*, 110–126.

and alchemical network to help transform one of the most socially aggressive and coercive aspects of Connecticut culture. Even as the witchcraft trials unfolded, he was employing these same personal attributes and connections internationally to help preserve the political integrity and relative autonomy of the colony itself.

(EIGHT)

"Matters of Present Utility"
John Winthrop, Jr., the Royal Society, and the Politics of Intelligence in Restoration New England

By the time John Winthrop, Jr., became governor of Connecticut in 1657, he had achieved an international reputation as an alchemist. During his travels to Europe he had met and made a lasting impression on members of the European republic of alchemy, several of whom he had sustained correspondence with once back in New England. Alchemists from New England who had known Winthrop and later relocated to England spread reports about him that further enhanced his reputation. In 1660, Samuel Hartlib wrote Winthrop, "You cannot believe what secret reports I have heard of you of which I would be so willingly informed." Hartlib indicated he had been told Winthrop had perfected, or nearly perfected, both the alkahest and the elixir that would effect transmutation. He hoped Winthrop would provide him with information about his alchemical abilities, so that Hartlib could "more freely provoke you to love and good works." The German physician and poet Johann Rist also remembered Winthrop. In his 1664 *Die alleredelste Tohrheit der gantzen Welt (The Most Noble Folly of the Whole World)* Rist described Winthrop as a reputed alchemical expert who could melt and form all kinds of stones, particularly diamonds and sapphires. The accuracy of the stories about Winthrop is less important than the impression they conveyed of him as an American natural philosopher with special alchemical knowledge and skill. This reputation was to open important doors for him in Restoration England and help make his 1661 mission to England to secure a charter for colonial Connecticut successful.[1]

---

1. Samuel Hartlib to John Winthrop, Jr., Mar. 16, 1660, Hartlib Papers [7/7/3b], University of Sheffield; Johann Rist, *Die alleredelste Tohrheit der gantzen Welt: vermittelst eines anmuhtigen und erbaulichen Gespräches welches ist diser Ahrt die dritte und zwahr eine Märtzens-Unterredung* (Hamburg, 1664), 238–241. Hartlib's *Ephemerides* for 1660 noted, "One of the Winthrops of New England Father or Son Hath the Art of Making Artificial Marble and of the Indian Pintadoes's or Painting of Indian Cloth." Hartlib Papers [29/8/11A].

The coronation of Charles II in 1660 raised serious issues for the Puritan colonies of New England. Connecticut, New Haven, Plymouth, and Rhode Island had supported the Puritan cause, and their lack of charters placed their current governments in jeopardy, at the will or whim of the new king. Connecticut's territorial claims rested on a dubious patent secured by the earl of Warwick in 1632, the rights to which Connecticut had purchased from George Fenwick of Saybrook, the presumed agent for the patentees, in 1645. The patent in question had apparently never been fully validated; moreover, no one seemed to possess an original copy of it. Without charter protection, the colony's political and territorial integrity was extremely vulnerable to challenge, since it had no defensible legal standing. This had not been of major concern in the early years of settlement or during the Interregnum, but with Charles II's accession it assumed prominence.[2]

The new monarch had immediately turned his attention to England's overseas plantations and instituted a series of reforms intended to rationalize colonial administration. The Navigation Acts of 1660 and 1662, the creation of the Board of Trade and the Council for Plantations, the formation of the Corporation for the Propagation of the Gospel in New England, and the chartering of the Royal Society—all within the first two years of Charles's reign—reflect a concerted effort by Whitehall to centralize control over the empire. Through better regulation of trade, increased oversight of activity, and concerted efforts to promote colonial economic and spiritual growth, the interests of both England and its plantations would be well served.[3]

The instructions Charles gave the newly organized Council for Foreign Plantations make explicit the imposition of central economic and political control informing the crown's new colonial policies.

> Wee have thought fitt ... to drawe those our distant dominions and the severall interests and governments thereof into a nearer prospect and consultacon ... they being now a greate and numerous people whose plentifull trade and commerce verie much imployes and increaseth the navigacon and expends the manufactures of our other dominions ... and bring a good accesse of treasure to our Excheqr. ... In consideracon whereof ... "Wee have judged it meete and necessary that soe many remote Colonies and Governments ... should now

---

2. Charles M. Andrews, *The Colonial Period in American History* (New Haven, Conn., 1934–1938), I, 354, 359, 365–368, 402–405; Charles J. Hoadly, *The Warwick Patent* ([Hartford, Conn.], 1902).

3. Margaret Ellen Newell, *From Dependency to Independence: Economic Revolution in Colonial New England* (Ithaca, N.Y., 1998), 78–81; Andrews, *Colonial Period of American History*, IV, 50–84; Bernard Bailyn, *The New England Merchants in the Seventeenth Century* (New York, 1964), 113.

no longer remaine in a loose and scattered [condition] but should be brought under such an uniforme inspeccon and conduct that Wee may the better apply our royal councells to their future regulacon securitie and improvement."[4]

A problem in instituting greater royal authority over the plantations was the scarcity of accurate information about them. Whitehall lacked sufficient knowledge about its overseas dependencies to administer them effectively. To address that problem, Charles ordered the council to institute aggressive policies to gather intelligence and establish uniform government. The council was to require that each plantation provide a "perticular and exact accompt of the state of their affaires," including the constitution of their laws and government, the number of men in the plantations, and the state of their fortifications and defenses. The council was to establish "a continuall correspondencie" with each plantation, so that it would be able to provide on demand an account of the current state of the colony and its government, its economic condition, and the commodities of every ship trading there.[5]

In addition to collecting useful information, the council was to take steps to standardize colonial governance, both internally and externally, and promote the spiritual well-being of the plantations through the provision of sufficient ministers and efforts to convert natives and servants. Charles envisioned an empire in which the crown actively regulated its colonies, which would exercise far less autonomy.[6]

The New England colonies were central to these plans for increasing the value of colonies through better regulation. Among the earliest, most populated, and most religiously inclined plantations, New England was seen as a laboratory for improving trade, commerce, and evangelization of the Indians. Its population growth, infrastructure, and religious orientations suggested that much of the groundwork for colonial economic expansion had already been laid. On the other hand, the Puritan colonies' reputation for religious intolerance, coupled with their traditions of relative political autonomy, made them a test case for the imposition of greater crown control.

Winthrop stepped into this new, assertive environment for colonial administration in September 1661 armed with a declaration of loyalty to the king and a petition to obtain a royal charter for Connecticut. To advise him in

---

4. "His Majesty's Commission for a Councill for Foreign Plantations," in E. B. O'Callaghan and Berthold Fernow, eds., *Documents relative to the Colonial History of the State of New-York* (Albany, N.Y., 1853–1887), III, 32–34 (hereafter *DCHNY*).

5. "Instructions," Dec. 1, 1660, ibid., 35.

6. Ibid., 32–36.

negotiating the rapids of Whitehall, Winthrop enlisted the support of Samuel Hartlib, whom he had first met on his 1641 journey and with whom he had been corresponding for two years. Hartlib was one of many scientifically oriented reformers who believed that aligning the scientific and technological gains of natural philosophy with the political power of the state offered the best approach to fulfilling the pansophic reform agenda he, John Dury, and Jan Comenius had been advancing since the 1640s.

During the Interregnum Hartlib and his circle had generated an array of proposed improvements for use in England and its plantations. Hartlib was the hub of an international network of philosopher-correspondents, including Winthrop, who believed that the prophesied premillennial instauration of knowledge was under way and that, in concert, they could simultaneously advance the cause of England, humankind, science, and Christ. By employing their alchemical, technological, and utilitarian advances in the Christian service of society Hartlib and his associates hoped, with state backing, to cure poverty, increase the wealth of the English nation, prolong and improve human life, and achieve dominion over nature. They developed schemes for cooperative scientific investigation, agricultural reform, and agrarian improvement. Winthrop and Hartlib shared a similar vision of the Christian call to employ natural philosophy in the service of society. Enlisting Hartlib's aid in understanding the political landscape at Whitehall was a natural and productive avenue for Winthrop to follow.[7]

Winthrop's arrival in London might have been fortunate for Hartlib as well. The small parliamentary pension on which he had lived during the Interregnum had been suspended at the Restoration. Hartlib, ill and in dire financial straits, had been forced the preceding fall to appeal for private support from Lord Herbert "till some other means of public Love and Encouragement maybe (if it maybe) determined." Given the goals of the new regime at Whitehall, the ability to guide a New England colonial governor into a proper relationship with the crown might have seemed to Hartlib an opportunity to reestablish needed patronage.[8]

One Hartlib associate who very much wanted to meet Winthrop was Benjamin Worsley. Former surveyor general of Ireland, a prime mover in the "Invisible College" that preceded the Royal Society and principal author of the Navigation Act of 1660, Worsley was a well-connected political figure and

---

7. Jim Bennett and Scott Mandelbrote, *The Garden, the Ark, the Tower, the Temple: Biblical Metaphors of Knowledge in Early Modern Europe* (Oxford, 1998), 33–43.

8. G. H. Turnbull, *Samuel Hartlib: A Sketch of His Life and His Relations to J. A. Comenius* (London, 1920), 57–58.

alchemical philosopher with a special interest in the American colonies and an immediate use for a New England ally such as he hoped to find in Winthrop. Hartlib characterized Worsley to Winthrop as "our Special Friend" and recommended that Winthrop contact him immediately on his arrival in London. Winthrop, however, for reasons that will become clear, was somewhat reluctant to do so. More than two weeks after his arrival, he had not made any effort to contact Worsley, and Worsley, eager for a meeting but reluctant to initiate it himself, urged Hartlib to be more insistent, giving him specific instructions on what to say to Winthrop to motivate him to seek him out.

> Pray Acquaint Mr. Wynthrop that you do heare they [at] Court are upon sending a Governor into new England and that there are some private Agitations on foote concerning that Countrey. That you being not able to learne any thing of perticulars . . . and knowing no busynesse can be done here at Court without <he hath> some Interest did againe the more earnestly desire that he and Mr Worsley were Acquainted and the rather seeing you know Mr Worsley hath much the eare of the Chancellor and you beleeve in reference to the Plantations <he is> privy to most Transactions.
>
> That he will finde mee every way a Civill man and one that you know will shew him a respect for the character Mr Worsley hath received from him. That in all things relating to publicke good, Just Lyberty of Conscience and any sort of ingenuus kinde of improvement, he will finde Mr. Worsley, as you beleeve, according to <his> owne hearts desire. . . . That you would be glad he would therefore finde Mr Worsley out as soone as he could.[9]

Worsley's instruction that Hartlib use the threat of the appointment of a governor general to gain Winthrop's attention was sure to garner a response, for it was fear of just such a move by the crown that had prompted Winthrop's urgent trip to London. For Winthrop and the colonists he represented, such a government posed a first-order threat. Consolidation of power in New England could come only at the cost of local autonomy, something the Connecticut charter journey was intended to preserve. To Worsley, Hartlib, and many of their scientific associates, however, unification of political authority

---

9. Extract, Benjamin Worsley to ———, with a Message for Winthrop, n.d., Hartlib Papers [33/2/27A-B:27B Blank]. Hartlib's message, sent to Winthrop by special messenger on October 9, essentially replicated Worsley's note (Massachusetts Historical Society, *Proceedings*, 1878 [1879], 215–216). The words enclosed in "< >" in quotations used from the Hartlib Papers represent words that Hartlib inserted in place of the original words in the texts, which he crossed out.

under a single figure reporting to the crown seemed the ideal administrative structure through which to implement colonial pansophic reform. This issue would pose a continuing series of challenges for Winthrop in his relations with the scientific and political communities of England. Although he shared the social, spiritual, and philosophical goals of many of the natural philosophers connected to the monarch, the nascent imperialist political agenda explicit in their program could, depending on their specific application, either aid or harm his interests. When the imperial objectives of the English cognoscenti and the colonial interests of Connecticut coincided, Winthrop could, and did, use them to advance mutually compatible programs. Other situations, however, and many of them, would attempt to direct Winthrop's interests as a natural philosopher against his interests as a colonial governor.

Such was not immediately the case, although, when Hartlib urged Winthrop to contact Worsley, he feared it might be. Winthrop's reluctance to reach out to Worsley was based on something Winthrop knew that Worsley apparently did not know he knew: that the person at court most active in seeking the appointment of a governor general in New England had been Worsley himself. Hartlib, apparently not realizing that Winthrop might object to the idea of a governor general, had written Winthrop the month before telling him that Worsley "is much dealing with his Maj. and some of his Privy Councel to bee sent over as an Agent or Resident of all the Plantations." Hartlib was excited at the prospects of such an appointment. "If it be granted, great numbers of honest People will replenish all English Plantations." Seeking control of colonial governments was not new to Worsley. In 1649, with the full support of the Hartlib circle, he had argued for removal of the royalist governor of Virginia in order to secure that colony's political loyalty and regulate its trade. Worsley sought to be the replacement governor, and he had intended to do in Virginia what Winthrop had intended to do in New London—establish a model plantation for development of trade and industry, educational reform, and the conversion and civilization of the native inhabitants. Winthrop, aware of Worsley's current interests, presumably had been reluctant to be introduced to the court by a man seeking to lead consolidation of the governments of New England. Despite his reluctance, however, the message Worsley dictated to Hartlib on October 9, which Hartlib had sent by special messenger the same night, prodded Winthrop into action.[10]

10. Samuel Hartlib to John Winthrop, Jr., Sept. 3, 1661, MHS, *Proceedings*, 1878, 212–214, 215–216; Charles Webster, "Benjamin Worsley: Engineering for Universal Reform from the Invisible College to the Navigation Act," in Mark Greengrass, Michael Leslie, and Timothy Raylor, eds., *Samuel Hartlib and Universal Reformation: Studies in Intellectual Communication* (Cambridge, 1994), 213–

Despite the differences in their views on colonial government, Winthrop found in Worseley, his junior by twelve years, a kindred spirit. Like Winthrop, Worsley had matriculated at Trinity College, Dublin. He was a Christian alchemist and alchemical physician and had lived and studied natural philosophy and emerging technologies in the Netherlands. Both men—Winthrop in New London, Worsley in the Old—had advanced ambitious but unsuccessful schemes to manufacture saltpeter for agricultural improvement and to meet the seemingly limitless early modern need for gunpowder. They shared an interest in optics, astronomy, and astrology. In the same way that Winthrop attempted to nurture the development of natural philosophers in New England, Worsley had sought, through his involvement in the group Robert Boyle named the Invisible College, to establish a charitable fraternity of natural philosophers who resided away from the communication networks in London, dedicated to universal reformation. Worsley believed in the powerful relationships between alchemy, medicine, theology, and astrology and in the ability of natural philosophers to play a leading role in the advancement of society. At the time Winthrop came to London, he, like Hartlib, had a strong sense that important millenarian transformations were under way.[11]

Before coming to England, Winthrop had asked Hartlib about his knowledge of the Rosicrucians, and Hartlib had informed him of a secret alchemical society that was "not onely a true possessor, but a reall dispenser of these Mysteries for the ends which God hath as steward entrusted them withal." Members of the society had come to England from all over the world, Hartlib reported, and were now hidden there but would make themselves known as soon as the political environment had stabilized. Hartlib sent Winthrop a book that would give him a sense of this society's mission, John Hall's *Model of Christian Society* and the *Right Hand of Christian Love Offered*. These were English translations of two Rosicrucian pamphlets, J. V. Andrae's *Christianae societatis imago* and *Christiani amoris dextera porrecta*. In Hartlib and Worsley, Winthrop found men who shared both his strong commitment to the utilitarian application of science and a vision of their age's special and godly mission to improve the world. He would at times diverge from them in the strategies

---

235. On Worsley's influence in the passage of the Navigation Act of 1651 and his role in shaping England's mercantilist policy, see Thomas Leng, "Commercial Conflict and Regulation in the Discourse of Trade in Seventeenth-Century England," *Historical Journal*, XLVIII (2005), 933–954.

11. Charles Webster, *The Great Instauration: Science, Medicine, and Reform, 1626–1660* (New York, 1976), 57–66, 377–380; Webster, "New Light on the Invisible College: The Social Relations of English Science in the Mid-Seventeenth Century," Royal Historical Society, *Transactions*, 5th Ser., XXIV (1974), 19–42; Benjamin Worsley to Samuel Hartlib, Feb. 4, 1659, Hartlib Papers [33/2/16a–17b], February 1656 [42/1/5a–6a], [42/1/16a–17b], [39/1/16a–20b], May 26, 1658 [33/2/9a–9b].

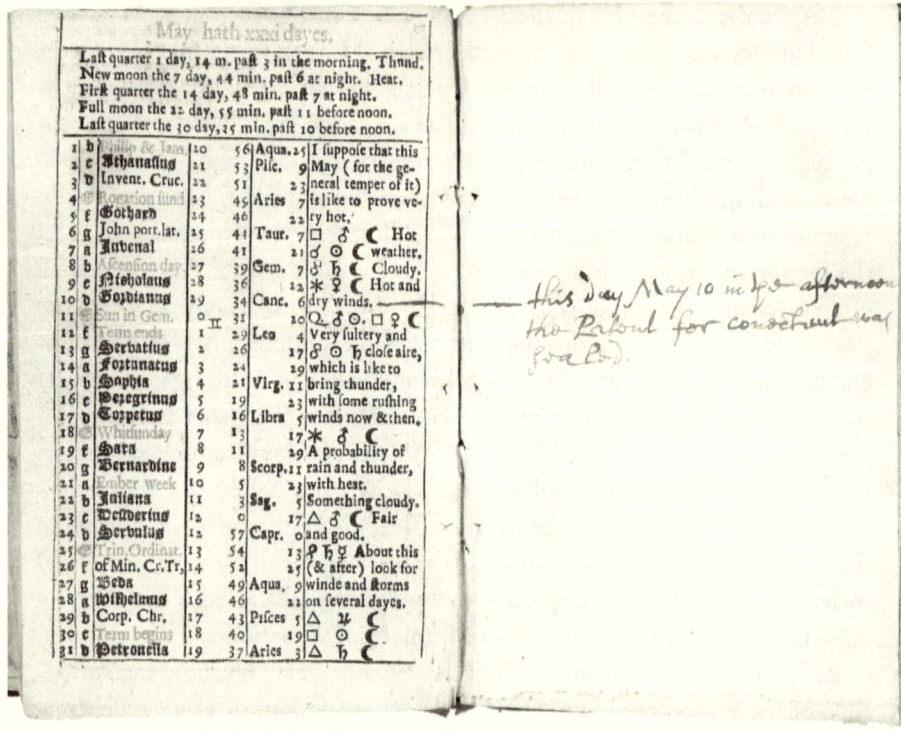

FIGURE 7. Winthrop Family Annotated Almanac. 1662. Almanac page for May 1662 and opposite page with manuscript notation. Manuscript, Winthrop Family Papers, Massachusetts Historical Society, Boston. *Courtesy of the Massachusetts Historical Society*

Winthrop specially noted the day Connecticut's charter received the great seal.

through which he believed such improvements could best be accomplished, but he shared their objectives.[12]

The affinities between Winthrop and Worsley led to a useful friendship. In spite of, or perhaps because of, his interest in consolidating New England's governments, Worsley put his considerable influence at Whitehall to work on Winthrop's behalf. When Winthrop submitted Connecticut's charter proposal to the Privy Council in February 1662, it was rushed through to approval. Customary steps, such as referring the petition to the Council for Plantations for further study, were dispensed with, and by the end of Febru-

12. Samuel Hartlib to John Winthrop, Jr., Mar. 13, 1660, Hartlib Papers [7/7/1a–8b]; Ronald Sterne Wilkinson, "John Winthrop, Jr. and the Origins of American Chemistry" (Ph.D. diss., Michigan State University, 1969), 272–273.

ary the decisive step of ordering the preparation of a provisional charter was achieved. Moreover, the charter that received the great seal on May 10, 1662, represented a major political coup for Winthrop and Connecticut. Under its provisions the size of Connecticut Colony was considerably expanded. The first act of the Privy Council toward centralizing authority in New England was one in which Winthrop himself participated, for his colony's new charter consolidated within its bounds all of New Haven Colony and half of the territory claimed by Rhode Island.

Connecticut's good fortune did not go unchallenged. Connecticut's charter caused turmoil in both New England and London. Dealing with the challenges raised in the metropole by agents of New Haven, Rhode Island, and the Atherton Company (a group of land speculators associated with Winthrop who had interests in the Rhode Island borderlands) kept Winthrop in England for another year. Winthrop was able to appease New Haven's agents until he could return to Connecticut and address the annexation question directly with that colony's leaders. Rhode Island, however, represented by Dr. John Clarke, raised an effective and vigorous opposition. Clarke's protest influenced Edward Hyde, first earl of Clarendon, to temporarily recall all copies of the Connecticut charter. He subsequently released them, however, and suggested when he did so that Winthrop send a copy to Connecticut. He also urged Winthrop and Clarke to come to a private resolution of their territorial dispute. Winthrop, seeking to capitalize on the royal interest in colonial consolidation, offered Rhode Island the opportunity to become part of Connecticut, which Clarke summarily declined. Subsequent negotiations, often acrimonious, dragged on throughout 1662 and into 1663. In the end, it took the assistance of Worsley and five other arbitrators to achieve even a temporary resolution. Along the way, Clarke managed to secure a charter for Rhode Island, but, with matters still unresolved, the crown decided to send a royal commission to New England to settle the boundary disputes once and for all and to bring further order to crown affairs.[13]

Winthrop used his extended stay in London to expand his network of court-connected patronage while pursuing his interests in alchemy, natural philosophy, and pansophic reforms, which he accomplished at the Gresham College meetings of the Royal Society. Shortly after his arrival in London, William Brereton, the future Lord Brereton, whom Winthrop met through Hartlib, began taking him to society meetings. On December 17, ninety days

13. Robert C. Black III, *The Younger John Winthrop* (New York, 1966), 212–226, 232–247; Sydney V. James, *John Clarke and His Legacies: Religion and Law in Colonial Rhode Island, 1638-1750*, ed. Theodore Dwight Bozeman (University Park, Pa., 1999), 59–83, esp. 76.

exactly after he had arrived in London, Brereton nominated Winthrop for membership. On January 1, 1662, Winthrop became the Royal Society's first colonial member, and during the next year and a half he was groomed to be a philosopher for and to the king.[14]

It is possible that Samuel Hartlib had planned Winthrop's election to the society even before Winthrop arrived in England. A letter to Hartlib from Winthrop's friend Johann Moraien, written from Arnhem the same month Winthrop was admitted to membership, inquired whether Winthrop had yet arrived in Europe and whether Gresham College, the home of the society, had "taken him in." Moraien might have been asking whether Winthrop had received an appointment to the staff of Gresham College. This was not an uncommon occupation for New England Puritan repatriates. The year prior to Winthrop's arrival, John Haynes, a former governor of Connecticut, had become lecturer in logic at Cambridge. No contemporary sources, however, allude to Winthrop's seeking or being considered for a position at Gresham College. Moreover, since the Royal Society had not as yet been formally chartered, Moraien's use of the institution's name as a shorthand label for the group of natural philosophers who met there makes sense. If that was indeed the case, then Winthrop's inclusion into the society must have been under consideration even before Winthrop reached England. Considering the access to cooperative scientific communication, patronage, and political influence the society could provide Winthrop and that group's great and growing interest in plantations, such efforts by Hartlib on Winthrop's behalf are understandable.[15]

The creation of the Royal Society was closely connected to the crown's attempt to gain greater understanding of, better authority over, and increased economic value from its plantations; and the society's millenarian reform program was wedded to a social vision of science, trade, empire, and reformation. The experimental philosophy promoted by the Royal Society was, among other things, a tool for exploiting the riches of the colonies; the information gathered from its international network of correspondents, a source of commercial and political intelligence. Even before its official incorporation in 1662, the king had urged the society to survey the riches of the empire,

---

14. Black, *Younger John Winthrop*, 218.

15. Johann Moraien to Hartlib, January 1662, Hartlib Papers [31/16a–b]; William L. Sachse, "The Migration of New Englanders to England," *American Historical Review*, LIII (1947–1948), 263. Hartlib had sought on other occasions to place scholars in academic positions in England. Stephen Clucas, "Samuel Hartlib and the Hamburg Scientific Community 1631–1660: A Study in Intellectual Communications," paper, 1998.

and much of its time after it was chartered was spent promoting industry, the empire, and trade.[16]

Robert Boyle, alchemist, millenarian, founding member of the society, and member of the Hartlib circle, embodies the interlocking directorates that were created among the crown's new institutions of colonial authority. Boyle was not only one of the leading minds of the Royal Society; he was an active member of the Council for Foreign Plantations and governor of the new Corporation for the Propagation of the Gospel in New England. His participation in all of these institutions enabled him to have strong influence in colonial policy creation and administration, which he readily employed on Winthrop's behalf. Boyle had, for a time, been the patron and admirer of George Starkey, who had been part of Winthrop's New England alchemical circle, and he had corresponded with Winthrop before his arrival in England. Clarendon assigned Boyle to be one of three arbitrators to help resolve the Connecticut–Rhode Island boundary dispute; Boyle had favored a resolution advanced by Winthrop. Moreover, when Winthrop's lengthy stay in London had created financial difficulties, Boyle arranged an introduction to Henry Ashurst, a wealthy promoter of the Corporation for the Propagation of the Gospel, who floated Winthrop a loan.[17]

Boyle's interest in the colonies was driven by a vision of their helping implement God's plan for perfecting human society. Boyle believed that man was inherently driven by covetousness, which he saw as divinely ordained: it sent men into the world to appease their appetites. Through this acquisitive effort came the discovery of the secrets of nature and productive output in trade and manufactures, all of which glorified God. The greatest satisfaction of human appetites came by means of production and exchange under market conditions and through shared information between philosophers and tradesmen. In Boyle's framework, natural philosophy was a divine tool to be exploited for knowledge and profit.

Boyle believed England's plantations were to play an important role in the capitalist process of social amelioration. Natural philosophers abroad, through collaboration with colonial agents, governors, traders, administrators, and natives, would advance the public good through the pursuit of new

---

16. J. R. Jacob, *Robert Boyle and the English Revolution: A Study in Social and Intellectual Change*, ed. L. Pearce Williams (New York, 1977), 144; Michael Hunter, *Science and Society in Restoration England* (Cambridge, 1981), 144–159.

17. William R. Newman, *Gehennical Fire: The Lives of George Starkey, an American Alchemist in the Scientific Revolution* (Cambridge, Mass., 1994), 70–72; James, *John Clarke and His Legacies,* ed. Bozeman, 70; Black, *Younger John Winthrop,* 227–228.

knowledge, practical advances in technology, economic exploitation of nature, and the careful observation and communication of information. These goals would fulfill God's plan to see man exploit his dominion over nature, even as they increased the strength of the British Empire. Perhaps because of Boyle's involvement in three of the institutions directly connected to colonial affairs, the plantation policy being elaborated at Whitehall was compatible with and to some degree an extension of Boyle's utilitarian and spiritual natural philosophy.[18]

Boyle was not the only natural philosopher connected with the Royal Society who had both an interest in the colonies and the king's ear. Sir Robert Moray, privy councillor, first president of the Royal Society, and an intimate of the king, was another influential link between the society and Whitehall. Lord William Brouncker, the charter president of the society, was another person closely connected to the monarch. A diary entry of Samuel Pepys in 1669 records a visit Pepys made with Moray and Brouncker to Charles II's private alchemical laboratory.

> Then down with Lord Brouncker to Sir R. Murray, into the King's little elaboratory, under his closet, a pretty place; and there saw a great many chymical glasses and things, but understood none of them.[19]

Members of the Royal Society continued to reflect the flourishing of ideas associated with alchemical, Hermetic, and Neoplatonic thought throughout the seventeenth century. During his attendance at society meetings, Winthrop associated with several like-minded members committed to Paracelsian cosmology, alchemy, and Christian scientific reform. In addition to Boyle, Moray, and Brouncker, all of whom became Winthrop correspondents, he met Elias Ashmole, who gave Winthrop a copy of his *Theatrum Chemicum Britannicum*, a defense of magic as a means of revealing the inner harmonies of the universe. Sir Kenelm Digby, the Catholic alchemical physician and pro-

---

18. This analysis was derived from Jacob's exegesis of Boyle's tract *Usefulness* in Jacob, *Robert Boyle and the English Revolution*, ed. Williams, 140–144. Other natural philosophers echoed Boyle's desire to deploy utilitarian scientific endeavors to achieve social betterment and attain divinely ordained goals. Jan Comenius was so enthusiastic about the Royal Society's potential for effecting the pansophic agenda he had outlined in England in 1642 that he dedicated the publication of a book version of "Via lucis" to the members of the society. Society member John Evelyn echoed this pansophic theme, declaring the mission of the society to be to "improve practical and Experimental knowledg, beyond all that has been hitherto attempted, for the Augmentation of Science and universal good of Man-kind." John Evelyn, *A Panegyric to Charles the Second* ... (London, 1661), 14; Charles Webster, *From Paracelsus to Newton: Magic and the Making of Modern Science* (Cambridge, 1982), 67.

19. H. B. Wheatley, ed., *The Diary of Samuel Pepys* (London, 1924), VIII, 201.

ponent of the sympathetic cure of the weapon salve, was a familiar face. Winthrop had become acquainted with him on his previous journey to England, and the two had corresponded after his return to America. Digby had, in 1655, attempted to lure Winthrop back to England, suggesting a colonial setting was insufficient for a person of Winthrop's abilities.[20]

Other society members sharing Winthrop's philosophical perspectives included the alchemical physicians Walter Charleton, Christopher Merrett, and Edmund Dickenson, the king's physician. Dickenson was another society member with whom Charles conducted alchemical experiments. Winthrop also met Thomas Henshaw, reputed by both Robert Child and Ashmole to be a remarkable alchemist. In 1650, after his return from America, Child had formed plans to create an alchemical society with Henshaw and the poet and translator of Rosicrucian tracts, Thomas Vaughan. This might have reflected an effort on Child's part to recast the "design of learning" he and Winthrop had envisioned in Connecticut.[21]

For Winthrop, membership in the Royal Society was immensely important. It offered high-level political patronage, potential access to private and public capital, elevated personal status, intellectual stimulation, and an opportunity to play a part in an organization that seemed on the verge of accomplishing at least some of the goals of world reformation. Winthrop sought to capitalize on the access to influence the society provided by advancing two schemes that he believed simultaneously served the practical interests of New England in colonial development, the new colonial focus at Whitehall and pansophic reform. The first was a proposal he submitted to Boyle in his capacity as governor of the Corporation for the Propagation of the Gospel. Winthrop asked that the corporation help raise money for a project to set the New England Indians to work in agriculture and processing. The Indians employed in the project—Winthrop intended to use a band under the Eastern Niantic sachem Ninigret, who resided in the disputed Rhode Island territory—would raise commodities that the proposed new company would vend commercially. The benefits to be gained were commensurate with both the society's reform agenda and the commercial imperatives of the Board of Trade and the Council for Foreign Plantations. The Indians would become more civilized and be

---

20. Webster, *From Paracelsus to Newton*, 10–11, 65; Elias Ashmole, *Theatrum Chemicum Britannicum* . . . (London, 1652); Wilkinson, "John Winthrop, Jr. and the Origins of American Chemistry," 292; Black, *Younger John Winthrop*, 172.

21. Wilkinson, "John Winthrop, Jr. and the Origins of American Chemistry," 295, 297–298; Samuel Hartlib, *Ephemerides*, 1649 [28/1/37a], [28/1/11a], 1650 [28/1/42b], [28/1/71B–84B]; Stephen Clucas, "Samuel Hartlib's *Ephemerides*, 1635–59, and the Pursuit of Scientific and Philosophical Manuscripts: The Religious Ethos of an Intelligencer," *Seventeenth Century*, VI (1991), 33–55.

prepared by work to receive the Christian religion. Simultaneously, through the adoption of English products and customs, their standard of living would increase. The empire's trade would be enhanced through the creation of a new market for English commodities, especially dry goods, which, Winthrop noted, the Indians were currently acquiring from the Dutch. "For there be many thousands which would willingly wear English apparel if they knew how to purchase it," Winthrop asserted, "which must easily be done by the improvement of their own labor in a due way; and besides, many other manufactures would be vended." In addition to creating a new and ready market for English goods, England would benefit by receiving substantial "hemp, flax . . . pitch, tar, wheat, prairie grass, and some other . . . commodities very proper to the country." It was to be no small-scale operation, but would quickly return a profit. The investors, from whom three thousand pounds would be needed in the first year and two thousand pounds the second, would receive a return on their investment within three to five years. Profits would come in the form of either interest or profit sharing. Surplus profits would be reinvested "in some purchase there" or some other investment whose returns would fund company operations."[22]

Winthrop's proposal reflected Boyle's belief in improvement of society through economic development and Whitehall's interests in pacification of the natives and increasing the direct commercial exchanges between the plantations and the home country. It set the stage for the Indian conversions that millenarians considered a prerequisite to the Second Coming of Christ. English people understood that reducing the Indians to Christ (as contemporaries termed the conversion process) involved, first, getting native people to adopt European customs and culture. Cultural colonization—shifting the Indians to English working practices and getting them to adopt English clothing and goods—was a necessary stage in a multistep process of bringing them in to Christ.[23]

Through Boyle, Winthrop presented his proposal to a meeting of the Corporation for the Propagation of the Gospel in New England, and, though they responded positively to it in principle, Winthrop had overestimated the new corporation's ability to secure capital. The company's finances were at the time in poor condition, and the labor scheme could not be funded. Winthrop's perception of how positive the corporation's reaction was was strong

---

22. "John Winthrop to the Governor of the Corporation for the Propagation of the Gospel in New England," MHS, *Collections*, 5th Ser., IX (1885), 45–47; William Kellaway, *The New England Company, 1649-1776: Missionary Society to the American Indians* (London, 1961), 52–57, 107–108.

23. Karen Ordahl Kupperman, ed., *America in European Consciousness, 1493-1750* (Chapel Hill, N.C., 1995), 10; Jacob, *Robert Boyle and the English Revolution*, ed. Williams, 152.

enough, though, that, two years later, upon hearing (incorrectly) that the corporation's funding had been replenished, he wrote requesting Boyle to have the corporation review his proposal again.[24]

Winthrop presented his second project directly to the Royal Society. Before his English journey, Samuel Hartlib had informed Winthrop that he had been promoting the idea of setting up a bank of lands and commodities in a plantation. Such a bank was of great interest to Winthrop, for land was New England's most abundant commodity, specie its scarcest. Collateralized land in return for specie or bills of exchange could be a powerful engine for economic development and social reform. Hartlib had sent Winthrop William Potter's *Key of Wealth* and "A Bank of Lands," two tracts elaborating the land bank concept. Reflecting perhaps the growing demands for landed inheritances caused by New England's expanding and unusually healthy creole population, Winthrop had responded that the concept of a land bank was very important but that New Englanders would be unwilling to engage their lands if the land behind the bank's capitalization had to be sequestered rather than remain available for improvement. Winthrop, however, had developed a proposal for a bank that did not require engaging the lands, and "thereby mony would flow in abundantly." Although he had considered promoting the scheme in New England, Winthrop had decided to defer doing so until he was in England, where it might be effected through the sponsorship of a single company. Winthrop, attuned to the society's imperial and pansophic agendas, declared that his land bank scheme "would greatly advance commerce and other publique concernments for the benefitt of poore and rich in great Brittaine and the good of these plantations." Invoking the alchemical code of nondisclosure, Winthrop further alluded to some secret aspect of his banking scheme. "Some matters that concerne the secretts of some waies of profitt in which the undertakers of such a banke would be invested . . . [may not] be conveniently intrusted in a letter." If he could explain it to Harltib in person, though, he would be able to "make it appeare really." Hartlib responded enthusiastically, suggesting he might be useful in helping advance Winthrop's project.[25]

---

24. After he returned to New England, Winthrop countered with a more modest proposal, that the corporation fund the conversion efforts of the New London minister William Thompson to the Pequot tribe. Frances Manwaring Caulkins, *History of New London, Connecticut: From the First Survey of the Coast in 1612 to 1860* (New London, Conn., 1895), 129; Thomas Birch, ed., *The Works of the Honourable Robert Boyle* (London, 1772), I, lxxi.

25. Samuel Hartlib to John Winthrop, Jr., Mar. 6, 1660, MHS, *Proceedings*, LXXII (1963), 48; William Potter, *The Key of Wealth* . . . (London, 1650), and "A Bank of Lands . . . ," in *Samuel Hartlib: His Legacy of Husbandry* . . . , 3d ed. (London, 1655), 289–299; Philip J. Greven, Jr., *Four Generations:*

Winthrop sought, probably near the time he was to return to Connecticut, to present this proposed banking venture to the consideration of the society membership. It did not, however, take priority on the society's meeting agenda, and Winthrop found that "their tyme was so short [and] that the discourse could not be read, only a little of the beginninge." Henry Oldenburg, secretary of the society, promised to have it transcribed and read at a future meeting. Once back in Connecticut, Winthrop wrote William Brereton, who had nominated him to membership, to determine whether the society collectively or Brereton individually had subsequently taken up the proposal. Winthrop was eager to know "whether it doth upon perusall appear to be a foundation of such use for the advance of trade, and setling a sure and easy way of a banke, that the honorable Society doe thinke fitt to owne, and to promote it to a way of practise." He was certain that, because of the Royal Society's reputation among the gentry, merchants, and citizens and the "insight that many of them [the members of the society] have into matters of trade and exchange," the society's sponsorship was the best way to capitalize the projected bank. He further offered the society an exclusive connection with the project. He had shown it to no one else; all the copies were in their hands. Should the society endorse the proposal, Winthrop wrote, "it may be quickly brought into a practical way, to the great advance of trade, and settlement of such a banke, as may answer all those ends that are attained in other pts of the world by bankes of ready money." Winthrop was, as it turned out, much more enthusiastic about his project's potential than his fellows. A letter from Henry Oldenburg, which predated Winthrop's letter to Brereton by three months but which Winthrop had not received before writing Brereton, informed Winthrop that his conceptions about the bank were in the hands of Brereton, who "doubts whether it be so fit to have them communicated to the person you named in your letter."[26]

Winthrop's project proposals demonstrate that he understood the Royal Society within the context of a web of relationships—political, economic, religious, and scientific—and not as a purely scientific entity. It operated from "an aggressive, acquisitive, materialistic, imperialistic ideology justified in the name of reformation." Winthrop sought to capitalize on that ideology by pro-

---

Population, Land, and Family in Colonial Andover, Massachusetts (Ithaca, N.Y., 1970); John Winthrop, Jr., to Samuel Hartlib, Jan. 7, 1661, MHS, *Proceedings*, LXXII, 65–66.

26. John Winthrop, Jr., to William Brereton, Nov. 6, 1663, MHS, *Proceedings*, 1878, 219–220. Winthrop did add a caveat to Brereton, asking that, if the society did not accept sponsorship, he hoped Brereton would suggest others to whom to present the concept. Henry Oldenburg to John Winthrop, Jr., Aug. 5, 1663, ibid., 216–218. The intended recipient of the proposal is not known.

moting programs of special benefit for himself and his colonial plantation that were commensurate with that ideology and served the cause of reform. The cool reception received by his banking proposal, however, suggests that at some level there was a clear distinction between his interests as a colonial projector and the interests of the society. It is possible, of course, that Winthrop's proposal was simply poorly conceived and was rejected solely on its lack of merit. It is also possible that the concept of banking itself seemed at the time simply too radical. Yet it is also likely that a scheme to transfer English capital into the control of colonial projectors where it could fund projects independently of central oversight or regulation did not fit the current objectives of either the society, the Board of Trade, or the Council for Plantations. Whatever its shortcomings, the society passed on its colonial member's bank scheme. He continued to beat the drum for the project and expressed neither surprise nor disappointment.[27]

THROUGHOUT THE TIME he was in England, Winthrop participated in the society's meetings. Based on presentations he was asked to make to the group, it is clear that the fellows appreciated his chemical and technological skills. He was asked, for instance, to join with Doctors Jonathan Goddard and Daniel Whistler to present the society an account of refining gold with antimony. He also displayed to the group an innovative, continuously fed tin lamp, called a bladder lamp, of which a drawing was made for the society archives.[28]

Despite the society's respect for Winthrop as a natural philosopher, the fundamental basis of its appreciation for him came from his association with and intimate knowledge of New England. Winthrop was himself a kind of colonial curiosity, of great interest and perhaps even greater value. The majority of his society presentations had explicit connections to New England's natural resources or technological operations. He reported on the history of making tar and pitch in New England, suggesting, in support of his proposal to Boyle, that it was a suitable employment for Indians as well as the English. He gave a detailed description of the cultivation of New England's Indian corn and was subsequently urged to provide a demonstration of brewing beer from maize. He displayed an earth brought from New England that, when stirred, stayed suspended in a water solution for a half hour. He read a paper recommending the suitability of building ships in "some of the northern parts of America,"

27. Jacob, *Robert Boyle and the English Revolution*, ed. Williams, 159.
28. "Active Members of the Royal Society," table 1, in Webster, *The Great Instauration*, 92; Wilkinson, "John Winthrop, Jr. and the Origins of American Chemistry," 309–320; Thomas Birch, *The History of the Royal Society of London* . . . , I (London, 1756), 77, 80.

and, when a ship arrived in London of New England construction, he was asked to report on it to the membership. A rattlesnake tail Winthrop brought to the society was investigated by Dr. Merret for its medicinal value. And, on the last day of 1662, during a discussion of black lead, Winthrop, referring to his mine site at Tantiusque, told the members that the only proper black lead in the world was found in England or New England.[29]

That Winthrop's membership in the society was intimately connected to his colonial residency does not mean he had a diminished status among the members. On the contrary, Winthrop represented to the fellows a type of natural philosopher who had been desired by reformers since Bacon's description of "Solomon's House" in *New Atlantis:* the traveling elders (one-third of Bacon's College) who gathered experimental information in other countries and returned, bringing their knowledge with them. Comenius's vision of universal reformation also had envisioned an international college of "learned and hard-working men, called from all nations," who would assemble the knowledge and skills necessary to regenerate the world and improve human living conditions.[30]

To members of the society such as Boyle, Moray, and Oldenburg, Winthrop represented the ideal on-the-ground informant—in what some believed was the most ideal location of all the colonies—for advancing the scientific agenda of the Royal Society: supporting the centralizing, imperial agenda of the Restoration monarchy. The society, through its effort to acquire systematic and comprehensive information about the natural, human, and technological resources of England's overseas plantations, served both as a scientific information clearinghouse and as an intelligence-gathering operation for Whitehall and English mercantile interests. Winthrop, because of his observational skills, scientific abilities, position of local authority, and shared belief in the pansophic agenda of society members such as Boyle, Evelyn, Digby, and Ashmole, was an ideal conduit through which to receive information useful in assessing the potential of and gaining control over the human and natural assets of the crown lands in North America.

Winthrop was being groomed by the society to become the natural historian of New England. This approach to information gathering and economic

---

29. Birch, *History of the Royal Society of London*, I, 112–113, 162, 166, 167, 206. The account of black lead was printed in the Royal Society's *Philosophical Transactions*, XI (1679), 1065–1069.

30. Frank E. Manuel and Fritzie P. Manuel, *Utopian Thought in the Western World* (Cambridge, Mass., 1979), 256–258; Johann Amos Comenius, *Opera Didactica Omnia* (1657) (Prague, 1957); Comenius, *The Way of Light* (1668), ed. and trans. E. T. Campagnac (Liverpool, 1938), 170–171; Robert Fitzgibbon Young, ed., *Comenius in England: The Visit of Jan Amos Komenský* . . . (London, 1932), 53.

development was integral to Bacon's proposed program for scientific advancement, and it had been promoted by the Hartlib circle. Bacon believed that the compilation of natural histories was crucial to the advancement of knowledge. He advocated a more systematic approach to collection of information than previously undertaken, one that paid minute attention to the processes of change and growth. The natural historian should observe weather, astronomy, and all the elements of the physical environment. Information on marvels, too, should be systematically recorded. Also, he added a new category, mechanical arts, to the subjects under scrutiny. Hartlib, expanding on the Baconian program, added trades, manufactures, and commerce to the subjects to be encompassed within natural histories. As the categories expanded, the concept of the documented natural history came to be synonymous with the concept of "systematic accounts"—carefully observed compilations of useful information, accurately compiled and distributed.[31]

Hartlib had proposed to Parliament that it set up an Office of Address, which would serve as a collection point and clearinghouse for comprehensive information on trade, employment, and property. Trade was broadly defined and conceived of as the collection of all information that would be of service in exploiting resources, improving commerce, and expanding England's economic power both at home and abroad. Ireland had formed the pilot site for the first attempt at a comprehensive natural history, and *Irelands Naturall History*, written in 1652 by Gerard and Arnold Boate, two Dutch physicians affiliated with Boyle, had provided a model upon which the society hoped to build. The utility of the Boates' natural history as both a source of information and a justification for conquest of the wild native inhabitants of Ireland had made it a powerful tool for natural philosophers, economic projectors, and English planters. The book explicitly anticipated the resettlement of Ireland after parliamentary conquest, a reminder of the imperialist component of the natural history projects.[32]

The practical and political utility of *Irelands Naturall History* prompted calls for additional natural histories and histories of trades. The future Royal Society member John Beale, who would become the most insistent champion of the need for a systematic accounting of New England, had urged Hartlib in 1658 to request the New Haven minister John Davenport to produce, or have produced, a natural history of his region:

---

31. Webster, *The Great Instauration*, 420–450; Wilkinson, "John Winthrop, Jr. and the Origins of American Chemistry," 327.

32. David Andrew Attis, "The Ascendancy of Mathematics: Mathematics and Irish Society from Cromwell to the Celtic Tiger" (Ph.D. diss., Princeton University, 2000), 11–13.

In your answere to Mr. Davenports enquiry you have an opportunity to move, That some happy pen would give us the history of Virginia, Newe England, and other neyboring plantations, as is Civilly exemplifyed in Mr Ligons Barbados. From Newe England wee may expect a more holy accompt of the demonstrations of divine providence in the afflictions, and also preserva[tion] of those good people who fled to that desert, as the Israelites out of Egypt through the red sea, To worship God. For I thinke it is, The only plantation, which went foorth upon the <apparent> accompt of God, to preserve their consciences from the Contagion of this wicked world, To vindicate the Solemnityes of the Sabbatical Worship of from the rude Minstrellsy of a corrupt courte, a conniving clergy, and a polluted people.

Tis fit the chiefe familyes, Their encrease and protection should bee faythfully bee recorded in a booke of Numbers. And the helpes of Naturall supplyes by Vegetables, animalls, Minerals; and soyle, Their arts and manufactures, Their governement, and Trade, their present attainements, and their hopeful enterprises, recomended to this nation and preserved for their owne posterity.[33]

Beale's combination of providentialism, utilitarianism, and desire for a systematic account of the human and natural resources of the colonies was typical of both the Hartlib circle's and the Royal Society's reform agendas. The natural history projects begun before 1660 built up into a substantial framework for the scientific effort of the Royal Society.[34]

The problem was not that there was a dearth of information in England about America. As early as 1600 an extensive amount of information circulated through Europe about the natural resources of the New World. The literature that accompanied the first waves of English colonization was massive. Neither the "catalogs of nature" nor the early settlement literature was adequate to the demands of the society. The scientific information was too generalized, often lacking adequate detail or specific references to locations. The colonization tracts were promotional and self-exculpatory. They worked diligently to justify colonization while building the case for the difficulty of achieving rapid returns for investors. Such information lacked the systematic, detailed accounting that was essential to the Royal Society's approach to his-

---

33. John Beale to Samuel Hartlib, Dec. 7, 1658, Hartlib Papers [51/39A–40B].

34. Webster, *From Paracelsus to Newton*, 62–64, and *The Great Instauration*, 420–446. On natural histories as a seventeenth- and eighteenth-century European phenomenon, see Alix Cooper, *Inventing the Indigenous: Local Knowledge and Natural History in Early Modern Europe* (Cambridge, 2007), 116–151.

tories. Moreover, they were dated; nearly a generation had passed since the wave of settlement literature had subsided. John Beale stressed this point to Oldenburg.

> Wee have their Annalls of their growth, and of all the removealls, wch have been made from them by dissenters in Religion. . . . Now there remaines 21 yeares, in which their growth is reported to be very great: Of which I have enquired, but can obtain no certain accompt.[35]

During his stay in England, Winthrop was requested repeatedly to make presentations to the society on subjects that would be suitable to a natural history of New England. Upon his return the call for him to produce a natural history would become a persistent and insistent demand from society correspondents. His willingness to provide detailed, accurate information about New England was explicitly connected to the continuation of patronage from the society's members. The problem this created for Winthrop was that, after he returned to New England, it became amply clear that the crown's desire to institute colonial consolidation could work both for and against his interests and Connecticut's. By providing intelligence through the natural histories that the society requested, he would increase and rationalize understanding about New England and enhance incentives for exploiting or taxing New England's resources. Given the society's "aggressive, acquisitive, materialistic, imperialistic ideology" and its links to Whitehall, such information would simultaneously encourage and facilitate the political desire to increase crown authority over New England. In every exchange with the society, Winthrop faced difficult choices: what kind of information to provide, and how much. In response, he developed strategies to maintain his good standing with the society while restricting the content of the information it received.[36]

WHILE HE WAS in England, the scientific patronage Winthrop enjoyed had allowed him to pursue aggressively and subsequently defend Connecticut's charter. Assured of the support of men such as Worsley, Boyle, Moray, and Brouncker, he had confidently held the line in his negotiations with John Clarke about the Rhode Island boundaries. Even after the Privy Council had tired of the affair and made a decision to send royal commissioners over to

---

35. John Beale to Henry Oldenburg, Aug. 29, 1668, in A. Rupert Hall and Marie Boas Hall, eds. and trans., *The Correspondence of Henry Oldenburg* (Madison, Wis., 1965–1971), V, 28–31; Henry Lowood, "The New World and the European Catalog of Nature," in Karen Ordahl Kupperman, ed., *America in European Consciousness*, 295–323, esp. 296–297, 302, 306–310; Kupperman, *Indians and English: Facing off in Early America* (Ithaca, N.Y., 2000), 3–4.

36. Jacob, *Robert Boyle and the English Revolution*, ed. Williiams, 159.

New England, Winthrop remained confident that his charter rights were fundamentally secure and that his patrons would effectively represent his rights at Whitehall.

He arrived back in America in early June 1663. Within days, he sent a letter to the society reporting on the failure of two experiments—one for collecting water from the ocean floor, another for determining the ocean's depth—that he had been requested to conduct during his Atlantic journey. He gave updates as well on other projects in which the society was interested. He was "necessitated to be up in the Countrey farre from the Sea," which would delay his report on a new salt-recovery experiment, but he already had workmen along the shore preparing the site. In a report that implicitly suggested to Boyle yet another employment for native people, he told of a mineral that an Indian had brought him that was a kind of marcasite but of which he had deferred making an assay until he could get a larger sample. Winthrop's Indian informant had told him about an additional site that appeared to promise copper, and Winthrop had engaged him to provide more information about it. In October, the members of the society sent Winthrop a thermometer and news that he had been assigned, in absentia, to the committee to "consider and improve all mechanical inventions" and to the committee "For Histories of Trade."[37]

Winthrop continued to nurture the scientific patronage network with which he was connected. He followed up with Brereton on the bank proposal he had submitted. He sent Boyle a request that the corporation provide a stipend for the widow of Thomas Mayhew, missionary to Martha's Vineyard, who had died at sea on a transatlantic crossing. With his letter he sent a "Rarity": Latin essays written by Indians who attended the college in Cambridge. Winthrop assured Boyle that he had "received them of those Indians out of their owne hands" and that, when questioned, they could provide answers in Latin and Greek.[38]

Henry Oldenburg wrote in the fall with the first of the many requests he would make to Winthrop to produce a natural history of New England. Hinting at existing commercial arrangements and greater "commissions" in the future, he informed Winthrop that the society expected that

> you will in time give them a better Account of the remarkables of your quarters than is any yet extant, concerning the mappe of your contry, the history of all its productions, and particularly of the sub-

---

37. Birch, *History of the Royal Society,* I, 280, 322, 406–407; Henry Oldenburg to Robert Boyle, Sept. 29, 1664, in Hall and Hall, eds., *The Correspondence of Henry Oldenburg,* II, 241.

38. John Winthrop, Jr., to Robert Boyle, Nov. 3, 1663, MHS, *Proceedings,* 1878, 218–219.

terraneous ones (concerning which you know what they look for from you, to the end that you may receave fuller instructions and ampler commission upon discoveries made made knowne to them), as also concerning your neighbors and their dealings with you, and your hopes of advancing further amongst them, likewise a relation of the Tides upon your coast, together with the course of the rivers; but, especially and above all, a full account of your successe in your new way of salt-making, whereof we could not compasse your experement here, as was much desired.[39]

Oldenburg requested cartographic information, details on economic production, accounts of intercolonial or English-native relations, reports on coastal tides and river courses, and information about mining and saltmaking activities that were certainly valuable to natural philosophers seeking to bring reformation to the whole world. The requests reflected a shift toward localization in intelligence gathering, away from the compendiums of the sixteenth and early seventeenth centuries, which treated the New World as a single category in opposition to the Old. But the same information could be more immediately useful to a monarch seeking to gain greater understanding of and control over his possessions abroad. In the same letter to Winthrop in which he issued the society's first request for a natural history, Oldenburg noted, "His Majesty presented his R. Society on munday last with a very noble mace of 60 lb. Sterl. which this very day will be the first time made use of, and be carried before our noble President at our meeting." Oldenburg hoped the king would soon visit and sign his name at the head of the list in the book of fellows and also "name something of his guift that may enable us to carry on our desseins." For Winthrop, this was another reminder of the power and patronage that society affiliation opened to him. In addition, if it were not already clear to Winthrop, it would soon be pointedly evident that the society served two masters: science and the imperial ambitions of Charles II.[40]

Winthrop never seems to have believed that the Privy Council would really send royal commissioners to New England or that, if it did, it might produce negative consequences for Connecticut. He continued to take a firm stance with Rhode Island's boundary negotiator even after the threat of send-

---

39. Henry Oldenburg to John Winthrop, Jr., Aug. 5, 1663, ibid., 216–218.
40. Ibid.; Lowood, "The New World and the European Catalog of Nature," in Kupperman, ed., *America in European Consciousness*, 302–303. For an account of how this crown–Royal Society collaboration for empire affected the implementation of slavery in colonial Jamaica, see M. Govier, "The Royal Society, Slavery, and the Island of Jamaica: 1660–1700," *Notes and Records of the Royal Society of London*, LIII (1999), 203–217.

ing a royal commission to judge the intercolonial dispute had been issued. When, before returning to New England, Winthop had been told that such a commission might feel the need to revise Connecticut's boundaries, he had stated with a flourish that Connecticut would accept any new boundary determination as if no prior charter existed. He might have been overconfident of the ability of his patrons to protect his interests, or he might have underestimated their dependence on the court. Nevertheless, soon after Winthrop left England, Whitehall decided to take a very different direction in American colonial affairs, and the members of the Royal Society supported the new imperial approach.[41]

Winthrop was on Long Island the following June, consolidating Connecticut's new charter gains there, when a note arrived from the colonial military leader John Underhill informing him that a fleet of British warships carrying five hundred soldiers and one thousand small arms was approaching New England "to settel government, and to reduse the Dutch." It took some time to learn the full implications of the fleet's arrival. For Winthrop and Connecticut, they were grave indeed. The king had granted his brother James, duke of York, a patent to a vast amount of territory: the present state of Maine between the Saint Croix and the Kennebec rivers, all the islands from Nantucket westward to Staten Island, and the entire mainland from the Connecticut River to the Delaware River. More than half of Winthrop's colony and all of the New Haven plantation that his charter had annexed the prior year were now to belong to York. Moreover, royal commissioners to New England had been appointed and were sailing with the fleet. Their stated intentions were to visit each colony, gather information for the king, institute an oath of allegiance, settle boundary disputes, and assure the accuracy of and enforce the terms of each colony's charter. Every New England colony had reasons to worry. One had only to examine a map—the type of map the society had requested from Winthrop the preceding fall—to realize that their charters were the only things preventing the creation of a single, unified royal colony from the Saint Croix to the Delaware.[42]

Knowledge of New England played an important part in the selection of at least one of the royal commissioners. Samuel Maverick, the only New Englander on the commission, had settled in Boston Harbor before the Winthrop fleet of 1630 and had lived among, but not with, the Massachusetts Puritans

---

41. "Instructions to Col. Nicolls and Commissioners to Connecticut," Apr. 23, 1664, *DCHNY*, III, 55.

42. Black, *Younger John Winthrop*, 269–278; John Underhill to John Winthrop, Jr., June 18, 1664, MHS, *Collections*, 4th Ser., VII (1865), 189.

ever since. A Royalist, Maverick had always opposed the Bay Colony's restrictive civil and ecclesiastical settlement. In 1647, he had joined with Robert Child and others in the effort to force Massachusetts to liberalize its qualifications for church membership and the franchise. He had old scores to settle, and his selection as a commissioner augured ill, especially for Massachusetts. Connecticut had reason for concern as well: Maverick had been one of the arbitrators assigned by Clarendon to attempt to resolve Winthrop's boundary dispute with Rhode Island. Maverick had favored Clarke, even helping him draft provisions for the new Rhode Island charter.[43]

Maverick had extensive knowledge of the New England plantations. He wrote, for either the Privy Council or the Council for Foreign Plantations, sometime after 1660, "A Briefe Discription of New England and the Severall Townes Therein, Together with the Present Government Thereof." "A Briefe Discription" was a kind of natural history. It provided useful information about the New England colonies and all the lands that were included in the duke of York's patent. While polemical and inconsistent, it represented a systematic effort to describe the political, religious, and physical landscape of New England. Maverick provided descriptions of eighty-six settlements extending from Nova Scotia to the Delaware Bay, including the Connecticut River towns and the Dutch and English settlements on Long Island. He gave the council a ship trader's view of each settlement. The reader, a passenger in Maverick's text, floated from settlement to settlement in a journey north to south, upriver and down, with an occasional inland digression to discuss a village in the interior. What concerned Maverick in his history were the same issues that would concern the royal commission: dates of settlement, patents under which lands were settled, and colonial loyalty. Maverick paid close attention to other details: the location and strength of New England's defensive fortifications (at the Isle of Shoals, Winisime, Strawberry Bank, Boston, and Saybrook). In some instances, he focused on a region's economic potential. Consider, for example, Maverick's description of the economic value of Winthrop's plantation at Pequot. While he chose not to call it by Winthrop's name of New London, he acknowledged the potential worth of the black lead mines and other mineral deposits and the industry and knowledge of the plantation founder:

> Before the Pequate River lyes Fishers Island, on which some people live, and there are store of Catle. This Pequat Plantation will in time

---

43. "Robert Child the Remonstrant," in Samuel Eliot Morison, *Builders of the Bay Colony* (Boston, 1930), 245–268; James, *John Clarke and His Legacies,* ed. Bozeman, 74.

produce Iron, And in the country about this is a Myne of Black Lead, and supposed there will be found better if not already by the industry of that ingenious Gentleman Mr. John Winthrop. It hath a very good Harbour, farr Surpassing all there about Connecticot River mouth to Pequate it is about eight Leagues.[44]

Maverick provided an overview of the history of each colony's founding, the charter under which it claimed legitimacy, the prevailing government and religious practices. He gave free rein to his resentment of the Puritans, especially of Massachusetts. In so doing, he built a fairly strong case against Massachusetts especially, as a colony in rebellion against the crown. He criticized the narrowness of its political franchise, its religious intolerance, and its refusal to grant nonconforming subjects traditional English rights. He cited the oath of allegiance to the colony instead of the king that freemen had to swear, and he railed against Massachusetts's illegal practice of coining money. Reaching back a quarter century to the early history of the Bay Colony, he also recounted the militant opposition that met previous rumors that a governor general was coming from England:

> All the Traine Bands in Boston, and Townes adjacent were in Armes in the streets and posts were sent to all other places to be in the same posture, in which they continued. . . . The generall and Publick report was that it was to oppose the landing of an Enemie a Governor sent from England, and with this they acquanted the Commanders.[45]

In addition to providing a textual map to the country and a motive for instituting greater controls over New England, Maverick's "Briefe Discription" provided a justification for England's claim to take New Netherlands: the Dutch had settled there in 1618, years after James I had granted the land to the New England Company of Plymouth. New Amsterdam lay a full degree within the New England patent, and Fort Orange (present-day Albany) two and one-half degrees. Furthermore, the Dutch had ceded any claims they might have had in 1629 or 1630 at Plymouth and again in 1632 or 1634, when an English trading ship had coursed the Hudson without challenge.

Maverick continued his narrative journey all the way to the mouth of the Delaware River, where "the earth brings forth plentifully all sorts of Graynes, also Hemp and fflax. The Woods affords store of good Timber for building of

---

44. Samuel Maverick, "A Briefe Discription of New England and the Severall Townes Therein, Together with the Present Government Therof," MHS, *Proceedings*, 2d Ser., I, 1860–1862 (Boston, 1862), 231–249.

45. Ibid., 240.

shipps Masts, Also Pitch and Tarre, The bowels of the earth yeilds excellent Iron Oare, and no doubt other Metalls if searched after." The Delaware, as depicted by Maverick, was a desirable place for English people to settle. In 1664, it would come to mark the southern boundary of the duke of York's patent.[46]

Maverick's natural history helped set the stage for a period of colonial consolidation that began with the issuing of the duke of York's patent and ended with the Glorious Revolution. Though his town descriptions were often overly succinct and the focus of his attention inconsistent, his "Briefe Discription" helped England decide that bold steps needed to be taken to bring New England into better alignment with an expanding British Empire. If Winthrop needed evidence that natural histories could produce political as well as scientific consequences, he would find it in the English fleet that anchored off Nantasket Point in July 1664.

Winthrop must have been shocked at the speed with which the gains of the Connecticut charter were evaporating and equally concerned about the apparent collapse of his patronage network at Whitehall. The arrival in Hartford of a packet of letters dispatched from the fleet helped provide clarification. In it was a letter from Robert Moray, first president of the Royal Society and privy councillor, fully endorsing the military commander of the expedition, Colonel Richard Nicolls, and recommending in the strongest terms that Winthrop lend total support to Nicolls's mission. Nicolls, a Royalist and long-term retainer to the duke of York, was a close friend of Moray. Like Winthrop, he was a natural philosopher and polymath, and, perhaps through Moray himself, Winthrop had become acquainted with Nicolls in England. Moray's endorsement of Nicolls was unqualified. He entreated Winthrop, "In all things apply and open your self to him, as you would do to the most vertuous person you know, if my strongest conjurations weigh with you." Such support was, Moray underscored, a question of loyalty to the crown. "For Hee being trusted so eminently by his Majesty in the affaires of these parts, I have not the least doubt of your respect to him in that regard, or that you will be [found] wanting in any thing wherein you can be usefull towards the advancement of His Majesties service." Moray did not mention the York charter's annexation of Connecticut lands directly but alluded pointedly to Nicolls's authority. "You know him to be a very worthy person, and may be perswaded of all the good offices he can do you." Moray wanted to triangulate a three-way alliance, with himself at the apex of the triangle, based on patronage, mutual self-interest, and support for the empire. "I would be glad that my interposi-

46. Ibid., 249.

tion might not onely encrease that confidence, but settle between you such a friendship as, upon further acquaintance, you may have the same kindness for one another that I pretend to from either of you."[47]

Winthrop understood perfectly the implications of Moray's letter. Although his colony's charter rights were in jeopardy, his best hope for securing a favorable outcome depended upon support for Nicolls's imperial mission. Winthrop's response showed his willingness to comply with the new rules of the game, but he could not keep from expressing a note of concern. "All your commands are strong obligations," he wrote Moray, "and shalbe attended with all due observance towards the honourable Colonell, to the greatest of my endeavours to acte a parte so eminently entrusted by his majestie as you were pleased to announce; and I hope is come for eminent good of these poore plantations."[48]

Winthrop, true to his word, joined the English commander at Gravesend, Long Island. There, accepting the realities of his new situation, he publicly announced that Connecticut had no claim to the English towns on Long Island, ignoring or repudiating his visit there the previous June to consolidate Connecticut's claims to jurisdiction over those towns. He then accompanied Nicolls to New Amsterdam, where he helped negotiate terms of capitulation with the Dutch. Winthrop was very much a presence when, on August 29, 1664, Fort Amsterdam became Fort James, and New Amsterdam became New York.[49]

As Moray had hoped and strongly advised, Winthrop and Nicolls did become friends, though perhaps not close. They were remarkably similar. Of the same latitudinarian temperament, both men were interested in the whole compass of natural philosophy, and both placed an emphasis on alchemy. Both were effective, charismatic rulers. Moray's hint that Nicolls might be able to provide good offices to Winthrop rang true. He and Winthrop agreed quickly on a provisional boundary between New York and Connecticut that preserved most of Connecticut's land west of the Connecticut River. Nicolls had no intention of ceding, or any portfolio to cede, the duke's charter rights, so his arrangement with Winthrop was unofficial. That unofficial status was surely intentional; it would and did provide useful background leverage for future intercolonial relations. For his part, Winthrop was pleased, officially at least, to get a glass half-full. He wrote Moray providing details of "the happy arrival of our noble friend Colonell Nicolls."[50]

47. Robert Moray to John Winthrop, Jr., Apr. 30, 1664, MHS, *Proceedings*, 1878, 222–223.
48. Ibid.
49. Black, *Younger John Winthrop*, 270–278.
50. Ibid., 270–282; John Winthrop, Jr., to Robert Moray, Sept. 20, 1664, MHS, *Proceedings*, 1878,

Despite the personal affinities between the two colonial leaders, at a fundamental level their interests differed. Winthrop was an elected New England governor; Nicolls was an official of the crown. Nicolls's primary focus of interest was Whitehall; Winthrop's was Hartford and Boston. Nicolls's interest was in the expansion of empire; Winthrop's was in maintaining colonial autonomy. Of course, the lines were not nearly that well defined. Winthrop's colony needed connections to the imperial bureaucracy, just as New York needed connections to the adjacent colonies. But at the core of the relationship between the two men there was a distinction of interests that could not be fully bridged. These differences were thrown into high relief by the visitation of the royal commissioners to New England.

An April 12, 1664, letter from Robert Boyle provided Winthrop assurances of what was and what was not at stake in the commissioners' investigations. It also demonstrated the continuing benefit for Winthrop and his colony of having society patronage at Whitehall. The letter was delivered to Winthrop by Dr. Sackville, physician to the commissioners. Boyle implicitly recognized that the people of New England (including Winthrop) might be inclined to view the commissioners as foreign agents, for he assured Winthrop there was not "any thing to be soe much apprehended in their embassy (as I may soe call it) into New England as the easily evitable want of a right understanding betwixt them and the planters." Boyle had met with the king and Clarendon that very day, and they had spoken of New England in a very favorable manner. Clarendon personally had given Boyle a commission to "assure some of your freinds in the Cyty, that the King intends not any injury to your charter, or the Dissolution of your sivil Government, or the infringment of your Liberty of Conscience and that the doeing of these things is none of the business of the Commissioners." Boyle's advice to Winthrop in dealing with the commissioners focused on the need for him to demonstrate allegiance.

> I have only time to add one word of freindly advice, which is that you would prevent the proposalls that you suspect may be made you by the Commissioners by doeing, as many of them [as] you think fitt to comply with of your own [ac]cord, And soe make those things the expressi[on of] your loyalty and affection, rather than barely of your obedience, such a course being that which would be much the most acceptable to the King.

222–223. For an analysis of the Nicolls administration in New York, see Donna Merwick, *Possessing Albany, 1630–1710: The Dutch and English Experiences* (Cambridge, 1990), 134–187.

Boyle closed by telling Winthrop he had had assisted the colony that day by showing the essays of the Indian scholars Winthrop previously sent him at court.[51]

The official instructions to the royal commissioners concerning Connecticut did specify the preservation of charter rights and free exercise of religion as long as that colony granted liberty of conscience to all its citizens. Other provisions, however, reflected the crown's desire to have a better understanding of and greater hand in colonial affairs. Intelligence gathering was a central part of the commissioners' public mission. They were, for example, to monitor public opinion in Connecticut, both before and after their visit, to learn the distinctions between that colony and Massachusetts, in terms of both civil and religious estates. They were to determine any encroachments made into the colony by the Dutch or the French and reduce them to obedience. They were to catalog all patents granted to individuals and determine whether such patented lands still lay dormant. Employing the land tenure principle of *vacuum domicilium*—that ownership depends on improvement of the land possessed—the commission announced the king's intention of voiding any land grants that had not yet been taken possession of and cultivated.

While in Connecticut, the commissioners were to ascertain what productive or economic benefits the crown might derive from local resources. Two of the commissioners' instructions seem closely tied to Maverick's description of the potential of Winthrop's plantation at New London, where the commissioners held their Connecticut hearing. One required that the commissioners identify all ironworks currently in operation and where others might productively be established. The other instructed them to gather information regarding "whether there have been at any tyme or yet are, any mines of Gold or Silver discovered and working there, and what hath arisen from thence; to the end that wee may receive an accompt of the fifth part thereof, which by their Charter is reserved to us." This is a particularly important provision, in light of Winthrop's continuing hopes for the mine at Tantiusque and the interest shown in minerals by the members of the society. It served as an explicit reminder that the crown took its charter rights in mineral wealth seriously and that, henceforward, it intended to be a silent partner and observer in any attempted exploitation of precious metals.[52]

Another set of instructions to Nicolls and the commissioners, about which Boyle apparently did not know, made the crown's real intentions with regard to the New England plantations abundantly clear. In the "Private Instruc-

---

51. Robert Boyle to John Winthrop, Jr., Apr. 12, 1664, MHS, *Proceedings*, 2d Ser., I, 376–377.
52. *DCHNY*, III, 55–56.

tions to Col. R. Nicolls, etc." the king's guarantee of charter protections was revealed to be what many New Englanders assumed it was: limited. What Charles wanted and what the commissioners were to obtain, or at least set the stage for, was control of the colonial executive and military. The actual mission Nicolls was to accomplish in New England was twofold: first, "to inform yourselves and us of the true and whole state of those severall Colonies," and, second, "by insinuateing yourselves by all kind and dextrous carriage into the good opinion of the principall persons there," to get the New England colonies to voluntarily yield up their charters for revision. The king intended that the New Englanders, brought of their own will to "an entyre submission and obedience to us and our Government," should yield him the right to appoint both their governors and the commanders of the colonial militias. Initially, Charles would be willing to select a governor from a list of three names presented by the colonial assemblies. For the present, despite the crown's large and growing financial needs, Nicolls was not to concern himself with revenues. His military mission was to take New Netherlands away from the Dutch, and his diplomatic mission to charm New England into subordinating itself to empire. The king was prepared to be patient; he might have recognized that achieving voluntary submission from New England immediately was naive. He had set a trajectory, however, that crown policy would follow for much of the next quarter century. It is in terms of this policy that Winthrop's complex relationship with both Nicolls and the Royal Society is best understood.[53]

53. "Private Instructions to Coll. R. Nicolls, etc.," Apr. 23, 1664, ibid., 57–61.
Charles II's handling of empire and colonial policy has been subject to many intrepretations. Charles M. Andrews's view of post-Restoration imperial policy was that there was no policy. Mercantilism shaped relations between colony and metropolis until a true imperial stance emerged before the American Revolution. Stephen Saunders Webb argued that imperial control was strong from the beginning of settlement, implemented and enforced through garrison government. J. M. Sosin thought Charles's imperial policy amateur, inept, and inconsistent. More recently, Brendan McConville has perceptively observed that the post-Restoration Stuart monarchs pursued a disjointed but increasingly threatening policy of instituting imperial controls that culminated in the imposition of the Dominion of New England. Andrews, *Colonial Period of American History*, IV, 126–127, 128; Stephen Saunders Webb, *The Governors-General: The English Army and the Definition of the Empire, 1569-1681* (Chapel Hill, N.C., 1979), and "Army and Empire: English Garrison Government in Britain and America, 1569–1763," *William and Mary Quarterly*, 3d Ser., XXXIV (1977), 1–31. For a challenge to the Webb thesis, see Richard R. Johnson, "The Imperial Webb: The Thesis of Garrison Government in Early America Reconsidered," *WMQ*, 3d Ser., XLIII (1986), 408–430; J. M. Sosin, *English America and the Restoration Monarchy of Charles II: Transatlantic Politics, Commerce, and Kinship* (Lincoln, Nebr., 1980), and *English America and the Revolution of 1688: Royal Administration and the Structure of Provincial Government* (Lincoln, Nebr., 1982); Brendan McConville, *The King's Three Faces: The Rise and Fall of Royal America, 1688-1776* (Chapel Hill, N.C., 2006). Carla Gardina Pestana's study of the divergence of interests between the metropolitan authorities and the English Atlantic

From the perspective of Winthrop and the colony of Connecticut, there was, as early as 1664, no question whether a new and threatening imperial policy was at work in New England. Moreover, it was a policy that had the potential to be backed by force at any time. The Navigation Acts, the creation of the Council for Foreign Plantations, the taking of New York from the Dutch, and the visitation of the royal commissioners in the company of a royal army all demonstrated a new intent by the crown to exercise its authority in colonial matters. To be sure, royal policy could seem capricious, but it could not be dismissed. The charter, granted the duke of York less than a year after Connecticut's charter passed the seals and annexing more than half of Connecticut into New York, might be considered a sign of ineptitude, a demonstration of the information vacuum at Whitehall, or a calculated implementation of imperial strategy. Whatever the cause, the newly negotiated unofficial boundary line between New York and Connecticut allowed Connecticut to retain its western lands only at the pleasure of the duke of York, and that left Connecticut truly vulnerable in its relations with the crown. Whether this state of affairs had been arrived at through caprice or calculation was immaterial; it still served imperial ends well. Nicolls's explanation to Whitehall for why he agreed to the provisional boundaries serves as another reminder of the ultimate goal of crown policy in New England. It would not be beneficial to annex Connecticut's former patent lands, Nicolls explained, unless all New England could be reduced to the duke's rule.[54]

Nicolls's and Winthrop's friendship should not blind one to the calculus of power affecting their interactions. Nicolls, as his private instructions commanded, sought to insinuate himself into Winthrop's good opinions. Winthrop sought through Nicolls to demonstrate his colony's loyalty to his majesty's interests while protecting it from royal imposition. Politics and policy permeated their personal encounters. Winthrop found that with Nicolls, as with the governors who would come after him, the imperial drive to bring Connecticut into alignment with crown policy was constant. Nicolls, like the Royal Society, applied firm but steady pressure on Winthrop to provide services of value to the empire. To the degree that he was able, Winthrop declined to be drawn into the greater web of imperial ambition yet continued to provide unceasing expressions of loyalty to the crown.

On the basis of Boyle's assurance that the the royal commissioners intended no interference with government or religion, Connecticut received the em-

---

colonies during the period prior to the Restoration provides interesting contextualization; see *The English Atlantic in an Age of Revolution, 1640–1661* (Cambridge, Mass., 2004).

54. Colonel Nicolls to the duke of York, n.d., *DCHNY*, III, 106.

bassy respectfully, even after it had rendered an unfavorable decision in the colony's continuing boundary dispute with Rhode Island. Before their arrival, Winthrop had, as Boyle recommended, written an effusive letter to Clarendon, assuring him that the commissioners would be most welcome. Upon the completion of their hearings in New London, Connecticut accepted four propositions: that all householders take an oath of allegiance to the king, that there be no religious qualification for the political franchise, that liberty of conscience in matters of religion be considered, and that any laws derogatory to the king (Connecticut got to decide what they might be) be repealed. Satisfied, the commissioners moved on, reporting favorably on Connecticut's positive response.[55]

Unlike Connecticut, Massachusetts elected to be unbending with the commissioners, rejecting their authority and refusing to meet with them. Massachusetts had become aware of Boyle's letter to Winthrop regarding the commissioners' intentions, and Governor John Endicott wrote Boyle urging him to intervene with the king on the colony's behalf to call the commissioners home. Boyle's response was curt, and it made clear that his assistance to Winthrop had not been based on an official relationship. "I did not imagine that what I occasionally writt to Mr. Winthrop, not so much as haveing been long Governor of so large a Colony, as upon the score of haveing been my particular acquaintence, should have been taken notice of by so considerable an assembly as yours." He rejected Massachusetts's "command" that he intercede for them: "I shall not successfully, and must confess that I canot very cheerfully, obey you."[56]

SUBSEQUENTLY, MASSACHUSETTS was charged by Commissioner Maverick with "standing out, or rather Rebellion" and was ordered to send representatives to England to answer for their actions. The benefits of Winthrop's Royal Society fellowship in providing access to patronage and the importance of his continuing to represent himself as a loyal rather than merely obedient subject could not have been clearer.[57]

---

55. John Winthrop, Jr., to Lord Clarendon, Winthrop Family Papers, Mss.5.51, Massachusetts Historical Society, Boston; Black, *Younger John Winthrop*, 284–285; MHS, *Collections*, 5th Ser., IX (1885), 72–73. To be sure, their visitation had helped produce a beneficial consequence for Connecticut. New Haven, facing the reality that it would be annexed into either Royalist New York or Puritan Connecticut, had chosen to become part of Connecticut, thus ending two years of contention.

56. Robert Boyle to John Endicott, Mar. 17, 1665, MHS, *Collections*, 5th Ser., I (1871), 400–403.

57. Samuel Maverick to John Winthrop, Jr., Aug. 9, 1666, ibid., 4th Ser., VII (Boston, 1865), 313–314.

Still, Connecticut's interests and those of the crown differed in important ways. There was much that Connecticut was not prepared to do for the king or his agents, but Winthrop would always find a suitable reason why Connecticut *could not,* rather than would not, comply with imperial requests. Most of the activities for which the royal governors sought colonial cooperation involved military matters. Connecticut did not want to be drawn into wars in which it had no immediate interest, yet, from the perspective of the royal governors, the crown's interests and Connecticut's interests were or should have been synonymous.

In the spring of 1665, concerned that the Dutch fleet of Admiral M. A. de Ruyter might attempt to recapture New York, Nicolls presented Winthrop with a formal motion that Connecticut and New York enter a mutual defense pact. Winthrop referred the proposal to the Connecticut assembly. The legislature, though "very zealous for his Majesty's interest," nevertheless "cold not see a clere way to act directly according to the proposall." The members had, Winthrop wrote Nicolls, "expressed a duty and willingness according to your capasitie to give assistance to his Majesties subjects in case of sudaine invasions about foreign enemies," but they feared sending troops to New York because their own plantations would be left vulnerable. Nicolls traveled to Hartford to insist that New York's safety and Connecticut's were inseparable. Winthrop ordered the Connecticut seacoast towns to stand in readiness for attack, but, after once more reciting his subjects' concern for his majesty's interest and their duty to provide aid, he again declined to join an alliance. This time he cited fears of local Indian assaults. Winthrop provided a good faith sign of his desire to support the crown by sending his son, Major Fitz-John Winthrop, to Albany to act as a liaison between the two colonies' military forces. But he did not send troops.[58]

A letter from the king in June 1666, announcing that England was at war with France and commanding Connecticut to supply troops for a march on Quebec, met with a similarly solicitous refusal. Winthrop told Nicolls that the danger from the Mohawks, who were at war with a local Connecticut tribe, made it impossible to comply with his request. He had, however, issued a special order putting both militia and cavalry units in a posture of defense. Though the route to Albany was difficult to travel, if occasion warranted, Connecticut would respond to the need. Winthrop had sent to Massachusetts to seek additional aid, and he informed Nicolls that he hoped all the colonies would be ready to attend their duty. In fact, the consultation with the colonies

---

58. John Winthrop, Jr., to Richard Nicolls, June 28, 1664, ibid., 5th Ser., VIII (Boston, 1882), 96–97.

was a delaying tactic. Three months later, the colonies unanimously agreed that, given the need for good ships to transport troops and the lateness of the year, "at present there could be nothing done by these colonies in reducing those places at or about Canada." Connecticut and Massachusetts did, however, send a scouting party out to find the French army, and Winthrop had personally interceded with some of the chief sachems of the Mohawks, he noted, persuading them not to ally with the French.[59]

Winthrop's resistance to imperial demands in opposition to his colony's parochial interests displayed a common pattern. Winthrop always expressed a great desire to comply with imperial commands, coupled them to a lengthy recital of the reasons compliance was impossible, and then offered some nominal response meant to demonstrate good faith without actual compliance. Winthrop's refusal to put troops into the field in winter to march against Canada or to send Connecticut forces to the aid of the English in the Caribbean was handled in the same manner. Nicolls found Connecticut's responses frustrating—but did not think them subversive. He wrote Henry Bennet, earl of Arlington, that Connecticut had been "so much tryall," but he was much more critical of Massachusetts, which was "too proud to be dealt with." Winthrop was adept at presenting refusal in a way that almost seemed like compliance.[60]

Winthrop and Nicolls remained on positive terms. Their shared interests in natural philosophy, alchemy, and minerals helped them gloss over political differences. When Nicolls returned to England in 1667, Winthrop continued to correspond with him and sought his aid with the king in resolving Connecticut's continuing boundary dispute with Rhode Island. Colonel Francis Lovelace, Nicolls's replacement, received similar loyal noncooperation from the Connecticut governor. Winthrop once left Lovelace waiting for five days at Milford, after the colonel arrived there with an official party and sent an express messenger urging Winthrop to come discuss various "affairs." Winthrop was very apologetic, pleaded inclement weather, the sickness of a friend, and miscommunications. But he intentionally avoided the meeting and later com-

---

59. John Winthrop, Jr., to Richard Nicolls, July 15, 1666, ibid., 99–101, John Winthrop, Jr., to Lord Arlington, Oct. 25, 1666, 101–103; Ian K. Steele, *Warpaths: Invasions of North America* (New York, 1994), 120–122. Fears of Indian wars were in fact warranted. The English conquest of the Dutch had raised concern among the Five Nations of the Iroquois, the Dutch traders' traditional allies. As the English had been allies of the Susquehannocks and the northern Algonkians, the Five Nations' traditional enemies, the Iroquois had entered into an alliance with the French. Fears among the English of a combination of all Indians against them received added impetus, when a declaration of war against France arrived in February 1666.

60. John Winthrop, Jr., to Lord Arlington, May 7, 1667, MHS, *Collections*, 5th Ser., VIII, 117–119; Colonel Nicolls to Lord Arlington, Apr. 9, 1666, *DCHNY*, III, 114–115, Dec. 12, 1667, 167.

mended Alexander Bryan, the leading Milford citizen, who had been pressed into service as the unofficial host to the group, for refusing to commit to one of the motions they presented. Winthrop did help Lovelace establish a regular monthly intercolonial post, so that "great conveniences of public importance" could be quickly communicated to all colonies, but in most instances Lovelace's efforts to entice Connecticut into a closer connection with the empire met with the usual pattern of positive noncompliance.[61]

WINTHROP'S APPROACH to dealing with the royal governors is interesting because it is similar to the way he dealt with some requests for information from the Royal Society. Winthrop's correspondence with the society has often been considered as an activity separate from his political life. Natural philosophy was (medical alchemy excluded) a diversion and relief for Winthrop from the pressures of colonial leadership, and membership in the society an honor. To be sure, the tone of Winthrop's correspondence changed upon his return from England in 1663. The once-ebullient promoter of New England's limitless potential became a cautious skeptic, doubtful about New England's present potential, and he often seemed unable or unwilling to comply with the society's requests for information. Arguably, the aging Winthrop was growing to feel a wilderness isolation and discouragement with New England's prospects. But such a reading fails to appreciate Winthrop's understanding of the context in which the generation and dissemination of information at the Royal Society took place.[62]

Winthrop's communications with the society were always in part political. The changed attitude in his correspondence regarding New England's development potential and the distinctions he made about the type of information he was willing to transmit to the society reflect both a personal and political agenda. Winthrop wanted to maintain his good standing as a fellow, both for personal and professional reasons. He needed the patronage and communication links with Whitehall that the society provided. Moreover, he fully supported the society's goal of pansophic reformation. He liked many of the members, with whom he shared natural philosophical interests and goals. But in light of the political realities in Connecticut, Winthrop was anxious to not provide information that would further fan or facilitate the king's or duke of York's interest in Connecticut's annexation into New York or colonial

---

61. Black, *Younger John Winthrop*, 300–302, 320–323; Francis Lovelace to John Winthrop, Jr., Dec. 27, 1672, MHS, *Collections*, 5th Ser., IX, 83–85, Jan. 22, 1673, 85–86.

62. John Canup, *Out of the Wilderness: The Emergence of an American Identity in Colonial New England* (Middletown, Conn., 1990), 208–210; Black, *Younger John Winthrop*, 311.

unification. Winthrop understood that the information he provided to the group at Gresham College would be circulated at Whitehall. Faced with royal commissioners, royal governors, and an uncertain charter, Winthrop avoided subjects that, despite their scientific value, had potential to undermine Connecticut's colonial autonomy. To the extent he was able, while appearing to meet his obligations as a fellow, Winthrop provided limited information and a negative interpretation of New England's mineral potential, and he simply refused to produce a natural history of New England. Widespread belief in New England's mineral wealth could furnish a strong incentive for annexation of Winthrop's colony. An up-to-date and detailed natural history would provide Whitehall a range of intelligence about New England that could prove detrimental to colonial autonomy. Winthrop continued to contribute to the goals of the society. He communicated freely on a wide range of subjects: astronomy, technology, husbandry, wonders, and medicine, but he was reluctant to provide any intelligence that could be used to advance an imperial policy directed against the New England colonies.

THE MOST BENEFICIAL product Winthrop could have produced for the Royal Society, and also indirectly for Whitehall, would have been a natural history of New England. Winthrop was the ideal person to undertake such a project. He had been positioned at the center of authority almost since the founding of Massachusetts, and he had participated in most of the important events that had taken place in New England. He had been a leader in two colonial governments and had founded three towns. He knew the country, and, because of his alchemical knowledge, he was in a better position to assess its resources than most other New Englanders. As a physician he had studied the botany of the region with a knowledgeable eye, and he had come to know the people and their diseases. As a projector and one of New England's most successful entrepreneurs, he had identified and implemented important approaches to exploiting the region's economic mineral potential. No one understood New England more thoroughly than Winthrop did in 1664, and the Royal Society hoped to capture and exploit that knowledge. Oldenburg's initial letter to Winthrop after he left England had put him on notice that the society would expect a natural history from him in time. His second letter, a year later, suggested that that time had come. Oldenburg connected the work he wanted Winthrop to compose to the "History of New England" that had been written by the senior Winthrop. Though their contents would by definition be different, Oldenburg implied that writing the natural history was Winthrop's filiopietistic duty, an intergenerational command performance. "Remember I entreat you, the History of New England, begun by your worthy Father, and continued by yourself," Oldenburg wrote. Boyle and Moray also urged Winthrop to work on a natural history.[63]

Without addressing the question of producing a natural history directly, Winthrop focused on other ways to provide the society with information of interest and value. He was more than willing to report on almost any subject whose content was, from the standpoint of New England, politically and economically neutral. Astronomy was particularly useful in this regard and an area in which Winthrop believed he had made an important discovery. He reported to Robert Moray that, while observing Jupiter in August 1664, he believed he had discovered a fifth moon circling that planet. He was cautious about the claim, and he hoped Moray would request other astronomers to attempt to verify his findings. Winthrop believed a larger telescope than he then possessed would be useful in making such a determination, and he subsequently attempted to build an eight-to-ten-foot telescope in Connecticut. He forwarded Boyle one of the "little tracts" about a comet, one of the brightest of the seventeenth century, that he had observed in November 1664. Another comet in 1678 prompted reports to Nicolls and Theodore Haak, and yet a third comet, which Winthrop himself had not seen but that his son Wait had reported to him, was the subject of a report to Oldenburg in 1672.[64]

Winthrop's medical interests also figured in his correspondence with the society. In response to an inquiry from Oldenburg, he gave a mostly negative opinion about the value of mercury-based medicines, although he did make use of purgative calomel. He sent roots of New England plants to be tried by medical acquaintances in England and wrote Doctors Jonathan Goddard, Christopher Merret, Benjamin Worsley, Johann Kuffler, and Daniel Whistler about a New England plant, probably nightshade, that resembled the purga-

---

63. Henry Oldenburg to John Winthrop, Jr., Mar. 26, 1664, in Hall and Hall, eds., *The Correspondence of Henry Oldenburg*, II, 149–150; Robert Moray to John Winthrop, Jr., Dec. 19, 1665, MHS, *Proceedings*, 1878, 224–225.

64. John Winthrop, Jr., to Robert Moray, Jan. 27, 1665, MHS, *Proceedings*, 1878, 220–222. Winthrop's small telescope, we now know, was actually incapable of making Jupiter's fifth moon visible. John W. Streeter, "John Winthrop, Junior, and the Fifth Satellite of Jupiter," *Isis*, XXXIX (1948), 159–163; Ronald Sterne Wilkinson, "John Winthrop, Jr., and America's First Telescopes," *New England Quarterly*, XXXV (1962), 520–523; John Winthrop, Jr., to Theodore Haak, 1668, Winthrop Family Papers, Mss.5.172; Wilkinson, "John Winthrop, Jr. and the Origins of American Chemistry," 337. Henry Oldenburg invited Winthrop to participate in an international effort to observe Mercury's transit across the sun on October 25, 1664, but the request failed to arrive in time: Henry Oldenburg to John Winthrop, Jr., Mar. 26, 1664, in Hall and Hall, eds., *The Correspondence of Henry Oldenburg*, II, 149–151; Black, *Younger John Winthrop*, 308–311; John Winthrop, Jr., to Robert Boyle, Oct. 29, 1666, MHS, *Proceedings*, 1878, 229; John Winthrop, Jr., to Henry Oldenburg, Sept. 25, 1672, Letter Book W3, Royal Society Library, cited in Black, *Younger John Winthrop*, 422.

tive jalap except that it also induced vomiting. Winthrop asked Boyle for information, provided it was not secret information, regarding a herbal remedy about which Boyle had written that was said to cure scrofula, a tuberculosis-related skin disease known as the King's Evil.[65]

Winthrop sent the society his thoughts on a way for vessels to determine longitude while at sea and reports on trials he had conducted on making potash, charcoal, and salt. He sent Robert Boyle supplements to the reports he had provided in London about making tar and about Indian corn. Robert Moray received information on the tides at the Bay of Fundy, along with a description of roiling tides of Hell Gate, the Dutch passage from Manhattan into the Long Island Sound. All but one of these reports was lost at sea, as was secret alchemical information that Winthrop had sent to the society. Winthrop was concerned the alchemical secrets might fall into unworthy hands and was relieved to get confirmation that they had not survived. "It was good satisfaction to my mind," Winthrop wrote Oldenburg, "to be certaine that the sea had those papers . . . I have thought it more expedient to reserve the mention of some hopefull considerations to an other tyme."[66]

The information Winthrop provided was always welcome, but the greater desire of and most insistent demand from the society was for the natural history. Time after time, Winthrop was implored to compose the natural history, and each time he either ignored the request or declined. "I cannot yet desist from recommending to you the Composure of a good History of New England," wrote Oldenburg, "from the beginning of the English arrival there, to this very time, containing the Geography, Natural Productions, and Civill Administration thereof, together wth the Notable progresse of that Plantation, and the remarkable occurences in thesame. An undertaking worthy of Mr Winthrop, and a member of the Royal Society!"[67]

To be sure, Winthrop was not alone in being asked to create natural histories for the society; it was part of a comprehensive program intended to encompass England, Ireland, and Scotland and to spread throughout the world. The natural history movement was completely in keeping with the society's pansophic reform agenda and resonated powerfully with the Baconian and

---

65. John Winthrop, Jr., to Henry Oldenburg, July 25, 1668, MHS, *Collections*, 5th Ser., VIII, 121–125, John Winthrop, Jr., to Henry Oldenburg, Nov. 12, 1668, 129–140, John Winthrop, Jr., to Robert Boyle, Oct. 29, 1666, 104; Wilkinson, "John Winthrop, Jr. and the Origins of American Chemistry," 336.

66. John Winthrop, Jr., to Henry Oldenburg, Nov. 12, 1668, in Hall and Hall, eds., *The Correspondence of Henry Oldenburg*, V, 152.

67. Henry Oldenburg to John Winthrop, Jr., Apr. 11, 1671, ibid., VII, 569.

Comenian reform agendas. In the case of New England, however, it could also be directly and immediately tied to the drive for imperial control and consolidation. Winthrop had ample reason to decline participation.[68]

A letter to Winthrop, in which Oldenburg again entreated Winthrop "not to forget the Annals of New Engl[and]," made the connection between the natural history project and colonial unification explicit. "How happy would it be," he wrote, "if there were an Union of all our English Colonies for free communications with mutuall assistances." Oldenburg estimated that, if Bermuda and the other English-inhabited islands were included in the union, their combined population would number more than a million persons. He envisioned "a numerous people there united, born and bred to agree with the Air and Soyle, and too strong to be supplanted by their Enemyes."[69]

Oldenburg applied current scientific theory to elaborate further on the benefits to be derived from uniting the colonies. And he advanced a mathematical formula through which the Atlantic colonies could unite rapidly.

> And if the English made it their busines to chuse their habitations about the heads of their rivers from New Engl. to Virginia, 'tis affirm'd, they would have a wholesomer Air, safer habitations, and the line of communication much shorter: I think, Sir, you told me once, that they had a foot passage from New England to Virginia through Maryland, and that it was not above 100 miles by that way. How easy then would it be to send forth 10 companies of Planters in 2. or 3. years from Virginia, Maryland, and N. England, to secure and maintain commerce through that passage, as at every 10 miles distance: And in 2. or 3. years more they may setle more plantations as at 5. miles distance."[70]

Oldenburg implied that rapid unification was not only desirable; it was the only way to avoid massive creolean degeneracy, the inversion of the widely accepted principle that civilizing the landscape civilizes the climate. Unimproved natural environments, it was believed, could strip away human civility, drawing people back into a savage state of nature. Oldenburg had heard this

---

68. Henry Oldenburg to Robert Boyle, Jan. 27, 1666, ibid., III, 32. On Oldenburg's special interest in compiling a universal natural history, see Marie Boas Hall, *Henry Oldenburg: Shaping the Royal Society* (Oxford, 2002), 150, 132–136; Susan Scott Parrish, *American Curiosity: Cultures of Natural History in the Colonial British Atlantic World* (Chapel Hill, N.C., 2006), 108–109.

69. Henry Oldenburg to John Winthrop, Jr., ca. Mar. 1, 1669, ibid., V, 425.

70. Ibid.; Karen Ordahl Kupperman, "Climate and Mastery of the Wilderness in Seventeenth-Century New England," in David D. Hall and David Grayson Allen, eds., *Seventeenth-Century New England* (Charlottesville, Va., 1984), 3–38.

was already happening on a wide scale in America. "But I am told, to my grief, that for want of due care of them (wch would fixe and setle them in convenient habitations) vast numbers of the English are become as wild as the Savages, and that they destroy all accomodations wherever they come, and so remove from place to place as disorderly as the wild Tartars."[71]

Colonial unification could reverse this degeneration and improve every aspect of life in the colonies. John Beale, who had first advanced the unification program to Oldenburg, understood colonies were divisive and might oppose unification. He urged it, however, as a national priority. "The English of these severall Colonyes are more apt to destroy each other, than to have thoughts of Uniteing: And the Governers, and Owners, or pretenders to Patternts, are not much otherwise minded. But this is the English Interest, and deserves to be carried by the overswaying power of England." Beale urged the appointment of a viceroy, or lieutenant, to oversee the entire continent, "for their Common safety and General Good."[72]

The information Oldenburg, Boyle, and Moray asked Winthrop to provide in the natural history was the intelligence needed to support such a colonial unification scheme. Oldenburg asked Winthrop to obtain "all the Mapps you can get, of New-England, New-Netherland, Mary-Land, and wherever the English are planted." He asked Winthrop to incorporate them into his "Annals of New England" and to add to them accounts of the neighboring plantations. Oldenburg believed such a volume "could not faile of a good Mart and Sale, as well amongst us as in all the colonies."[73]

AS HE HAD DONE with the royal governors, Winthrop continually expressed willingness to comply with the society's desire for information about New England but never actually provided useful intelligence. Beginning in 1669, the society began to receive carefully packed boxes from Winthrop containing a wide-ranging selection of New England curiosities, intended for display at society meetings and possible inclusion in the society's collections. Among the items included were stuffed rattlesnakes, native wampum, a flying fish, a starfish, twelve-row Indian corn, miniature oaks, barnacles, a piece of limestone, sand, tree bark from Nova Scotia, a deformed deer head, winter and summer wheat, native baskets and bowls, horseshoe crab, a hummingbird

---

71. On the pervasive seventeenth- and eighteenth-century angst over the possibility of such creolean degeneracy, see Parrish, *American Curiosity*, 90–105; Henry Oldenburg to John Winthrop, Jr., ca. Mar. 1, 1669, in Hall and Hall, eds., *The Correspondence of Henry Oldenburg*, V, 425.

72. John Beale to Henry Oldenburg, Aug. 29, 1668, in Hall and Hall, eds., *The Correspondence of Henry Oldenburg*, V, 28–31.

73. Henry Oldenburg to John Winthrop, Jr., Mar. 1, 1669, ibid., 422–426.

with nest and eggs, a fly with feathers, artifacts from a hill that "jumped into a river," snakeweed (a sure cure for snakebite), butternuts, American hazelnuts, a native thunderstone, a bow, arrows with a variety of points, a dogskin quiver. Along with the boxes, Winthrop sent long letters describing and contextualizing the items. They were, as Winthrop described them, either novelties (amusements) or curiosities (rare and interesting); none of them had significant value or potential value, and none of them provided information of use to Whitehall. They allowed, however, the members of the society to become witnesses to the uniqueness of New England, and they were very well received.[74]

Winthrop's shift of focus was both strategically and intellectually astute. Early modern natural philosophers and scientific institutions such as the Royal Society were possessed by an insatiable curiosity about natural "curiosities," especially those from exotic and distant locations. A newly awakened interest in discovery and description of the natural world, impelled by the early modern voyages of exploration, expanding trade, and an exploding consumer revolution, had produced an epistemological revolution. The shock of discovering so much that was new and exotic in the New World simultaneously had prompted a reexamination of the Old, with the result that early modern European men and women observed the world they could see through a new analytical lens of "objectivity" while simultaneously developing unquenchable curiosity about those parts of the world they could not see. A desire for owning, understanding, comparing, and defining nature in all its variety placed an intellectual emphasis on the accurate measurement and description of all objects and put a premium on the information that could be provided by men like Winthrop, who had, through their access to exotic goods, the ability to satisfy this burgeoning and insatiable curiosity about the nature of places distant and unknown.[75]

74. John Winthrop, Jr., to Henry Oldenburg, Oct. 4, 1669, ibid., VI, 253–256, Aug. 26, 1670, VII, 142–145, Oct. 11, 1670, VII, 201–203, Oct. 26, 1670, VII, 221–224, Wait Winthrop to Henry Oldenburg, Oct. 17, 1671, VIII, 305–306; Henry Oldenburg to John Winthrop, Jr., Mar. 26, 1670, MHS, *Proceedings*, 1878, 244–245. Matthew Underwood sees the Winthrop curiosities as advertisements by Winthrop of the great economic potential of New England. Given Winthrop's reluctance to furnish information about the economic potential of the projects that really interested the society's members and the limited commercial value of most of the curiosities presented, I am skeptical of this interpretation. Matthew Underwood, "Unpacking Winthrop's Boxes," *Common-Place*, VII, no. 4 (July 2007), www.common-place.org (last accessed May 28, 2008); Steven Shapin and Simon Schaffer, *Leviathan and the Air-Pump: Hobbes, Boyle, and the Experimental Life* . . . (Princeton, N.J., 1985), 55–60.

75. Harold J. Cook, *Matters of Exchange: Commerce, Medicine, and Science in the Dutch Golden Age* (New Haven, Conn., 2007); Cooper, *Inventing the Indigenous*; Ken Arnold, *Cabinets for the Curious: Looking Back at Early English Museums* (Burlington, Vt., 2006); Parrish, *American Curiosity*, 103–

Oldenburg wrote on two occasions to tell Winthrop that not only the society but also the king himself had been taken with his submissions. "The King saw them with no common satisfaction, expressing his desire in particular to have that Stellar fish engraven and printed." An engraving of the fish appeared in the society's *Philosophical Transactions*, as did an excerpt from one of Winthrop's descriptive letters. The king was, according to Oldenburg, equally taken with the hummingbird's nest. Winthrop sent Moray fibers from the silkgrass plant for a pillow to be given to the king.[76]

The curiosities increased Winthrop's standing among the members of the society and helped relieve him of the pressure to produce a natural history. In 1670, rather than ignoring Oldenburg's request, he formally rejected it, citing the recency with which the country had been settled. "I thinke it may be too soone to undertake that worke, there having beene but little tyme of experience since our beginnings heere and the remote Inland partes little discovered." He added that such a history might, however, be possible, with a little more time. The next year Winthrop declined again, citing the same grounds.[77]

Finally, in 1672, Oldenburg put the full weight of the society's patronage network behind a final request that Winthrop compose the natural history of New England. "Your noble friends here, My Lord Brereton, Mr. Boyle, Sir Robert Moray etc. . . . continue wth me their earnest request, that you would not delay to put into writing what you know of the constitution and productions etc. of New England. Though it cannot be perfect, yet it will be very welcome, as much as can be said of it by you. What remains, and what shall be discover'd hereafter, will be the work of those, that shall survive us." Oldenburg asked Winthrop to pardon the importunity of someone who was "somewhat impatient of all delays in matters of present utility."[78]

---

135. Parrish details how the Royal Society's communication with its American correspondents throughout the seventeenth and eighteenth centuries came to be shaped largely by a desire for information about American curiosities and how the Americans came to both take pride in and recoil against this emphasis on their status as distant observers.

76. Henry Oldenburg to John Winthrop, Jr., Mar. 26, 1670, MHS, *Proceedings*, 1878, 244-245; "An Extract of a Letter, Written by John Winthrop, Esq; Governour . . . ," Royal Society, *Philosophical Transactions*, V (Mar. 25, 1670), 1151-1153, "A Further Accompt of the Stellar Fish . . . , VI (Jan. 1, 1671), 2221-2224; Henry Oldenburg to John Winthrop, Jr., Apr. 11, 1671, in Hall and Hall. eds., *The Correspondence of Henry Oldenburg*, VII, 568. On the cachet that came from royal attention to a correspondent's curiosities in the eighteenth century, see Parrish, *American Curiosity*, 123-124; John Winthrop, Jr., to Henry Oldenburg, September 1671, in Hall and Hall, eds., *The Correspondence of Henry Oldenburg*, VIII, 265-267.

77. John Winthrop, Jr., to Henry Oldenburg, Aug. 26, 1670, in Hall and Hall, eds., *The Correspondence of Henry Oldenburg*, VII, 143, September 1671, VIII, 267.

78. Henry Oldenburg to John Winthrop, Jr., Mar. 18, 1672, ibid., VIII, 595.

It was the "present utility" of the natural history that had always kept Winthrop away from the project. And he would not change now.

THE OTHER AREA of interest on which the members of the Royal Society focused their attention was mining and minerals. The demand for Winthrop to provide information regarding New England's mineral potential was almost as great as the pressure to produce a natural history. Here, too, Winthrop elected to work against an imperial program. While he was in England, at the height of his charter success, Winthrop had made much of New England's mineral prospects. He had returned home with specific instructions from the society to report on his colony's "subteranneous" productions: "You know what they look for from you," Oldenburg reminded him in his first letter to New England, "to the end that you may receave fuller instruction and ampler commission upon discoveries made knowne to them." Winthrop had responded to Oldenburg's request quickly, sending accounts to both him and William Brereton of a piece of marcasite he had obtained from an Indian, which differed from ordinary marcasite in being whiter and clearer. The native had also told Winthrop of a potential copper mine, about which Winthrop hoped to provide more information in a future report.[79]

News of the duke of York's patent and the royal commissioners arrived the following spring, however, and with it Winthrop's concern about the consequences of building a strong economic case for New England's mineral wealth intensified. To divert attention from New England, Winthrop shifted his focus to the mineral discoveries in the newly acquired Dutch territories. Writing to Robert Moray, Winthrop described the same piece of marcasite about which he had written to Oldenburg the preceding November and said that the Indian wars had prevented him from traveling to investigate that mine site. The real opportunity, however, lay in the former Dutch territory. He had heard there would be peace among those Indians shortly. "Then there will be oportunity to search that part of the Country, which before the Dutch suffered not whilst the land was in their power. I should be glad," he added, "there could be found any minerall matters of reall worth."[80]

Generalized reports of mineral prospecting were not what members of the Royal Society hoped to receive from Winthrop, however. In a letter to Governor Nicolls, which Moray explicitly invited Nicolls to share with Winthrop,

---

79. Henry Oldenburg to John Winthrop, Jr., Aug. 5, 1663, MHS, *Proceedings*, 1878, 216–218. Marcasite is iron sulfide, a brassy yellow mineral with a greenish tint often confused with pyrite. Henry Oldenburg to Robert Boyle, Sep. 29, 1664, in Hall and Hall, eds., *The Correspondence of Henry Oldenburg*, II, 241.

80. John Winthrop, Jr., to Robert Moray, Sept. 20, 1664, MHS, *Proceedings*, 1878, 223–224.

Moray asked Nicolls to send "parcells of every kind of minerall you meet with, be it earth, clay, sand, stones, or what else soever, and in good quantity." Moray was eager for Nicolls to look for quicksilver and to assay some discoveries he had made of golden earth. Moray was himself experimenting with extracting silver from lead ore and was considering investing in silver mines in Cardiganshire.[81]

Letters Winthrop wrote to Moray about mineral matters were lost in transit during the second Anglo-Dutch War (1665–1667). Moray, unaware that Winthrop had written, prodded him for information. Adopting a tone that belied his position as a privy councillor and former society president, Moray pouted: "You do not acquaint none of your friends hereaway with any thing you have don, found out, or do designe nor give them any account of such matters as you are very well able to do. . . . If you would not be chid, you must be at som more trouble to correspond with friends here."[82]

Moray was a well-connected patron, and his criticism drew a lengthy explanation from Winthrop, whose earlier letters had been lost at sea. Winthrop, however, declined to repeat what he had written in those letters, except to say that he had been very diligent in his search for all sorts of metals, but constant Indian wars had made discovery impossible. The places that had the best hopes for silver and quicksilver, he repeated, were in New York. Colonel Nicolls had knowledge of a mine site "that possibly would have beene of better worth then hath beene before knowne in these pts of America," but Nicolls, too, would attest to the difficulties of making full discovery while the Indian wars continued. Winthrop further declined to provide details about New England's iron ore and ironworks. "I forbeare to mention any thing againe now about these matters, having written largely formerly, as also concerning lead and great probabilities of lead mines, and something about copper and some considerable expences bestowed rashly upon trialls of a stone that holdeth (as is supposed) some small quantity of that mettall . . . of all which I may hope againe to recollect my thoughts about those perticulars, of which I have formerly written."[83]

The pattern of withholding information is similar to the pattern Winthrop used in declining requests from the royal governors: expressing the desire to comply, explaining the inability to comply, citing the possibility of future compliance, and providing an alternative to show loyalty without compliance.

---

81. Robert Moray to Richard Nicolls, Dec. 19, 1655, ibid., 225–229.
82. Robert Moray to John Winthrop, Jr., Dec. 19, 1665, ibid., 224–225.
83. John Winthrop, Jr., to Robert Moray, Aug. 18, 1668 (Postscript Aug. 26), ibid., 232–234.

With Oldenburg, Winthrop became patently pessimistic. Winthrop, who had long been New England's greatest champion of mineral schemes, seemed to fall into despair.

> I wish I could tell you some certainty of any good mines in this North America: I have made, as carefull and diligent inquiry, as I could, and might have travailed further hopefully therein, had not the continued warres amongst the Indians wholy hitherto disapointed all such discoveries . . . There lieth this no small discouragement about inquiry after mines heere: we may suppose that if Rich they lie usually deepe in the bowells of the earth. And although some pregnant signes upon the superficies may give hopes, and probabilities, and possibly scattering pieces may be causally found of metalline substance, yet there may be great uncertainty to find a continuing veine, great summes may be expended, and yet misse thereof, as hath beene in England and other parts in knowne minerall grounds.

Mining success, Winthrop implied, was a matter of chance and the risk unacceptable to most planters.

> Some (I have heard) spend much and misse, others hitt upon a profitable discovery. It would not be likely to induce persons, especially our planters, to adventure much upon such probabilities, which they cannot looke into by their owne judgments. There have beene attempts, but profitt not presently appearing soone discouraged, and given over.

Successful mining operations in young colonies were simply not part of the providential plan, Winthrop suggested: God had reserved such success for future generations. Almost as an afterthought, he hoped that massive public investment "would not be all fruitlesse," but the overall tone of his correspondence made it unmistakably clear that intelligent investors would seek returns elsewhere.[84]

Winthrop's pessimism regarding New England's mineral potential is so strikingly uncharacteristic of his normally ebullient personality that almost every biographer has reflected on it. Most agree that it simply reflected Winthrop's growing cynicism about the difficulties of mineral projecting. Winthrop might indeed have been discouraged, but the emphatic nature of his skepticism, especially after decades of promotion, suggests additional intent. Dissuading interest in mineral development provided, in the environment of

---

84. John Winthrop, Jr., to Henry Oldenburg, Nov. 12, 1668, in Hall and Hall, eds., *The Correspondence of Henry Oldenburg*, V, 150–157.

the late 1660s and the 1670s, political advantage. I do not suggest that Winthrop *fabricated* pessimism just to dissuade interest in New England. Rather, I suggest that Winthrop gave free rein to skepticism, in the same way that he had formerly been New England's greatest mineral promoter and projector.[85]

Not even a letter from Henry Oldenburg, mentioning a proposal being advanced to save English timber stocks by transporting all iron production to New England, "where is both store of Iron and a superabundance of wood," drew a positive comment from Winthrop. The man who had gone to England to raise the funds for the first iron furnace in New England, who now saw an infant iron industry growing in several New England towns, was discouraging. He summarized for Oldenburg what he had already asserted: "There is yet no certainty, of what is underground, there are some appearances upon the surface in some places of lead and other mineralls, but there have yet none wrought but Ironston, which hath beene also mostly of that sort which is called the Bog mine, which is found only in low grounds not deepe nor of any great thicknesse."[86]

Moray seems not to have been persuaded by Winthrop's pessimism. He accepted that Winthrop might be delayed until the Indian wars were over, but then he wrote, "I do presume you will prosecute the designe you have to enquire after Mineralls." He also sought to acquire the mineral information that Winthrop had said was lost at sea. "It had been too much trouble to you to have repeated what was in your former letter, but it would doubtless have been not onely satisfactory to me, but usefull to know everything you wrote." Oldenburg supported Moray's request. He wrote Winthrop "to sollicite you, that you would not think it a trouble, to recover again such particulars, as you intimate in yr letters to be lost, wch were likewise intended for the R. Society; especially of Minerals."[87]

Winthrop did not provide the lost information, nor did he explain his reluctance to. To the persistent Moray, who wrote him describing a new process for turning cast iron into steel, Winthrop continued to downplay local opportunities, directing Moray's interest away from New England and toward the

---

85. Parrish has noted how the Royal Society's later American correspondents frequently downplayed the significance of their reports, possibly as a strategic device. Winthrop's despair over New England's mineral potential, however, runs counter to a long-standing pattern of optimism about and promotion of investment in New England mining ventures. *American Curiosity,* 116–118.

86. Henry Oldenburg to John Winthrop, Jr., 1669, MHS, *Proceedings,* 1878, 422–426; John Winthrop, Jr., to Henry Oldenburg, Aug. 26, 1670, in Hall and Hall, eds., *The Correspondence of Henry Oldenburg,* VII, 143.

87. Robert Moray to John Winthrop, Jr., July 17, 1669, MHS, *Proceedings,* 1878, 243; Henry Oldenburg to John Winthrop, Jr., Mar. 26, 1670, in Hall and Hall, eds., *The Correspondence of Henry Oldenburg,* VI, 594–596.

interior. Describing a number of failed attempts at mining in Massachusetts and near the seacoast, he wrote, "These parts affoard little of novelty worth your notice . . . but it seemes very probable there may be such mineralls in the inland countries."[88]

ON MAY 1, 1675, the Connecticut General Court and John Winthrop, Jr., read a letter from the newly appointed governor general of New York, Edmund Andros. It demanded that Connecticut immediately surrender "in his Royall Highness behalfe, that part of his Territoryes, as yet under your Jurisdiction." Connecticut was to hand over to Andros all its land west of the Connecticut River. The imperial imposition Winthrop had worked for years to prevent had been suddenly set in motion. An angry exchange of letters between Connecticut authorities and Andros was followed, on July 8, by the arrival of ships bearing royal soldiers at the mouth of the Connecticut River. That the confrontation came to a stalemate and Andros withdrew his troops was due largely to the demands of a larger confrontation, the outbreak of King Philip's War, whose imperatives were to occupy Winthrop for the remaining nine months of his life.[89]

Ever since his charter trip to London in 1661, Winthrop had understood the direction in which the British Empire was moving. Then he had used his alchemical reputation, membership in the Royal Society, and the access to court patronage those provided to manipulate the new monarch's goal of consolidating New England's colonies to Connecticut's advantage. The royal charter Winthrop secured in 1662, which annexed into Connecticut both New Haven Colony and half the territory of Rhode Island, can, in this sense, be seen as a faltering first step by Charles's government toward what would ultimately become the imperial Dominion of New England. That the Connecticut charter was challenged almost immediately by a conflicting patent given to the king's brother in 1664 is a sign of both the disorganized state of Charles II's emerging imperial administration and the crown's realization that Winthrop's colonial interests and the imperial agenda might be more at odds than had at first seemed.

Winthrop's response to the threat to Connecticut's newly gained autonomy was both subtle and effective. For more than a decade he used the patronage he

---

88. John Winthrop, Jr., to Henry Oldenburg, Aug. 26, 1670, in Hall and Hall, eds., *The Correspondence of Henry Oldenburg*, VII, 143; Robert Moray to John Winthrop, Jr., Apr. 8, 1671, MHS, *Proceedings*, 1878, 249; John Winthrop, Jr., to Robert Moray, 1671, MHS, *Collections*, 5th Ser., VIII, 140–142.

89. E. Andros to the Connecticut General Court, May 1, 1675, *DCHNY*, XIV, 689; Stephen Saunders Webb, *1676: The End of American Independence* (New York, 1984), 342–353.

received through the Royal Society membership to stay informed of imperial plans and assert his colony's unwavering loyalty to Whitehall while simultaneously discouraging interest in the region's economic potential and ignoring requests to provide the kind of intelligence that might facilitate royal control. Winthrop's communications with society members, when read against the political situation in which Winthrop found himself after 1664, underscore how the quest for scientific knowledge and the quest for political authority were closely interrelated in at least one part of the early British Empire.

Viewed through the longer lens of later United States history, it is ironic that Whitehall and the natural philosophers of the Royal Society were the first to believe that unification of the English American colonies was a desirable goal. It is equally ironic that colonial leaders like Winthrop, in defense of their autonomy, used science and diplomacy to keep America divided. A century later, American colonial leaders—some of whom shared the same combination of scientific interest and political sophistication—would create the United States of America to counteract a British government that sought to divide and conquer them.

Afterword

John Winthrop, Jr., died on April 5, 1676, in Boston, where, despite failing health, he had been helping frame the colonial response to the continuing devastation of King Philip's War. The preceding month had brought fierce Indian attacks to seven New England towns; colonial prospects for winning the war were in doubt. Nevertheless, the anxious and beleaguered colonists paused to ceremonially mourn the passing of a leader they had come to revere as much for knowledge and compassion as for service. "The Blaze of Towns was up like Torches light, To guide him to his Grave," eulogized the Boston poet Benjamin Tompson. Forty-four years previously, Winthrop had been welcomed to Boston with the sound of artillery company musket volleys as he came ashore from the English ship *Lyon*. On Monday, April 10, Winthrop was carried to the sound of drums, horns, and minute gun salutes from the same artillery company to the "old Burying place." There, he was laid to rest in the tomb of his father.[1]

In the wake of his passing, contemporaries who summed up Winthrop's contributions to New England consistently noted three things: the depth and usefulness of his alchemical knowledge; his commitment to tolerance, especially in religious matters; and his political acumen. Each of these traits was deeply intertwined with the alchemical culture that had been a central feature of his own life and that had, through Winthrop and his associates, put a lasting imprint on New England's emerging culture.[2]

---

1. In March Indians had attacked Groton, Longmeadow, Northampton, Marlboro, and Rehoboth, Massachusetts, as well as destroyed Simsbury, Connecticut, and Providence, Rhode Island. [Benjamin Tompson], *New-Englands Tears for Her Present Miseries: or, A Late and True Relation of the Calamities* . . . (London, 1676), 7; J. Hammond Trumbull, ed., *The Public Records of the Colony of Connecticut*, II (Hartford, Conn., 1852), 432–433, 452–453; Samuel A. Green, ed., *Diary by Increase Mather, March, 1675-December, 1676* (Cambridge, 1900), 27–28; Samuel Eliot Morison, *Builders of the Bay Colony* (Boston, 1930), 287; M. Halsey Thomas, ed., *The Diary of Samuel Sewall, 1674-1729* (New York, 1973), I, 15.

2. Benjamin Tompson, the Boston poet who eulogized Winthrop twice — in a poetic account of King Philip's War and an elegiac broadside — praised Winthrop as a charitable alchemical physician, an incomparable mineralogist, a leader of "Meekness and Justice" committed to "modest

As a Christian alchemist in Puritan New England, John Winthrop, Jr., was a stellar man of his times, though not a man for all times. He embodied the notion of a fully integrated life, one in which a scientific spirit, social spirit, medical spirit, political spirit, and religious belief were complementary and symbiotic. He also embodied a set of ideals and aspirations that resulted from the unique confluence of confessional conflict and scientific discovery particular to the early modern period. He and many of his contemporaries believed that religion and science were deeply intertwined, that through the successful pursuit of divinely inspired alchemical discovery the world could be improved and the ultimate goal of Christian endeavor—the return of Christ to rule the earth—accelerated. These ideals and aspirations, and the theories of scientific discovery they produced, did not survive the centuries following Winthrop's death any better than Puritanism. Important aspects of the alchemical culture Winthrop helped bring to New England, however, did. These became foundational threads in the complex tapestry of New England's culture, and, like Puritanism, their influence still resonates. Yankee convictions about the importance of knowledge and discovery as tools for social improvement, the idea that technology can solve most of society's problems, and the notion still prevalent in many circles that America has a duty to lead the world in scientific discovery echo in secular and nationalist form the more prayerful and ecumenical aspirations that ascended with the smoke from Winthrop's alchemical furnaces.

Despite the stresses of war that preoccupied his final days, Winthrop could, at the conclusion of his life, look back upon much with satisfaction. His medical service to his generation had made him a beloved and revered patriarch. Moreover, through his support and example, a younger generation of healers—men such as James Noyes, Gershom Bulkeley, Thomas Palmer, and his own son Wait Winthrop—were continuing to provide New Englanders greatly appreciated alchemical medical services. A plan had also been advanced to build an alchemical laboratory at Harvard College, and the number of people studying the spagyric arts in New England seems to have been in-

---

bounds, in Church and Commonwealth," and a true adept. Stephen Chester, of Wethersfield, Connecticut, who also penned a funeral broadside, echoed the same themes, praising Winthrop's "Judgement into Chimistry," his medical ability, his religious tolerance, and his political acumen. Benjamin Tompson, *A Funeral Tribute to the Honourablle Dust of the Most Charitable Christian, Unbiassed Politician, and Unimitable Pyrotechnist John Winthrop, Esq* . . . [Boston, 1676], broadside reprinted in Thomas Franklin Waters, *A Sketch of the Life of John Winthrop the Younger* . . . , Ipswich Historical Society, Publications, VII (Cambridge, Mass., 1899), 75–77; [Stephen Chester], *A Funeral Elegy upon the Death of That Excellent and Most Worthy Gentleman John Winthrop Esq.* . . . [Boston, 1676], broadside at Massachusetts Historical Society; Harold S. Jantz, "The First Century of New England Verse," American Antiquarian Society, *Proceedings*, LIII (1944), 282, 287–290, 403, 407, 480–485.

A FUNERAL TRIBUTE

To the Honourable Dust of that most Charitable Christian, Unbiassed Politician, And unimitable Pyrotechnist

John Winthrope esq:

A Member of the Royal Society, & Governour of Conecticut Colony in

NEW-ENGLAND.

Who expired in his Countreys Service, April. 6th. 1676.

ANother Black Parenthesis of woe
The *Printer* wills that all the World should know
Sage *Winthrop* prest with publick sorrow Dies
As the Sum total of our Miseries:
A Man of worth who well may ranked be
Not with the thirty but the peerless three
Of *Western Worthies*, Heir to all the Stock
Of praise his sire received from his Flock:
GREAT *WINTHROPS* Name shall never be forgotten
Till all *NEW-ENGLANDS* Race be dead and rotten;
That Common Stock of all his Countries weal
Whom Grave and Tomb-stone never can conceal.
Three Colonies his *PATIENTS* bleeding lie
Deserted by their great *PHYSICIANS* eye;
Whose common sluce is pozed for their tears,
And Gates fly open to a Sea of fears.
His Christian Modesty would never let
His Name be near unto his *SAVIOURS* set:
Yet Miracles set by, hee'd act his part
Better to LIFE then Doctors of his Art,
Projections various by fire he made
Where Nature had her common Treasure laid,
Some thought the tincture *Philosophick* lay
Hatcht by the Mineral Sun in *WINTHROPS* way;
And clear it shines to me he had a Stone
Grav'd with his Name which he could read alone.
To say how like a *SCEVOLA* in Court
Or ancient *CONSULS* Histories report
I here forbear, hoping some learned Tongue
Will quaintly write, and not his Honour wrong,
His common Acts with brightest lustre shone,
But in *Apollo's* Art he was alone.
Sometimes Earths veins creeping from endless holes
Would stop his plodding eyes: anon the Coals
Must search his Treasure, conversant in use
Not of the Mettals only but the juice.
Sometimes his wary steps, but wandring too
Would carry him the Christal Mountains to
Where Nature locks her Gems, each costly spark
Mocking the Stars, spher'd in their Cloisters dark,
Sometimes the Hough, anon the Gardners Spade
He deign'd to use, and tools of th' Chymick trade.
His fruit of Toyl Hermetically done
Stream to the poor as light doth from the Sun.

The lavish Garb of silks, Rich Plush and Rings
Physicians Livery, at his feet he flings.
One hand the Bellows hold, by t'other Coals
Disposes he to hatch the health of Souls;
Which Mysteries this *Chiron* was more wise
Then unto Ideots to Anatomize.
But in a second person hopes I have
His Art will live though he possess the Grave.
To treat the *MORALS* of this Healer *Luke*
Were to essay to write a *PENTATUKE*,
Since all the Law as to the *MORAL* part
Had its impression in his spotless heart:
The vertues shining brightest in his Crown
Were self depression, scorning all renown;
Meekness and Justice were together laid
When any Subject from good order straid.
Neither did ever Artificial fire
Boyle up the Choler of his temper higher
Then modest bounds. In Church and Common-wealth
Who was the Balsome of his Countries Health,
Europe sure knew his worth who fixt his Name
Among its glorious Stars of present fame. (there
Here Royal *CHARLES* leads up, stands *WINTHROPE*
Amongst the *Virtuosi* in the Rear:
But for his Art with hundreds of the rest
He might be be plac'd in Front and come a Breast.
What Soul in fouldings 'tother side the Scene
With Souls turn'd Angels guess we to have been
When first his Chariot wheels the threshold felt (dwelt?
Where *WINTHROPS, DUDLYS, COTTONS* Spirits
VVhat melting joys are there? Sorrows below,
Should adequately from *New-England* flow:
If Saints be intercessors, heres our hope
VVe need not be beholding to the Pope.
VVe have as good our selves, an honest Brother
Outvies their Saintship, there or any other.
Now *Helmonts* lines so learned and abstruse
Are laid aside and quite cast out of use:
And Authors which such vast expenses spent
Lye like his Corps; his Ear is only lent
To Heavenly Harmonies, all things his Eye
Views in the platforme whence all forms did fly;
His labours cease for ever, but the fruit
He reaps at Fountain head without dispute.

B. Thompson.

FIGURE 8. *A Funeral Tribute to the Honourable Dust of . . . John Winthrope.* Broadside by Benjamin Tompson. [Boston, 1676]. *Courtesy of the Massachusetts Historical Society*

creasing. Alchemy continued to remain a vital interest in New England well into the eighteenth century. In 1773, Judge Samuel Danforth of Massachusetts offered to show Benjamin Franklin a sample of the philosopher's stone, and during the same decade Yale president Ezra Stiles displayed a great interest in and knowledge of alchemical authors, although he claimed not to have personally conducted alchemical experiments.

The quest for mines and mineral wealth continued throughout New England and would be an engine of economic activity well into the eighteenth century. The fledgling ironworks industry was expanding, and by the end of the century thriving ironworks had been established in several New England towns. The black lead mine site at Tantiusque would continue to be a major focus of activity within the Winthrop family itself. Although Winthrop's second attempt to extract ore from the site in the 1650s had been unsuccessful, it is clear that belief in the mine's possibilities was transferred to succeeding generations. Winthrop's grandson and namesake, who also became a member of the Royal Society, promoted the mine site's potential in England in the early decades of the eighteenth century. Another eighteenth-century Connectican, the minister-physician Jared Eliot of Guilford and later Killingworth, would receive a medal from the Royal Society for his research and subsequent treatise on ironmaking.

Winthrop could also take pride in his role in bringing to a close the first chapter in witchcraft prosecution in New England. Thanks in no small part to his efforts, Connecticut would never again execute a witch and would not try another witchcraft case until 1692. The Katherine Harrison case had echoed throughout New England. Massachusetts did not execute a witch again until the case of Goody Glover in 1688, and then only reluctantly, after it had sought a medical determination of Glover's case from no fewer than six Boston physicians. Through Winthrop's and Gershom Bulkeley's actions in the Harrison case, we gain a new insight into the relationship between elite and popular magic in colonial New England. Where scholars once held that all magical practices were proscribed by Puritan leaders, we now see that no small number of them were students of natural magic and used their knowledge of the occult to scrutinize the accusations made against less knowledgeable people accused of diabolical magic practices. The cultural framework in which magic was practiced was much more complex and the boundaries far more permeable than a simple binary opposition will allow. Both elite and rank-and-file New Englanders understood magical practices to occur along a continuum that ranged from diabolical magic to divinely granted insights into occult operations within the natural world. Moreover, the community opinion regarding where any practitioner's magic was located along this continuum was dynamic, subject to continual assessment and review. A world of wonders is almost by definition a world of magical possibilities, and this study shows that certain types of elite magic practice were integral and fully accepted aspects of early New England's culture. The philosophical ideas underpinning Winthrop's occult philosophies continued to exert influence well beyond the age of witchcraft, becoming part of what Catherine Albanese has called "the magico-religious repertory that helped to shape American metaphysics," influencing the subsequent elaboration of Freemasonry, Mormonism, Universalism, and Transcendentalism.[3]

Connecticut's 1662 charter, which Winthrop had secured in no small part because of his alchemical reputation and the well-placed patronage it gave him access to, would survive a series of challenges, including the short-lived Dominion of New England, and continue to provide Connecticut a degree of political autonomy unique among the mainstream Puritan colonies. Connecticut's relative self-rule was so favorable to its interests that it was continually challenged in the eighteenth century by the royal administrations of both New York and Massachusetts. Connecticut's elected governors, taking a page from the Winthrop playbook, resisted these Royalists' insistence that Connecticut subordinate itself to crown authority by following the strategy of positive noncompliance Winthrop himself had used successfully in his dealings with New York governors Nicolls and Lovelace. Connecticut's charter political autonomy was the critical factor in Connecticut's early adoption of whig government, which in turn helped Connecticut be among the first colonies to mobilize for the patriot cause in the run up to the American Revolution. Significantly, when the Revolution came, Connecticut saw no need to create a new constitution. It simply changed the authorizing agent in the 1662 charter to the people of Connecticut and continued to rule under that document until 1818.

This study also underscores the problematic nature of the lingering conception that Puritanism and profit were opposed. The heart of the pansophic enterprise was belief in world improvement in preparation for the return of Christ. Those improvements, which depended on the successful implementation of a broad-ranging series of economic development schemes, were intended to produce profits that would in turn produce additional improvements as well as benefit those who had made the improvements possible.

---

3. Catherine L. Albanese, *A Republic of Mind and Spirit: A Cultural History of American Metaphysical Religion* (New Haven, Conn., 2007), 66–82, esp. 67.

Every major project Winthrop participated in, from the ironworks at Braintree to the saltworks in Barbados, was intended to generate a positive economic return to investors. This does not mean, however, that Winthrop was New England's first capitalist entrepreneur, unfettered by the restraints of a moral economy. Rather, profit was seen as the indispensable means to achieving godly ends, and personal gain was to be a by-product of that effort, not an end in itself. Puritan projectors placed great importance on the providential implications of their endeavors. God's guiding hand was seen as influencing the outcome of most ventures, and economic success, far from being antithetical to Puritan values, confirmed providential support, God's "blessing on the means." Had Winthrop's mine at Tantiusque actually been a rich silver mine, he and his associates would undoubtedly have read it as a providential sign from God that New England's people and the pansophic program there were, quite literally, richly blessed.

All Winthrop's projects were initiated with the understanding that they were being undertaken in the service of God and that, if they were in accord with God's present designs, they would prosper accordingly. Of course, in any endeavor in which the goals of personal profit and godly service are inseparable, the relative influence of self-interest to Christian service as a motivating force is ultimately impossible to determine. Like the alchemical quest for gold, seen as a striving for purity by some and a demonstration of unbridled greed by others, the perception of what motivated Winthrop's, or anyone else's, pansophic agenda depends largely on who's doing the looking.

To his contemporaries and those immediately after him, Winthrop appeared to be the alchemical model of Christian charity. When Cotton Mather, who was only twelve at Winthrop's passing, penned Winthrop's biography for his *Magnalia Christi Americana,* the central themes in Winthrop's life of alchemical service, religious tolerance, and political savvy remained as strong as they had been a quarter century before. Praising Winthrop, whom he dubbed Hermes Christianus, for his medical abilities, "Genius and Faculty for *Experimental Philosophy,*" and his political achievements in both Old and New England, Mather depicted Winthrop as characterizing all the qualities of the true adept.

> The Description which the most Sober and Solid Writers of the *Great Philosophick* Work do give of those Persons, who alone are qualified for the Smiles of Heaven upon their Enterprizes, would have exactly fitted him . . . And he had herewithal a certain *Extension of Soul,* which disposed him to a *Generous Behaviour* towards those, who by Learning, Breeding and Virtue, deserve Respects, though of a Perswasion and

Profession in Religion very different from *his own;* which was *that* of a Reformed *Protestant,* and a *New-English Puritan.*[4]

An alchemical adept, a practitioner of religious toleration, and a New English Puritan. This was a combination of traits highly commendable and completely compatible to Cotton Mather. To many of the historians of New England colonial history who followed him, however, these concepts became separated and distorted. Tolerance as a strain of Puritanism was all but forgotten, alchemy became a secretive and marginalized occult practice, and Puritanism itself was transformed into a stern and forbidding monolithic religion of command and control. Along the way, Winthrop, too, was transformed, into an indecisive and inferior but good-natured foil to his father, a son who could never quite decide what he wanted to be when he grew up. Today, thanks to newer generations of historians, New England Puritanism has been reconsidered. We now understand it as a multistranded and multitextured religious movement that included among its members many proponents of religious toleration. Early modern alchemy has also been reconceptualized as a serious intellectual pursuit and an important contributor to the development of empirical scientific practice. The goal of this study has been to recast the older view of John Winthrop, Jr., and the importance of his alchemical practice to New England culture by helping the reader better understand him and his work in the context of the world in which he lived and hoped.

---

4. Cotton Mather, *Magnalia Christi Americana: or, The Ecclesiastical History of New-England* . . . (London, 1702), book 2, 31–32.

Index

Adamic knowledge, 1, 20, 40, 178–179
Agricultural improvement, 66, 68, 89–91, 122, 140; and climate, 141
Agrippa, Heinrich Cornelius, 17–18, 150, 203, 216, 223
Alchemical networks, European, 3, 27, 42, 49, 65, 69–70, 91, 259
Alchemical networks, New England, 5, 143–146, 151, 200–209, 259, 263; and status of practitioners, 200–201; value of, to community, 201
Alchemical physicians in New England, 200–209; status of, 200; list of, 201
Alchemical texts: obscurity of, 25, 26
Alchemists: repression of, in New England, 5, 94, 122
Alcock, George, 206
Alcock, John, 145, 205–206
Alkahest, 2, 157, 179, 181, 196, 200–201, 243, 253
Allyn, John, 235
Allyn, Matthew, 214
Amsterdam, 4, 63, 66, 68
Andrews, William, 190
Andros, Edmund, 300
Antilia, 71–72
Antimonial medicines, 194–195
Antimony, 66, 174, 194, 269; dangers of, 194
Antinomianism, 46, 50–52, 124. *See also* Free grace controvery
Ashmole, Elias, 264, 270
Ashurst, Henry, 263
Astrology, 19, 38, 217, 259; and astral influences, 205; condemnation of judicial, 208, 217, 239, 242, 249–250; natural, 208, 217, 247; and witchcraft, 239, 246–247, 249–250
Astronomy, 290

Atherton, Humphrey, 144–146
Augur, Nicholas, 222
Aurum potabile, 68
Avery, Jonathan, 207
Avery, William, 206–207
Ayres, Goody, 230–232, 235

Bacon, Francis, 1, 2, 26, 270–271, 292; affinities of, with alchemists, 26; and universal knowledge, 28, 40; and magic, 216
Bailey, Nicholas, and wife, 225–226
Barnes, Mary, 234
Beale, John, 155–156, 271–273, 293
Bennet, Henry, earl of Arlington, 287
Berkeley, William, 144
Bernard, Richard, 233
Bishop, Mary, 219–221
Bismuth, 82, 148, 149; as *mater argenti,* 87
Blackleach, John, 241
Black lead mining, 72, 80–82, 86–89, 270
Blinman, Richard, 146, 191
Boate, Gerard and Arnold, 8, 271
Boswell, William, 70, 183
Boyle, Robert, 7, 91–92, 123, 203, 205, 206, 263–264, 270; as Winthrop's patron, 263, 265–267, 274, 281–282, 284–285, 290–291, 295
Bracy, Thomas, 239
Bradstreet, Anne, 206
Brereton, William, 261–262, 268, 274, 295
Brewster, Betty, 220, 221
Brewster, Jonathan, 139, 145, 146, 157–160, 200
Broughton Thomas, 151–152
Brouncker, William, 2d Viscount Brouncker, 7, 264
Bryan, Alexander, 288
Bulkeley, Gershom, 7, 145, 200, 204–205, 212,

243, 303, 305; opposition of, to witchcraft prosecution, 243–244
Bulkeley, John, 205
Burton, Robert, 172

Cabala, 18, 35, 38, 44, 203
Cassacinamon. *See* Robin Cassacinamon
Catholics, 92
Charivari, 240
Charles II, 229, 237, 254, 264, 295; and efforts to rationalize colonial administration, 254–255; imperial agenda of, 254–255, 262, 273, 275, 279–284, 286, 300; and lack of knowledge about colonies, 255, 272–273, 282
Charleton, Walter, 265
Charter of 1662 of Connecticut, 7, 229–230, 237, 253, 254, 260–261, 273; challenges to, 261, 276, 280–282, 284–285, 300, 306
Chauncey, Charles, 204
Chauncey, Elnathan, 204
Chauncey, Ichabod, 204
Chauncey, Isaac, 204
Chauncey, Israel, 204
Child, Ephraim, 198
Child, Robert, 5, 13, 52, 65, 77, 81, 84, 87–90, 121, 140, 143, 150–151, 203, 265; remonstrance of, 5, 93–94, 121–123, 138–139, 182, 277; branding of, as Jesuit, 123
Christian alchemy, 2, 11–13, 20, 44–45, 47, 61–62, 66, 157–159, 204, 256, 258, 302; and profit motive, 20–22, 26–27, 66, 68, 263, 306–307; motivation to practice, 157–159; and medicine, 161–162, 178, 183, 191, 199
Chrysopoeia. *See* Transmutation
Clark, Daniel, 190
Clarke, John, 261, 273, 277
Clarke, Thomas, 154
Coddington, William, 52
Codes, 83–85, 144, 150
Cole, Ann, 232, 234, 235
College of Light, 63, 64, 73, 83, 270; in New Haven, 154–156
Cologne, 66, 67
Colonial consolidation, 279, 288–289, 292–293, 300, 306

Colonial economic development, 45; and alchemy, 44, 53, 146, 152–154
Colonization, 3
Comenius, Jan, 3, 4, 49, 54–63, 70, 256; and presidency of Harvard, 71, 154, 204
Comets, 290
Connecticut: witch prosecution in, 7, 147–148, 211–213, 227; as focus of European attention, 43, 91–92; conflicts of, with Massachusetts, 92, 93, 96–98, 100–102, 108; and concern about alchemical plantation, 94, 115; alliance of, with Mohegans, 98–99, 115; and purchase of Warwick patent rights, 101, 254; and suspicion of Winthrop's arrival, 102, 115, 117–118; and extension of New London boundaries, 146; and Winthrop's mining proposal, 149; change in attitudes of, toward Winthrop, 213–214. *See also* Charter of 1662 of Connecticut; Winthrop, John, Jr.: and Connecticut
Corporation for the Propagation of the Gospel in New England, 254, 263, 265–267
Cotta, John, 174
Cotton, John, 164
Council for Foreign Plantations, 254, 263, 265, 269, 277
Countermagic, 227
Court of Assistants, Connecticut, 237, 240
Creolean degeneracy, 292–293
Croll, Osvald, 177, 201, 204
Cullick, John, 238
Culpeper, Cheney, 147
Curiosities, 293–295

Danforth, Samuel, 305
Davenport, Elizabeth, 198–199
Davenport, John, 72, 154, 156, 176, 191, 196, 198, 214, 219, 225, 271–272
Day, Stephen, 86
Dee, John, 3, 22, 29, 33–35, 45, 65, 67, 70, 218, 248, 249
Delaware, 278–279
Descartes, René, 62, 67
Dickenson, Edmund, 265
Digby, Sir Kenelm, 91–92, 196, 264, 270

Disease theory, 164–165, 177–178
Divination, 250
Doctresses, 187. *See also* Healers, female
Downing, Emmanuel, 77, 87, 140
Downing, Lucy, 54, 140–141
Doxey, Katherine, 113, 184–185
Drebbel, Cornelius, 32
Dury, John, 49, 59, 63, 72, 154, 256
Dyer, Mary, 171

Eaton, Theophilus, 214
Eliot, Jared, 305
Eliot, John, 142, 168
Endecott, John, 139, 285
Epidemics, 168–170
Everard, John, 46, 48–49

Familiars, 221, 223
Familism, 4, 46–48, 51, 124
Fenwick, George, 50, 101, 254; and sale of Warwick patent to Connecticut, 101
Ficino, Marsilio, 16–17
Fishers Island, 85, 90–91, 145, 151, 277; assault of, by Mohegans, 125; and design of learning, 151–152
Fludd, Robert, 3, 44–45, 54, 196, 247, 248, 249
Fones, Thomas, 30
Fowle, Thomas, 81
Foxon, 126
Frances, Joan, 239
Free grace controversy, 50–52, 99, 123; unresolved questions of, 5, 93. *See also* Antinomianism

Galen, 23, 161, 172, 176
Garlick, Elizabeth, 226–228
Garlick, Joshua, 227
Garrett, Margaret, 235
Glassmaking, 139
Glauber, Johann, 54, 67, 68, 79, 90, 143, 148, 150, 183, 204
Glorious Revolution, 279
Glover, Goody, 305
Goddard, Jonathan, 269, 290
Godman, Elizabeth, 219–226

Golius, Jacob, 32–33, 54
Goodwin, William, 184, 190
Goodyear, Mr. and Mrs., 220
Gorton, Samuel, 121, 123, 198
Graphite, 88
Graves, William, 238
Greensmith, Goody, 232–235
Gresham College, 261–262, 289
Greville, Robert, Lord Brooke, 50, 60, 63, 72
Griswold, Michael, 240

Hague, The, 70
Hale, Mary, 239
Hale, Samuel, 239
Hall, John, 259
Hamburg, 4, 63, 65, 82
Harriot, Thomas, 104
Harrison, Katherine: witchcraft trial of, 212, 238–251, 305
Hart, James, 174
Hartford witch-hunt, 7, 226–235
Hartlib, Samuel, 3, 4, 13, 59–60, 63–64, 70, 81, 121, 150, 154–155, 253, 256, 258, 259, 262, 267, 271; interest of, in America, 71–72; and criticism of New England Puritans, 73
Hartlib circle, 60, 66, 69–70, 154–155, 256, 258, 263, 271–272
Haynes, Edward, 148
Haynes, John, 214, 262; wife of, 113, 185, 186, 191; daughter of, 187
Haynes, Joseph, 234, 235
Healers, female, 186–187; and distribution of Winthrop's medicines, 198–199
Helia Arista, 150
Helmont, Jan Baptista van, 143, 203–206, 223, 243
Henshaw, Thomas, 265
Herbal remedies, 186, 196–197, 290
Herbert, George, 174
Hermes Christianus, 183–199, 209, 216, 307
Hermes Trismegistus, 16–17, 46
Hermeticism, 18, 19, 40, 54, 68, 216, 223, 264
Hippocrates, 161
Hoar, Leonard, 201, 203–204
Holyoke, Elizur, 190–191

Hooke, William, 219–220
Hooker, Samuel, 234
Hooker, Susannah, 186, 187
Hopkins, Edward, 50, 129, 156; wife of, 184
Hopkins, Matthew, 233
Howell, Betty, 226
Howes, Edward, 3, 30, 44–46, 48, 53, 83; advice of, on relations with Indians, 103–105, 142
Hutchinson, Anne, 50–52, 93, 134, 171
Hyde, Edward, earl of Clarendon, 261, 281, 285

Intercolonial competition, 5, 94–95
Intercultural alliances: importance of, 95, 105, 108, 125; dominant English view of, 136
Ipswich, Mass., 50, 53
Ironworks, New England, 65, 70, 76–78, 122, 141, 149–150, 282, 297, 299, 305; at New London, 86; and New Haven, 153–154

James I, 248
Jennings, Nicholas and Margaret, 230
Jesuits: alchemists as, 123
Johnson, Edward, 169
Johnson, Jacob, 239

Kelly, Elizabeth, 230–231
Kerckringe, Theodore, 194
Kercum, Mary, 239
King, Thomas, 87, 139
King Philip's War, 300, 302
Kuffler, Abraham and Johann, 32, 53, 66, 83, 84, 290

Labor problems, 141
Lake, Margaret, 113
Lamberton, Hannah, 220, 221
Land bank, 267–269
La Rochelle expedition, 32
Leader, Richard, 78, 139, 143, 152
Lead mines: importance of, 80, 148–150, 297; and silver, 80–81. *See also* Black lead mining
Lee, Samuel, 206–207
Leete, William, 198, 231
Leveritch, William, 155

Lilly, William, 239, 246–248
Liquor distilling, 140–141
Lodowick, Christian, 206
London, 64; as alchemical center, 24
Long Island, 139, 226–228, 280
Lovelace, Francis, 287–288, 306
Loyal noncooperation, 287, 289, 293, 297

Macaria (Plattes), 4, 64, 68
Macrocosm and microcosm, 19, 23, 61, 68, 216, 247
Magic, 150, 196, 201, 203–204, 208; and science, 34, 38, 112; natural and diabolical, 35, 203, 215, 217, 220–221, 246–250, 305; study of, by Puritan elite, 201, 203, 204, 208, 211, 215–217, 251, 305; and alchemy, 209; and continuum of acceptability, 211, 217–218, 221, 244; and status, 218, 249; authority over interpretation of, 221, 249; boundary between intellectual and folk, 239–240, 247–250
Manitou, 112
Maps, 274–276, 278, 293
Martin, Samuel, 239
Mason, John, 106, 129–131, 199
Massachusetts: as limiting Puritan improvement, 49, 73; religious intolerance in, 93–94, 122, 278, 285; alliance of, with Narragansetts, 97–98; and Pequot lands, 97–98, 100–101, 108; resistance of, to expansion, 121; concern of, about United Colonies, 121
Mathematics, 45, 54
Mather, Cotton, 71, 154, 160, 164–165, 175, 193, 207–209, 216, 307
Maverick, Samuel, 8, 276–277, 285; natural history by, 8, 277–279
Mayhew, Thomas, 141–142, 274
Medical charity, 191, 193, 198–199; as drain on resources, 191
Medical correspondence, 171, 184–187, 198–199; from women, 184, 187
Medical culture of England, 162, 166, 171, 173–175, 178, 180; and opposition to alchemical medicine, 181
Medical culture of New England, 162–164,

166, 173, 175–176, 180, 182, 187, 191, 198–199, 303
Medical hierarchy of New England, 187
Medical maladies, 185–190; gendered dimensions of, 185; nature of, 185; pervasive, 189
Medical practice by ministers, 173–175, 217; opposition to, in England, 174–175
Medical providentialism, 162–182; and diseases of settlement, 166–167; and Indian epidemics, 168–170; as contested in England, 171–173; influence of, on New England medical practice, 173, 175–176, 178
Medications, alchemical, 194–196
Medicine, alchemical, 6, 66, 143, 148; and medical providentialism, 6, 87, 171, 176–179; and Indians, 110, 142; importance of, in New England, 161–162, 175–176, 178, 183, 200–209, 214; and spiritual linkage, 162, 170, 176, 178–179, 198; practical advantages of, 179–180; criticism of, as dangerous, 181–182; and reputation, 193–194, 214; and purity, 197; and status, 200
Medicine, Galenic, 176–179, 181–182
Medicine, household, 182, 186–187
Mede, Joseph, 59
Mercury, 144
Merrett, Christopher, 265, 270, 290
Miantonomi, 98–100, 111, 113, 118
Millenialism and science, 1–3, 27–29, 41, 61, 64, 150, 178, 263–264, 303
Mineral prospecting, 79–80, 148–149
Mining, 68, 139; and New England as mineral storehouse, 78, 139–140, 148–149; English preoccupation with, 78, 296–300, 305. *See also* Black lead mining
Mirandola, Pico della, 17
Mohawks, 286
Mohegans, 95, 97–100, 102, 111, 115; and assaults on Winthrop plantation, 115–117, 125–126, 129–131
Monas hieroglyphica of John Dee, 34–36; adopted by Winthrop, 35, 37
Moody, Sir Henry, 196
Moraien, Johann, 13, 66–67, 69, 70, 75, 143, 183, 194, 262

Moray, Sir Robert, 7, 264, 270, 279–280, 290, 295–297, 299
Mortality of settlers, 167–168
Mortimer, Cromwell, 91

Nameaug (Pequot plantation), 91–92, 105–108, 126, 129, 146, 156, 277
Napier, Richard, 172
Narragansetts, 95, 97–99, 103, 109, 111, 118, 125; outrage of, at Miantonomi's death, 100; viewed as enemy by English, 119
Natural histories: value of, 8, 9, 270–272, 289, 291–292; Gerard Boate's, 8, 271; Samuel Maverick's, 8, 276–279; pressure on Winthrop to produce, 9, 270, 273–275, 289–295
Neoplatonism, 4, 16–17, 49, 52, 59, 123, 134, 223, 264
New England: alchemical potential of, 40–41, 44; and economic collapse and alchemical response, 55–56, 64–65, 76; as pansophic laboratory, 71–75, 83–85, 122; constraints on research in, 144; as laboratory for English imperial consolidation, 255
New Haven, 7, 153–156, 300; Comenian college in, 154–156
New London, 277; as alchemical research center, 4, 5, 82–92, 145, 157; as hospital town, 6, 112–113, 188–191; Mohegan harassment of, 114–116, 125–131, 138; English resistance to, 115, 124–131; suspicions of, 123; reconceptualization of, 138–139; and development, 146, 152; name of, confirmed, 156
Newton, Roger, 189
Niantics, 95, 103, 109, 118, 125, 128, 131, 142, 265; viewed as enemies by English, 119
Nicolls, Richard, 8, 279–287, 290, 296–297, 306
Nightshade, 290
Ninigret, 142, 265
Nipmucks, 86, 87, 102, 109, 125; attack on, by Mohegans, 125, 139
Nitre. *See* Saltpeter
Nonmonshot, 86–87
Nowequa, 125–127
Noyes, James, 145, 303

Objectivity, 294
Occult forces, 215, 217
Occult knowledge, 2, 7, 44, 61, 211–212, 215–216, 243; and witchcraft, 212, 215–218, 239, 241–242, 249
O'Dell, Lydia, 187–188
Office of Address, 271
Olcott, Mary, 239
Oldenburg, Henry, 268, 270, 274–275, 289–293, 295, 299
Oldham, John, 80
Oliver, James, 206
Optics, 54, 66, 67, 156, 259
Osborne, William, 149–150

Paine, William, 154
Palgreave, Richard, 206
Palmer, Thomas, 180, 197, 303
Pansophism, 2, 4, 49, 56–63, 151, 258, 263–264, 288, 291, 306; and conversion of Indians, 58–59, 70, 76, 141–142, 265–266, 274; and Comenian agenda, 62–63, 292; and pansophic moment, 63, 69, 70, 74; and transmutation, 66–67; and pragmatic reform, 71, 147, 306–307; and internal regulation, 76; threats to, 122; and educational reform, 154–156; and Christian service, 191
Paracelsianism, 20–23, 264; problems with, 24–25, 177; and metals, 79
Paracelsian medicine, 18–19, 23, 66, 176–177, 183, 196; and syncretism, 177–178
Paracelsus, 3, 18, 68, 150, 161, 176–177, 181, 201, 204, 223, 243
Paramount sachem, 108, 114–115
Particular Court of Connecticut, 227; role of magistrates in, 227; role of governor in, 228, 237; and Court of Assistants, 237
Patronage, alchemical and political, 256–258, 261, 265, 274–275, 279–285, 287–288, 297, 300
Peck, William, 186, 187
Pepys, Samuel, 264
Pequot Indians, 95, 103, 106, 125, 126; as tributaries of Uncas, 115, 120, 126–127; intimidation of, by Uncas, 126, 130–132; and release from Uncas, 135–136
Pequot War: aftermath of, 5, 93, 103; and competition for Pequot lands, 94–98, 100–101
Pequot/Thames River watershed, 102
Perkins, William, 233
Pessicus, 100
Peter, Hugh, 4, 50, 53, 229
Peters, Thomas, 97, 101, 111, 116, 117, 183
Pierson, Abraham, 184
Philosopher's stone, 179
Pitkin, William, 241
Plattes, Gabriel, 3, 4, 63–65, 78, 83
Polygyny, 113–114
Potter, William, 267
Providence Island Company, 50
Pseudoscience: alchemy as, 14
Purgatives: as measure of a medicine's effectiveness, 197; semiotic significance of, 197
Puritanism, 4; and science, 14–15; and Christian alchemy, 38–41; opposition of, to alchemy, 47
Pynchon, John, 191
Pynchon, Mary, 113
Pynchon, William, 86, 109

Quakers, 52

Regicides, 229
Religion and economy, 2–4
Religion and magic, 15, 110–111, 215, 217, 233
Religious toleration, 4, 5, 49, 51–53, 67, 278, 282, 285, 308; Puritan concerns about, 93
Remonstrance of Robert Child, 5, 93–94, 121–123, 138–139, 182, 187
Republic of alchemy, 65, 69
Richardson, Amos, 109
Rist, Johann von, 65, 253
Robin Cassacinamon: alliance of, with Winthrop, 103, 105–108, 114–115, 136; as agent for Uncas, 106; Pequot sachem, 106; as ahtaskoaog, 110
Rood, Sarah, 184

Rosicrucians, 1, 2, 28–29, 31, 46–48, 74, 259, 265
Rossiter, Bray, 198, 231
Royal College of Physicians, 172, 174, 181, 206
Royal commissioners, 8, 273–277, 281–285, 296
Royal Society, 3, 193, 207, 261–262, 264, 272, 305; and imperial agenda, 7, 261–263, 266, 268, 270, 273, 276, 279–280, 283, 292–293, 296–299; and intelligence gathering, 8, 262, 270, 273, 275, 288; potential origin of, in Connecticut, 91–92; and political patronage, 265, 269, 279–281, 285, 288–293, 297, 300–301; reports to, by Winthrop, 269–270, 290–291, 293–299; and natural histories, 272–273, 289–293; and mining and minerals, 296–300
Rubila, 66, 195
Rudolf II, 23

Sackville, Dr., 281
Salt: philosophical importance of, 54–55, 150
Saltpeter, 147–148, 194, 259; medical properties of, 195
Saltworks, 53–55, 140, 148, 152, 274–275
Sanford, Andrew, 234
Sanford, Mary, 234
Sassacus, 113
Saybrook plantation, 50, 108, 128, 277; and claims of Pequot lands, 101
Schlegel, Paul Marquart, 65, 149
Scholasticism, 16, 23
Seager, Elizabeth, 234–238, 241, 242
Secrecy, 11, 25, 75, 83–84, 144–145, 205, 208, 215, 247–248, 253, 267, 283, 291; and status, 25; and sacred knowledge, 39, 157–159; and intellectual property, 152, 248–250
Sendivogious, Michael, 177
Sergeant, Jonathan, Jr., 184
Settlement, literature of, 272, 275
Shantok, 95, 111, 118
Shepard, Thomas, 142
Silver-bearing mines, 4, 78, 80–82, 86–89, 100, 139, 149, 282, 297
Smith, Elizabeth, 239

Snelling, William, 191
Sodium sulfate, 68
Spectral apparitions, 215, 239–240, 242, 245
Spectral evidence, 245–246
Spiritual alchemy, 11–12
Spiritus mundi, 66, 68, 112
Stanton, Thomas, 128
Starkey, George, 123, 143, 178, 181–182, 200, 205–206, 263
Stiles, Ezra, 305
Stone, Samuel, 186, 191, 213, 234, 243
Swift, Joanna, 187
Sympathies and antipathies, 19, 186, 207, 216
Systematic accounts, 271

Tanckmarus, Augustus, 142–143, 149, 183
Tanckmarus, Johannes, 65
Tantiusque, 80, 86–87, 109, 138, 270, 278, 282, 305, 307; and recording of land sale, 109–110. *See also* Black lead mining; Silver-bearing mines
Tar and turpentine production, 140, 269, 291
Taylor, Edward, 39, 201
Telescope, 290
Thames / Pequot River, 4
Thomson, George, 181
Tompson, Benjamin, 302
Tradescant, John, 89
Transmutation, 2, 20–21, 66, 68, 159, 253; and greed, 25
Travel and alchemical knowledge, 30–31
Treat, Dorothy Bulkeley, 205
Treaty of Hartford, 95, 99, 108, 115, 127

Uncas, 98, 100, 102, 111, 118; efforts of, to assume regional dominance, 98–100, 106, 113, 114; and alliance with Connecticut, 98–99, 106, 115; and welcome of Winthrop, 111, 114; Winthrop as rival of, 115; and campaign against Winthrop, 115–116, 125–126, 138; political savvy of, 118; and conversion, 142
Underhill, John, 276
United Colonies of New England, 5, 99; and support of Uncas, 100, 118, 120; and

Pequot lands, 101–102; and criticism of New London settlers, 119; and Winthrop guardianship of Pequots, 119–120, 126; and representation and voting, 120; and Child's remonstrance, 121; and mistrust of Winthrop's plantation, 121, 124; hostility of, to Winthrop, 124, 126, 127–128; and Uncas's seizure of Pequots, 129–132; shifting position of, on Pequots, 135; and release of Pequots from Uncas's control, 135–136
Universal College. *See* College of Light
Universal reformation, 28–29, 62–64, 68, 256, 259
Utilitarian alchemy, 12, 20–24, 35, 55, 259
Utopian tracts, 64

Valentine, Basil, 38, 177, 180, 194, 204
Vaughan, Thomas, 204, 265
Venetian laboratory, 83, 151

Waples, Thomas, 239
Warren, William, 239
Warwick patent, 101
Weapon salve, 196, 201, 215, 218, 265
Webster, John, 79, 81, 82, 201
Webuckasham and Wascomo, 86–87, 109
Wequashcook, 113, 118–119
Whistler, Daniel, 269, 290
White, William, 139, 145, 152
Whiting, John, 232, 234
Wigglesworth, Edward, 185–186
Wigglesworth, Michael, 165–166, 185
Wilkins, John, 91–92
Willard, Josiah, 239
Williams, Nathaniel, 206
Williams, Robert, 140
Williams, Roger, 98–99, 106, 131, 139, 183
Wilson, John, 191
Winslow, Edward, 123
Winthrop, Adam, 131
Winthrop, Elizabeth Reade, 50, 113, 125
Winthrop, Fitz John, 286
Winthrop, John, Jr.: overview of, 11, 33, 43, 53, 193, 213–214, 253, 302–305; and father, 51–52, 94, 131–135; and religious toleration, 52–53, 67, 73, 93, 134, 257, 308; and pansophic program for New England, 71–76, 85–86, 105, 151, 191; and New England ironworks, 76–78; and black lead–silver mine, 78–79, 81–82, 86–89, 100, 102, 109; and design of learning, 151–152, 265; and New Haven ventures, 153–156; as Hermes Christianus, 183–199, 209, 216, 307; competition for services of, 213–214; and natural magic, 216
—and alchemy: Christian, 30; and pursuit of knowledge of, in Middle East, 31, 33; and New London as research center in, 83–86, 100, 102, 146; and recruiting European experts in, 84, 91–93; and New England network in, 143–146
—and Connecticut: and Saybrook, 50; as governor of, 156, 193, 199, 226, 229; and royal charter of, 236, 253, 255–257, 260–262; and English imperial agenda, 258, 281, 284–301
—and Indians: cultural sensitivity of, to relations with, 102–105, 108–110, 114, 125, 136; and alliance with Robin Cassacinamon, 103, 105–108, 114–115, 136; and Sagamore of Agawam, 103; and settling among Pequots, 105; and defense of Pequot interests, 108, 115, 128–129, 131, 135–136; as paramount sachem among, 108–114; as powaw, 110, 112; and Uncas, 111, 114–117, 125–132, 135; and guardianship of Pequots, 117–118, 126, 146
—and medicine: as practitioner of, 6, 30, 44, 111–112, 183–199; and reputation in, 112, 143, 160, 182–183, 191–193, 214; and medicines employed by, 183, 193–196; and extent of medical practice of, 184, 189–191, 195; and demand for medicines of, 198; as referral physician, 186–187; and appreciation of, for women healers, 198–199; and commitment of, to medical charity, 191, 193, 198–199
—and Royal Society: 9, 193, 236, 262–265, 267–269, 274, 280, 290–291; and strategies for dealing with requests of, 273, 280, 288–291, 293–300
—and witchcraft: and skepticism regarding charges of, 212, 223, 228; and refusal to enforce witchcraft convictions of, 212, 237;

and undermining legal foundations for convictions of, 212, 243–252, 305–306; and authority in cases of, 221–222, 228, 229, 237–238, 240–241; and strategies in cases of, 224–225, 228, 229, 236, 242, 249; as target of criticism for position on, 241, 248–249

Winthrop, John, Sr., 33, 51–52, 80, 94, 106, 108, 139, 170, 289; concern of, about son's pansophic project, 94, 123–124; on assault by Uncas, 131; last request of, to son, 131, 135; relations of, with son, 132–135

Winthrop, Stephen, 87

Winthrop, Wait Still, 208, 290, 303

Witchcraft, 6, 134, 204, 210–252; transformation of prosecution of, in Connecticut, 210–213, 229, 237, 242–252, 305; and Enlightenment skepticism, 210–211, 251; and occult philosophers' authority over interpretation, 211–212, 218–219, 222, 228, 238–241, 249; and Puritan elite, 212, 221–222, 242; and social pathology, 212, 224, 232; and evidentiary standards, 212, 233, 242, 244–252; and cause of illness, 219–221, 223, 226–227, 231, 232, 238–239; and foreknowledge of events, 219–221, 239, 249–252; association of, with the devil, 220–221, 225, 231–233, 235, 249–252; and harm to animals, 227; attitudes of prosecutors toward, 233–234, 241–242; and ordeal by water, 233, 235, 236, 241; and women healers, 238–239; and two-witness rule, 244–245; at Salem, Mass., 245

Witch's teat, 223

Worsley, Benjamin, 7, 39, 147, 256–259, 261, 290

York patent, 8, 276–277, 279, 280, 284, 296, 300

Yale, Mrs., 224

www.ingramcontent.com/pod-product-compliance
Lightning Source LLC
Chambersburg PA
CBHW030106010526
44116CB00005B/122